S0-CPS-635

THE PSYCHOLOGY OF ALGEBRA

THE MACMILLAN COMPANY
NEW YORK · BOSTON · CHICAGO · DALLAS
ATLANTA · SAN FRANCISCO

MACMILLAN & CO., LIMITED
LONDON · BOMBAY · CALCUTTA
MELBOURNE

THE MACMILLAN CO. OF CANADA, LTD.
TORONTO

THE PSYCHOLOGY OF ALGEBRA

BY

EDWARD L. THORNDIKE, MARGARET V. COBB,
JACOB S. ORLEANS, PERCIVAL M. SYMONDS,
ELVA WALD AND ELLA WOODYARD

OF THE

INSTITUTE OF EDUCATIONAL RESEARCH
TEACHERS COLLEGE, COLUMBIA UNIVERSITY

The investigations reported in this volume were made possible by a grant from the Commonwealth Fund.

New York
THE MACMILLAN COMPANY
1924

PRINTED IN THE UNITED STATES OF AMERICA

Set up and electrotyped. Published, June, 1923.

PREFACE

Nunn, Smith, Young, and other men of notable educational insight and wisdom have written manuals of the pedagogy of algebra. Rugg and Clark and their co-workers have applied psychology and psychological experimentation especially to the problems of the selection of subject matter in algebra. The National Committee on Mathematical Requirements has made important studies of many matters relating to pupils, subject matter, and methods.

In this work we have been in part gleaners after all these, checking their insights by experimentation, and filling out needed details. In part, however, our work has been very different, inasmuch as we have resolutely applied to the pedagogy of algebra the facts and principles which recent work in the psychology of learning has established. The nature of this recent work will appear more suitably in the actual uses of it in this volume than in any brief verbal account. Suffice it to say here that it emphasizes the dynamic aspect of the mind as a system of connections between situations and responses; treats learning as the formation of such connections or bonds or elementary habits; and finds that thought and reasoning—the so-called higher powers—are not forces opposing these habits but are these habits organized to work together and selectively.

There has been general coöperation in the work reported here, but special credit may be assigned as follows: To Miss Cobb, for the last section of Chapter I, the work on the degree of intellect required for success in high-school studies; to Mr. Orleans, for Chapters XV and XVI; to Mr. Symonds for the section of Chapter I on the careers of high-school pupils and for Chapter XVII; to Miss Wald for Chapter XIV; to Miss Woodyard for Chapter XI and the second section of Chapter II; to Mr. Orleans and Miss Woodyard for Chapter X. For the execution of the remainder of the work Mr. Thorndike has been responsible, and also for the general planning and supervision. The investigations owe much to the intelligent and careful work of Miss Eleanor Robinson and Miss Theresa Shulkin.

We wish to acknowledge here our indebtedness to all those who have helped, some repeatedly, in our investigations. The list is far too long to print, over two hundred teachers and scientific workers having given information or advice or expert opinion. We thank also those who have offered us facilities for experiment which lack of time prevented us from using.

Our work was made possible by a grant from the Commonwealth Fund and by the provision made for the Institute of Educational Research by the Trustees of Teachers College, Columbia University.

Teachers College
Columbia University
July, 1922

CONTENTS

LIST OF TABLES

THE PSYCHOLOGY OF ALGEBRA

CHAPTER I

The High School Pupil

In this chapter we shall not rehearse the facts concerning boys and girls of fourteen or fifteen which are available in books on the psychology of childhood and adolescence, but shall report three studies which add new information concerning the select group of boys and girls who enter American high schools.

CHANGES IN THE QUALITY OF THE PUPILS ENTERING HIGH SCHOOL [1]

The mathematician or college teacher who concerns himself with the teaching of algebra tends to think of the high school and its pupils as it was when he himself attended high school. There is thus a tendency for, say, the textbook which he writes or the selection of subject-matter and methods which he advocates to be adapted to the pupils of twenty-five to thirty years ago. This tendency persists even when he visits high schools frequently and teaches their classes occasionally.

[1] This section is reprinted, with additions, from *The School Review*, May, 1922.

The pupils of today, however, are different from those of twenty-five years ago, not only in their experiences and interests, but also in their inborn abilities. The data available for 1918 may be compared with the figures for 1890, since the interval between a person's study of algebra as a pupil and his opinion about it as an educational expert would probably average approximately twenty-eight years.

The essential available facts are given on this page, according to bulletins[1] of the Bureau of Education and the census report[2] of 1910. All numbers are to the nearest thousand.

Thus, the number of high-school pupils in 1918 was six times that in 1890, while the number of children of high-school age in 1918 was less than one and two-thirds times that in 1890. The number graduated, which is in some respects a better measure, was eight times as large in 1918 as in 1890.

	1890	1918
Number of students in all secondary schools	298,000	1,804,000
Number graduated	30,000	248,000
Total population	62,622,000	105,253,000
Population, ages 10-14	7,034,000	10,400,000*
Population, ages 15-19	6,558,000	10,400,000*

*Estimated from the total population for 1918, allowing for the general change in the distribution of the population by age groups. In 1900 there were 8,080,000 persons ten to fourteen years of age and 7,556,000 fifteen to nineteen; in 1910 there were 9,107,000 ten to fourteen and 9,064,000 fifteen to nineteen. By 1918 the numbers in these two age groups were probably approximately equal.

Neither of these, however, is just the comparison we wish. We are concerned here primarily with the number of pupils in the first year of high school (minus repeaters,

[1] "Private High Schools and Academies, 1917-18," *Bureau of Education Bulletin No. 3*, 1920, p. 4. "Statistics of Public High Schools, 1917-18," *Bureau of Education Bulletin No. 19*, 1920, pp. 11 f.

[2] *Thirteenth Census of the United States*, I (1910), 306.

since we do not wish to count a child twice because he takes the work of the grade a second time). This we would preferably compare with the number of children who passed a certain age mark, say fourteen, during the year in question.

On the basis of the primary data we may estimate the number of children who became fourteen years old during 1890 and 1918 as 1,365,000 and 2,080,000, respectively. The numbers who entered high schools and remained for four months or more in the school years 1890-91 and 1918-19 are more difficult to determine. The students in the first year of high school in recent years are about 40 per cent of the total high-school enrollment.[1] Some of these are repeaters and must be subtracted to leave the actual number entering high school that year. On the other hand, the enrollment figures omit some of the students who enter high school, study algebra for a few months, and then leave school. The percentage of repeaters is not known. About 1905 in certain cities one pupil in five failed of promotion in the first year of high school.[2] The pupils who failed of promotion would, of course, not all repeat the grade, the tendency to leave school being notably strong in those who fail. Considering these and other factors, we may set 90 percent of 40 percent of 1,804,000 as an approximate count of the number entering the first year of high school annually. This is almost 650,000; that is, almost one in three of the children reaching their teens in the United States enters high school.

This is a fact worth remembering. Nothing like it has ever occurred before in the world's history. The corresponding figure for 1890 is almost certainly not over one in ten.

Neither the percentage which the number of first-year

[1] *Report of the United States Commissioner of Education for 1916*, I, 448.

[2] G. D. Strayer and E. L. Thorndike, *Educational Administration, Quantitative Studies*, pp. 29 f. New York: Macmillan Co., 1913.

high-school pupils was of the total high-school enrollment in 1890, nor the proportion of repeaters in the first year at that time is known. Since four-year high schools have been replacing schools with shorter courses, and since also the gain from 1890 to 1918 has been greater for graduates than for total enrollment, it is probable that the percentage put at 40 for 1918 should be put at 43 or more for 1890. Even if it is put as high as 46, and if we use 90 percent of 46 percent of 298,000 as our high-school inflow in 1890, we have an enormous change from 1890 to 1918. For every one hundred children who reached fourteen there were about three and one-half times as many beginning high school in 1918 as in 1890! [1]

We lack measures of the inborn capacities of the one in ten or eleven of a generation ago and have only very scanty measures of the capacities of the one in three of today. We

[1] At the same time that we were making this investigation, Byrne [1922] was, entirely independently, studying the same matter, but more extensively. He does not make just the same allowances for distribution among the grades within the high school or for over-enrollment and repeaters that we have made, and secures estimates of ratios of enrollment to age populations which are a little higher than our estimates of the ratios of "school flow" to "age flow," both at 1890 and at 1918. His ratios are 35.9 for 1918 and 11.5 for 1890, or 3.1 times as high for 1918; ours give 3.5 times as high. We quote his table as Table 1.

TABLE 1

Ratios of Enrollment to Age Population at Different Dates

Grade	U. S. 1870	U. S. 1890	U. S. 1904	U. S. 1918
6	68.6	77.9	79.8	92.1
7	60.2	68.4	70.8	78.8
8	44.5	50.5	53.6	72.0
H. S. I	5.0	11.5	21.3	35.9
H. S. II	2.9	6.8	13.0	24.2
H. S. III	1.8	4.2	8.3	16.9
H. S. IV	1.0	2.4	5.2	13.7
College	....	1.1	1.1	3.5

have, however, excellent reasons for believing that the one in ten had greater capacities for algebra and for intellectual tasks generally than the one in three of today.

We know that education is selective, that the correlation between native capacity and continuance in school to higher and higher grades is positive. Let us consider the effect of

TABLE 2

DISTRIBUTION OF INTELLECT IN 1000 PUPILS OF THE FIRST YEAR OF HIGH SCHOOL ACCORDING AS ONE IN TEN OR ONE IN THREE ENTERS HIGH SCHOOL

	Lowest Tenth	Second Tenth	Third Tenth	Fourth Tenth	Fifth Tenth	Sixth Tenth	Seventh Tenth	Eighth Tenth	Ninth Tenth	Highest Tenth
Selective force: $r = .70$										
One in ten..	0	2	7	15	28	48	80	130	223	467
One in three	4	17	32	47	71	95	124	157	198	255
Selective force: $r = .60$										
One in ten..	2	9	18	29	46	66	97	137	204	392
One in three	12	29	45	61	80	98	121	146	177	230
Selective force: $r = .80$										
One in ten..	0	0	1	4	11	26	54	111	231	562
One in three	1	6	17	35	57	89	126	170	222	277

what seems a reasonable degree of this selective force, say that represented by a correlation of .70.[1] If the correlation is .70 and one-tenth of each oncoming age group is selected for entrance to high school and for the study of algebra, we shall have approximately the results shown in the first line

[1] The justification of this estimate would lead into highly technical arguments, unsuitable for presentation here. The actual data available are complicated by various factors; and their interpretation depends upon one's opinion of the relative shares of native capacity and the amount of education in determining the score made in such tests as the Stanford-Binet, the Army Alpha, and the Army Beta. The correlation between Alpha score and grade reached, as reported by the recruit, may be taken as .75; that for the Nonverbal Beta Test, .65; and that for the Stanford-Binet, .65 ("Psychological Exmining in the United States Army," *Memoirs of the National Academy of Sciences*, XV, 779, 783, 805. Washington: Government Printing office, 1920).

of entries of Table 2. If, the correlation being .70 as before, the selection is widened to one-third, we shall have approximately the results shown in the second line of entries of the table. In the former case, 95 percent of the pupils studying algebra will be above average in native intellectual capacity; in the latter case, only 83 percent. In the former case seven-tenths of the pupils will be in the top fifth of human beings for intellect; in the latter case, only four and one-half tenths. If we set a forty-percentile human intellect as the minimum able to profit by the study of algebra, there would be only $2\frac{1}{2}$ percent of pupils unable to profit in the one case as against 10 percent in the other. If we set the median intellect as the minimum able to profit by the study of algebra, the corresponding percentages would be 5 and 17. For readers who would estimate the selective force as greater or less than that denoted by $r = .70$, the results of similar calculations are included in Table 2.

It may be well to note here that the different selection of pupils in the European high schools must be considered in connection with any plans to adopt in whole or in part their subject-matter or methods in the teaching of mathematics. Their selection was very much narrower, and unless they were foolish in their control of it, as by rigidity of systems of caste, wealth, and the like, they should have obtained a much higher fraction of their youth in respect to intellect than we obtain of our youth.

Just how much narrower it was, it is not important for our purpose to measure, but one case may be presented. In Prussia, in 1910, the numbers of pupils in the different grades of *Gymnasien*, *Progymnasien*, *Realgymnasien*, *Oberrealschulen*, and *Realschulen* altogether were as follows: upper first, 8,788; lower first, 10,993; upper second, 14,369; lower second, 25,060; upper third, 27,966; lower third,

32,172; fourth, 35,044; fifth, 35,230; sixth, 37,071.[1] According to the *Statesman's Year Book*, the total population of Prussia in 1910 was 40,165,219. Estimating the number reaching a stage corresponding to our first year of high school as 30,000 and the number reaching fourteen years of age as 800,000, we have about one child in twenty-seven being selected for continuance to that educational level. The selection in 1913 was only a little wider, and in 1905 it was not very much narrower, no such increase of secondary education having been in progress there as in the United States.

Byrne [1922] has made the best estimates he could for all the leading European nations and Japan, for the ratio of the school enrollment corresponding to that in our four-year high-school to the population of four corresponding years. This is an extremely difficult task if one tries to secure the best data in the original foreign reports; and is a very risky enterprise if one works, as Byrne does, only with secondary accounts in English. We quote his results below. Experts for any particular country may be able to show flaws in his procedure, and some of his estimates (as of Ireland's being 10 per cent better off than England and Wales in the proportion continuing to high school, or of Norway's being over three times as well off as Sweden, or of Japan's being better off than Germany, or of Russia's being only a little below Germany) seem unlikely. It is, however, a most useful enterprise, and, regardless of particular flaws, it abundantly verifies the unique and unprecedented character of high-school enrollment in the United States.

[1] Data from Ergänzungsheft 27 of the *Centralblatt für die gesamte Unterrichtsverwaltung in Preussen*, 1910.

TABLE 3

Ratios of Enrollments in Upper Four Years (High School) to Estimated Four-Year Age Populations. In Percents
(Byrne, 1922, p. 5)

Nation	Ratio Four-Year Enrollment to Four-Year Population
England and Wales	3.9
Scotland	9.1
Ireland	4.3
France	2.4
Belgium	2.5
Germany	1.7
Austria	2.0
Hungary	1.5
Switzerland	2.2
Netherlands	2.6
Denmark	5.4
Norway	4.5
Sweden	1.4
Italy	.7
Spain	1.1
Russia	1.3
Japan	1.8

THE OCCUPATIONS OF HIGH-SCHOOL GRADUATES AND NON-GRADUATES [1]

High-school graduates make their life-careers in occupations that are among the more intellectual and more refined. Any observing high-school principal or teacher knows that his pupils very rarely become farm laborers, factory hands, cabmen, domestic servants, or salesgirls. Just how they are distributed, however, is not known for the country at large or for even any large geographical unit or large city.

Investigations have been made by Shallies [1913], Mitchell [1914], Counts [1915], Inglis [1915], Koons [1917], and in connection with the Elyria Survey [1918], with the following results.

[1] This section is reprinted from the *School Review*, June, 1922.

Shallies reports that of 734 students graduating from seventy-five high schools in New York State in June, 1908, 33.6 percent went to college, 16.6 percent to normal school, 5.4 percent to professional school; 16.0 percent were engaged in teaching, 11.7 percent in business, 8.5 percent in trade; 5.6 percent were at home, and 2.6 percent were unknown as to their careers.

The letter of inquiry which secured these facts was sent in March, 1911, to the principals of certain high schools, one or more of whose graduates in 1908 had graduated from a normal school in 1910. This procedure served to select high-school graduates who went on into normal schools in excess of their representation in the state in general. How many of the principals who were asked to report the facts did so is not stated. Probably not all reported, and those who did probably represented schools somewhat above the average in the dignity of their graduates' careers.

Mitchell reports the distribution of 845[1] graduates from forty-eight high schools in Iowa in 1908 as follows:

	Percentage
Agriculture	5.8
Professions	4.5
Business	3.8
Commercial employees	16.1
Teaching	30.0
Studying in liberal arts colleges	16.7
Girls married	17.0
Musicians	3.2
At home	3.0

Counts, studying the immediate futures of 20,389 graduates in 1913 from high schools in the North Central Association, reports the following distribution:

[1] The numbers for the groups specified total 841, and this total has been used in computing the percentages.

	Percentage
College	26.9
Commercial school	3.7
Trades	3.4
Farming	2.9
Normal school	7.3
Business	10.1
At home	15.1
Other occupations	14.3
Professions*	3.3
Domestic economy and agriculture	2.4
Teaching	4.3
Unknown	6.3

* This presumably means, at least to some extent, students in professional schools.

The primary data used by Counts were voluntary returns sent in by the principals of 1,000 high schools during the first half of the school year 1913-14. They were asked to report the number of the boys and girls in the previous year's graduating class belonging in each of the following groups: college, commercial school, trades, farming, normal school, business, at home, other occupations, medicine, dentistry, engineering, pharmacy, law, domestic economy, agriculture, and unknown. This request is somewhat ambiguous, and, oddly enough, omits teaching, the one occupation most widely undertaken by high-school girls in the year following graduation. Counts interprets medicine, dentistry, etc., as professions, but they may oftener mean students preparing for these professions.

Inglis computes, from the data of the United States Bureau of Education for 1912, that the central tendency by states is for a trifle over one-half of the high-school graduates to continue in colleges, normal schools, or other institutions of higher education.

Koons studied the graduates of a single high school (Murphysboro, Illinois) which was established in 1901. Up to January, 1916, there were 269 graduates, 117 boys and

152 girls. The occupations of these are reported by Koons as shown in the following table. For comparison similar facts are given for 117 boys and 152 girls selected at random from those who left school before graduation. The entries are expressed in percentages.

	Graduates		Non-Graduates	
	Boys	Girls	Boys	Girls
Agriculture	3.4	0.0	9.4	0.0
Business	23.0	15.1	23.1	7.2
Industries	14.5	0.0	26.5	5.3
Housewives	0.0	46.7	0.0	46.1
Teaching	6.9	21.0	0.9	3.3
Professions	12.0	0.0	1.7	0.0
Students	23.1	2.6	0.9	1.3
Miscellaneous	7.7	2.0	12.0	4.6
At home	0.0	9.2	0.0	14.5
No occupation	2.6	0.0	2.6	0.0
Occupation not known	3.4	2.0	21.4	13.2
Deceased	3.4	1.3	1.7	4.0

The Elyria Survey, reporting on 51 boys and 97 girls graduating in 1915 and 1916, finds that the percentage distribution on the basis of occupation within a year or two after graduation is as follows:

	Boys	Girls
Engaged in further study	35.3	34.0
Teaching	0.0	22.7
Trade (all in banks)	7.8	0.0
Manufacturing and mechanical industries	37.3	2.1
Milliner	0.0	2.1
Clerical occupations	11.8	18.6
Married	0.0	5.2
At home	2.0	13.5
Miscellaneous	5.9	2.1

In the spring of 1917 forty-six boys who entered the Elyria High School in the fall of 1913 but had left school were distributed as follows:

	Percentage
Agriculture	10.9
Trade	26.1
Industry	39.1
Transportation	10.9
Bell boy	2.2
Unknown	8.7
Loafing	2.2

Thirty girls of the same group who had left school were distributed as follows:

	Percentage
Industry	10.0
Telephone operator	26.6
Clerical work	30.0
Nurse	3.3
At home	20.0
Married	6.7
Unknown	3.3

These studies are valuable in helping to define our knowledge of the sort of life-work in which high-school students will engage, but they are limited in two important ways. They do not keep track of the individuals long enough for us to establish their eventual status; and the classifications employed (agriculture, business, industries) are too general. Agriculture may mean that a person operates a farm worth $50,000 and employs ten men or that he is a farm laborer; business may mean an errand boy or a bank president.

An attempt has been made to overcome these limitations by searching for representative high schools which have published or unpublished lists of their alumni for many years back and which report occupations in detail. Information from the following institutions seemed especially suited to our purpose: The Ottawa Township High School, Ottawa, Illinois; data for the classes from 1878 to 1914. The Labette County High School, Altamont, Kansas; data for the classes from 1896 to 1915. The Atchison County High School,

Effingham, Kansas; data for the classes from 1892 to 1916. The Pontiac Township High School, Pontiac, Illinois; data for the classes from 1895 to 1916. The Gouverneur Schools, Gouverneur, New York; data for the classes from 1888 to 1915. The Auburn Academic High School, Auburn, New York; data for the classes from 1868 to 1906.

We have collated the facts for graduates from 1892 to 1901 (1895 to 1901 for Pontiac and 1896 to 1901 for Labette County) and also for graduates from 1902 to 1911 (1902 to 1906 for Auburn). The former group represents graduates most of whom had more than ten years after graduation in which to establish themselves in life and deserves special attention.

The occupations mentioned are given in full, partly because these details give a concrete sense of the work of high-school graduates and partly to permit the reader to classify them as he wishes. They are also classified into groups. The detailed lists presented in Tables 4 and 5 show that the high-school graduates of 1892 to 1901 engaged in the main in the top quarter of the country's work as rated on the basis of desirability and importance. In Table 4 the ninety-nine cases labeled "business" are all from two schools, Ottawa and Auburn; and the more detailed records of the other four schools show that only a very small percentage of these ninety-nine are engaged as porters, drivers, and the like. Even if we regard the "unknowns" as much inferior to the "knowns," and estimate the number of "farmers" who are really farm laborers very generously, we have out of 466 males, only 41 who may be doing work below the level of a stenographer, salesman, or electrician (3 in Army and Navy, 6 in the factory group, 9 out of 33 farmers, 1 lumberman, 2 out of 7 in the manufacturing group, 8 out of 99 in business, and 12 out of 24 unknowns).

Let the reader arrange the occupations of the 1892 to 1901 male graduates along a scale of seven units in which one represents an unskilled day laborer; four, a blacksmith, carpenter, mason, or plumber; seven, a doctor, lawyer, engineer, or operator of an industrial or commercial plant with an income of $3,500 or more. Two and three represent equal steps between one and four. Five and six represent equal steps between four and seven. Distribute the "farmers," "unknowns," and other doubtful cases as seems just. It will probably be found that more than one-half of the male high-school graduates will be placed in the two highest compartments, and more than four-fifths of them in the three highest. The occupations of the women may be similarly distributed by using a scale in which one represents an unskilled laborer, a dish washer or factory hand of low

TABLE 4[1]

PERCENTAGE DISTRIBUTION, ON BASIS OF OCCUPATION, OF MALE GRADUATES FROM SIX HIGH SCHOOLS

	1892–1901	1902–1911
Artist, 1, 0; architecture, 2, 1; cartoonist, 1, 0; chorister, 0, 1; music, 1, 2; music teacher, 0, 1; stage, 1, 0	1.3	0.8
Accountant, 1, 1; auditor, 2, 0; banking, 14, 2; bank cashier, 3, 4; bank solicitor, 1, 0; city treasurer, 1, 0; post-office inspector, 1, 0; postmaster, 2, 2; registrar, 2, 0; county treasurer, 0, 1	5.8	1.5
Army, 1, 0; Navy, 2, 1	0.6	0.2
Advertising agent, 0, 1; business, 99, 131; book business, 2, 0; business manager, 0, 1; druggist, 4, 3; hardware, 0, 3; insurance, 0, 1; jeweler, 0, 1; lumber dealer, 1, 0; music dealer, 1, 0; manager of coal office, 0, 2; manager of grocery depot, 2, 2; manager farmer's elevator, 0. 1; manager of show, 0, 1; real estate, 6, 4; merchant, 4, 7; salesman, 6, 4; shoe dealer, 0, 1; shoe salesman, 0, 1; traveling, 1, 0	27.0	24.7

[1] The wording of the original is followed in the list of occupations. The numbers after each occupation are the actual occurrences, the first being for the decade 1892 to 1901, the second for the decade 1902 to 1911. The totals are: 1892 to 1901, 466; 1902 to 1911, 664. The numbers in the columns at the right are percentages of the total. Thus, of the graduates from 1892 to 1901, 1.3 percent are artists, architects, etc.; 5.8 percent are accountants, auditors, etc.

TABLE 4—*Continued.*

	1892–1901	1902–1911
Clergyman, 11, 3; dentistry, 6, 2; editor, 5, 0; education, 18, 4; entomology, 0, 1; journalism, 5, 3; law, 57, 14; medicine, 19, 5; missionary, 1, 0; pastor, 1, 0; pathology, 1, 0; principal of school, 0, 1; physical director, 0, 1; physician, 17, 5; professor, 1, 1; student, 1, 105; superintendent of schools, 2, 2; teaching, 7, 37; Y.M.C.A., 3, 1	33.3	27.9
Chemistry, 2, 5; civil engineer, 7, 7; engineering, 14, 9; electrical engineer, 1, 1; mechanical engineer, 1, 1; mining engineer, 0, 3; architectural engineer, 0, 1	5.4	4.1
Draughting, 2, 2; electricity, 2, 9; forestry service, 0, 1; inspector (meat and dairy), 2, 0; machinist, 2, 2; painter, 0, 1; pharmacy, 2, 5; plumbing, 1, 1; photography, 3, 0; telegraphy, 1, 0; surveyor, 0, 1	3.2	3.3
Bookkeeping, 7, 9; billing clerk, 0, 1; cashier, 1, 2; clerking, 11, 20; drug clerk, 0, 1; government clerk, 0, 1; groceryman, 0, 1; mail clerk, 3, 8; postal service, 1, 0; Santa Fe office, 0, 1; stenographer, 1, 14; secretary, 1, 0; railroad employee, 4, 1	6.2	8.9
Factory, 1, 1; chauffeur, 0, 1; deliverer, 0, 1; mail carrier, 1, 3; mechanic, 4, 1; work in lumberyard, 0, 1; working for coal company, 0, 1	1.3	1.4
Farming, 33, 77	7.1	11.6
Lumberman, 1, 3	0.2	0.5
Manufacturing, 7, 2	1.5	0.3
Agricultural agent, 4, 2; internal revenue officer, 1, 0; osteopath, 1, 0; tailor, 1, 0; United States mail, 2, 4; undertaker, 0, 1; veterinary, 0, 1	1.9	1.2
At home, 0, 3	0.0	0.5
Not stated, 23, 83; unknown, 1, 6	5.1	13.4

grade; four, a clerk or typist; seven, a teacher in high school or an operator of a store or shop with an income of $2,000 or more. This procedure will probably result in placing nearly three-fifths of the unmarried women graduates in the two highest compartments and nearly nine-tenths of them in the three highest.

High-school graduates (male) have over ten times their quota in the professions. According to the census of 1900, in the general male population twenty-five to thirty-four years of age there were just about as many draymen, hackmen and teamsters as there were in the four professions plus journalism and dentistry; but among the high-school graduates there were possibly one-sixtieth as many. Among women in general of the ages twenty-five to thirty-four, dressmakers and milliners outnumber teachers considerably, but among the high-school graduates teachers are about fifty times as numerous as milliners and dressmakers together.

The life-careers of non-graduates cannot be pictured with surety. We have been unable to find records save those quoted from Koons and the Elyria Survey and those given

TABLE 5[1]

PERCENTAGE DISTRIBUTION, ON BASIS OF OCCUPATION, OF FEMALE GRADUATES FROM SIX HIGH SCHOOLS

	All Women		Unmarried Women	
	1892–1901	1902–1911	1892–1901	1902–1911
Artist, 0, 2; music, 7, 10; musician, 1, 0; music teacher, 6, 4; stage, 1, 0; supervisor of music, 1, 0	1.9	1.5	3.8	2.5
Business, 80, 63; abstract office, 1, 0; music dealer, 1, 0; deputy register of deeds, 0, 1; superintendent, 0, 1	9.5	6.2	19.4	10.3
Bookkeeper, 13, 16; clerk, 13, 15; office work, 0, 1; private secretary, 2, 1; postal clerk, 0, 2; saleslady, 1, 2; secretary, 0, 1; stenography, 24, 37; assistant postmistress, 1, 0	6.3	7.1	12.8	11.9
Chautauqua system, 0, 1; education, 107, 22; librarian, 5, 5; married teacher, 3, 4; student, 3, 70; settlement work, 0, 1; teaching, 44, 204	18.9	29.3	38.4	48.7
Nursing, 6, 12; married nurse, 1, 0; osteopathy, 1, 0; pharmacy, 1, 0	1.0	1.1	2.1	1.9
Dressmaking, 2, 1; millinery, 1, 1	0.3	0.2	0.7	0.3
Domestic, 0, 1; housekeeping, 2, 0	0.2	0.1	0.5	0.2
Machine operator, 1, 0	0.1	0.0	0.2	0.0
Married, 437, 417	50.9	39.8	......	......
At home, 36, 94; in Japan, 1, 0; unemployed, 34, 18	8.3	10.7	16.8	17.7
Not stated, 18, 34; unknown, 4, 7	2.6	3.9	5.2	6.5

[1] The numbers after each occupation are the actual occurrences when all women are included, the first being for the decade 1892 to 1901, the second for the decade 1902 to 1911. The totals are: all women, 1892 to 1901, 859; 1902 to 1911, 1,048; unmarried women, 1892 to 1901, 421; 1902 to 1911, 631.

in Table 6 comparing graduates and non-graduates of the Northeast High School of Philadelphia, a school of rather special nature and history. The one fact which is certain is that the non-graduate takes, in respect to occupation, a position intermediate between the graduate and the person who has not attended high school.

TABLE 6[1]

PERCENTAGE DISTRIBUTION, ON BASIS OF OCCUPATION, OF GRADUATES AND NON-GRADUATES OF THE NORTHEAST HIGH SCHOOL, PHILADELPHIA, 1893–1911

	GRADUATES	NON-GRADUATES
Actors, 0.0, 0.2; artists and designers, 3.2, 0.7; engravers, 0.2, 0.4; musicians, 0.3; 0.4	3.7	1.7
Accountants, 0.6, 0.4; bankers and stockbrokers, 1.1, 0.5; builders, 0.9, 0.4; contractors, 0.8, 0.2	3.4	1.5
Architects and architectural draughtsmen	3.2	.9
Advertising solicitors and writers, 0.5, 0.7; commercial travelers, etc., 3.0, 7.3; insurance agents, etc., 1.4, 1.1	4.9	9.1
Brewers, 0.3, 0.2; druggists, 0.6, 0.9; manufacturers, 2.7, 2.0; merchants, 1.7, 2.4	5.3	5.5
Bookkeepers, 2.7, 2.0; clerks, etc., 5.3, 8.2; private secretaries, 0.3, 0.5; stenographers, 0.8, 0.0	9.1	10.7
Chemists, 1.7, 0.4; engineers, 8.8, 2.4	10.5	2.8
Clergymen, 0.4, 0.2; dentists, 1.1, 0.4; journalists, 1.5, 0.4; physicians and surgeons, 2.7, 0.9; teachers, 2.4, 0.4; consuls and lawyers, 2.3, 0.7	10.4	3.0
Draughtsmen, 8.6, 3.5; decorators, 0.8, 0.4; electricians, 5.3, 0.2; instrument makers, 0.8, 0.0; jewelers, 0.1, 0.2; machinists and pattern makers, 1.5, 0.9; photographers, 0.6, 0.0; surveyors, rodmen and transitmen, 2.1, 0.7	19.8	5.9
Osteopaths, 0.0, 0.4; veterinary surgeons, 0.1, 0.4	0.1	0.8
Plumbers, 0.4, 0.7; printers, 0.5, 0.9; tailors, 0.1, 0.0	1.0	1.6
Students	15.7	1.5
Supervisors, 0.4, 0.0; superintendents, foreman, and managers, 1.9, 5.1	2.3	5.1
Other occupations and not reported	10.6	50.4

From both the details and the grouped results shown in Table 6[2] it is obvious that, with any reasonable distribution

[1] The first number after each occupation is the percentage of living graduates; the second, the percentage of living non-graduates.

[2] These facts are compiled from the *Catalog of the Northeast High School*, Philadelphia, 1912–13, and the *Handbook of the Alumni Association of the Northeast Manual Training High School*, Philadelphia, 1913.

of the "unknowns," the non-graduates will be relatively less frequent than the graduates in the more dignified and important occupations and more frequent in those less dignified and important. This is shown also by the data quoted from Koons and the Elyria Survey.

We have shown in the previous section that from 1890 to 1918 the percentage of children reaching a given age who go to high school has trebled and that the percentage of children reaching a given age who graduate from high school has increased even more. It seems unlikely that the enviable status shown for graduates in 1892 to 1901 in respect to occupations can be fully maintained now and in the future. To maintain it would require that the favored occupations be practically closed to all but high-school graduates. This may perhaps be taking place. The supply of high-school graduates is increasing so fast that any profession or reputable semiprofession may demand such. Even if it is not fully maintained — indeed, even if there is a considerable movement downward — the high-school graduates will still have a notably high occupational status; the correlation between amount of education and dignity of occupation will still be close.

THE INTELLIGENCE OF HIGH-SCHOOL PUPILS [1]

The "intelligence" or "amount of intelligence" or "degree of intelligence" or "intelligence score" of a pupil, as used in this section, will mean the score he obtains in a first trial with the Army Alpha test. This test, presented on pages 38 to 46, contains 212 items. A correct response to

[1] The evidence for the statements of this section, and some critical discussion of it, may be found in an article entitled "The Limits Set to Educational Achievement by Limited Intelligence," in the *Journal of Educational Psychology* for November and December, 1922.

any one of them counts one. An incorrect response counts zero, except where the pupil has an even chance of being right by guessing; then it counts minus one. About half of the native white adults of the United States would score 65 or higher. About a third of them would score above 85. If the reader will go through the test, checking the 65 items that seem easiest, he will gain a rough idea of the degree of intelligence signified by a score of 65; in the same way he may obtain a rough sense of the meaning of a score of 85.

The Intelligence of High-School Pupils

If pupils in high school are given the Army Alpha test,[1] three facts appear very clearly. First, the average is much above that of the general population; second, the average rises frcm first to second, second to third and third to fourth year pupils; third, there is a very wide range of variation.

Tables 7, 8, 9 and 10 show the facts, in the form of the number of pupils who obtained scores from 0 to 4, 5 to 9, 10 to 14, 15 to 19, and so on. Thus Table 7 reads: "Of the high-school freshmen in Mt. Clemens, one obtained a score from 165 to 169; one a score from 145 to 149, one a score from 140 to 144, and so on. The median score (which is approximately the score such that 50 per cent of the pupils fall below it and 50 per cent exceed it) for the 85 Mt. Clemens freshmen was 96."

The scores in general average high. See Table 11, totals for Ill., Ia., Wis. and Mich. The median freshman score is 96. The median sophomore score is 111. The median junior score is 123. The median senior score is 126. They increase from year to year as shown above and in Table 11. The difference between freshmen and seniors is shown graphi-

[1]A representative form of Army Alpha is reprinted in full at the end of this chapter.

TABLE 7

ALPHA DISTRIBUTION OF HIGH-SCHOOL FRESHMEN[1]

Score on Alpha	Mt. Clemens	Milan	Mt. Pleasant	Alma	Mich. Outside Detroit	Ill., Wis. and Iowa	Total Ill., Wis., Iowa and Mich.	
							No. Cases	Per-cent
180–184	......	......	......	......	....	3	3	.17
175–179	......	......	......	......	....	0	0	....
170	......	......	......	......	....	2	2	.12
165	1	......	......	......	1	2	3	.17
160	......	......	1	1	2	5	7	.41
155	......	......	......	......	....	7	7	.41
150	......	1	......	2	3	9	12	.70
145	1	2	......	......	3	13	16	.93
140	1	......	......	2	3	17	20	1.16
135	1	2	2	4	9	31	40	2.32
130	1	2	2	3	8	50	58	3.37
125	4	1	2	6	13	67	80	4.65
120	3	3	2	7	15	65	80	4.65
115	6	2	3	11	22	72	94	5.46
110	6	4	2	10	22	73	95	5.52
105	8	2	8	5	23	97	120	6.97
100	5	6	3	12	26	87	113	6.57
95	7	6	5	27	45	112	157	9.13
90	4	7	10	11	32	98	130	7.55
85	8	11	6	18	43	113	156	9.07
80	3	7	11	10	31	88	119	6.92
75	5	7	9	18	39	73	109	6.34
70	6	14	8	11	39	61	99	5.75
65	5	8	6	11	30	37	71	4.15
60	4	4	7	12	27	23	50	2.90
55	3	1	3	6	13	16	29	1.68
50	1	4	......	8	13	15	28	1.63
45	2	3	1	3	9	5	14	.81
40	......	1	......	......	1	0	1	.06
35–39	......	1	......	1	2	4	6	.35
30–34	......	......	......	......	....	0	0	....
25–29	......	......	......	......	....	2	2	.12
No. Cases	85	99	91	199	474	1247	1721	100.02
Median..	96	85	85	91	89	99	96	....

[1] These results are typical for the country as a whole. There is very great variation among communities. For example, the median score for recruits from the highest ranking State in the Army data was almost twice that of the lowest ranking State (79 and 41).

TABLE 8

ALPHA DISTRIBUTION OF HIGH-SCHOOL SOPHOMORES

Score on Alpha	Mt. Clemens	Milan	Mt. Pleasant	Alma	Mich. Outside Detroit	Ill., Wis. and Iowa	Total Ill., Wis., Iowa and Mich.	
							No. Cases	Per cent
180–184	......	......	......	......	....	2	2	.16
175–179	......	......	......	......	....	1	1	.08
170	......	......	......	......	....	6	6	.48
165	2	......	......	......	2	7	9	.79
160	1	......	1	......	2	7	9	.79
155	......	......	......	2	2	15	17	1.36
150	......	1	......	1	2	20	22	1.76
145	4	......	1	4	9	30	39	3.11
140	4	......	......	......	4	47	51	4.07
135	......	......	4	......	4	34	38	3.03
130	6	2	4	......	12	63	75	5.90
125	6	5	2	2	15	67	82	6.54
120	3	0	8	4	15	76	91	7.26
115	7	3	2	9	21	72	93	7.42
110	5	1	5	5	16	94	110	8.78
105	7	6	3	7	23	70	93	7.42
100	3	3	5	6	17	83	100	7.98
95	2	5	4	9	20	65	85	6.78
90	5	5	9	9	28	67	95	7.58
85	1	5	5	7	18	50	68	5.42
80	4	3	3	7	17	30	47	3.75
75	1	2	5	5	13	31	44	3.51
70	......	1	2	3	6	22	28	2.23
65	......	......	3	2	5	16	21	1.68
60	......	......	2	3	5	8	13	1.04
55	......	2	......	2	4	1	5	.40
50	......	1	1	......	2	3	5	.40
45	......	......	......	......	....	2	2	.16
40	......	......	......	......	....	0	0	....
35	......	......	......	......	....	1	1	.08
30	......	......	......	......	....	0	0	....
25–29	......	......	......	......	....	0	0	....
20–24	......	......	......	......	....	0	0	....
15–19	......	......	......	......	....	1	1	.08
No. Cases	61	45	69	87	262	991	1253	100.03
Median..	117	99	101	98	104	112	111	....

TABLE 9

Alpha Distribution of High-School Juniors

Score on Alpha	Mt. Clem-ens	Milan	Mt. Pleas-ant	Alma	Mich. Outside Detroit	Ill., Wis. and Iowa	Total Ill., Wis., Iowa and Mich.	
							No. Cases	Per-cent
185–189	......	......	......	......	......	2	2	.20
180–184	......	......	1	......	1	3	4	.41
175	......	......	......	2	2	5	7	.72
170	2	......	1	......	3	8	11	1.13
165	1	......	2	......	3	15	18	1.84
160	1	......	......	......	1	15	16	1.64
155	4	......	2	1	7	32	39	3.99
150	2	......	2	......	4	38	42	4.30
145	2	......	2	3	7	34	41	4.20
140	2	......	4	4	10	51	61	6.24
135	1	......	2	5	8	60	68	6.96
130	3	......	......	3	6	62	68	6.96
125	4	1	4	1	10	66	76	7.78
120	5	......	3	2	10	74	84	8.60
115	5	......	7	4	16	76	92	9.41
110	3	......	2	6	11	68	79	8.09
105	3	......	3	2	8	52	60	6.14
100	3	1	3	3	10	48	58	5.94
95	1	1	2	7	11	42	53	5.42
90	1	......	3	2	6	23	29	2.97
85	1	......	1	4	6	19	25	2.56
80	......	......	1	3	4	13	17	1.74
75	1	......	1	2	4	7	11	1.13
70	......	......	......	3	3	3	6	.61
65	......	......	......	1	1	3	4	.41
60	......	......	......	2	2	0	2	.20
55–59	1	......	......	2	3	0	3	.31
50–54	......	......	......	......	......	0	0	......
45–49	......	......	......	......	......	1	1	.10
No. Cases	46	3	46	62	157	820	977	100.00
Median..	124	......	120	110	118	124	123	......

TABLE 10

ALPHA DISTRIBUTION OF HIGH-SCHOOL SENIORS

Score on Alpha	Mt. Clemens	Milan	Mt. Pleasant	Alma	Mich. Outside Detroit	Ill., Wis. and Iowa	Total Ill., Wis., Iowa and Mich. Exc. Detroit		Detroit	
							No. Cases	Per cent	No. Cases	Per cent
200–204	.....	.....	.....	.....	.....	1	1	.13	.....	.....
195–199	.....	.....	.....	.....	.....	0	0	.....	.....	.....
190–194	.....	.....	.....	.....	.....	2	2	.26	.....	.....
185	.....	.....	.....	.....	.....	0	0	.....	.....	.....
180	1	.....	.....	1	2	2	4	.52	3	.48
175	1	.....	.....	.....	1	5	6	.78	3	.48
170	1	.....	.....	1	2	12	14	1.83	11	1.77
165	.....	.....	1	2	3	15	18	2.35	6	.97
160	.....	.....	2	1	3	18	21	2.74	17	2.74
155	1	.....	3	3	7	31	38	4.96	19	3.06
150	2	.....	2	3	7	26	33	4.30	30	4.83
145	.....	.....	3	2	5	36	41	5.35	41	6.60
140	2	.....	4	3	9	51	60	7.83	34	5.47
135	1	.....	4	1	6	41	47	6.13	41	6.60
130	1	.....	3	1	5	45	50	6.52	41	6.60
125	1	.....	4	3	8	59	67	8.75	50	8.05
120	4	1	4	7	16	44	60	7.83	44	7.08
115	2	.....	2	2	6	55	61	7.96	48	7.72
110	2	.....	5	4	11	49	60	7.83	46	7.40
105	2	.....	1	6	9	42	51	6.66	49	7.89
100	1	.....	1	6	8	34	42	5.48	32	5.15
95	1	.....	2	1	4	18	22	2.87	33	5.31
90	2	.....	2	1	5	15	20	2.61	19	3.06
85	1	.....	3	1	5	4	9	1.17	22	3.54
80	.....	.....	.....	1	1	7	8	1.04	8	1.29
75	.....	.....	2	3	5	15	20	2.61	11	1.77
70	1	.....	.....	.....	1	3	4	.52	4	.64
65	.....	.....	.....	1	1	3	4	.52	3	.48
60–64	.....	.....	.....	.....	0	1	1	.13	2	.32
55–59	1	.....	.....	.....	1	1	2	.26	3	.48
50–54	.....	.....	.....	.....	.....	.....	.....	.....	1	.16
No. Cases	28	1	48	54	131	635	766	99.94	621	99.94
Median..	121	.....	128	121	123	127	126	.....	123	.....

cally in Fig. 1. They vary as shown in the tables, and (for the freshmen and seniors) in Fig. 1.

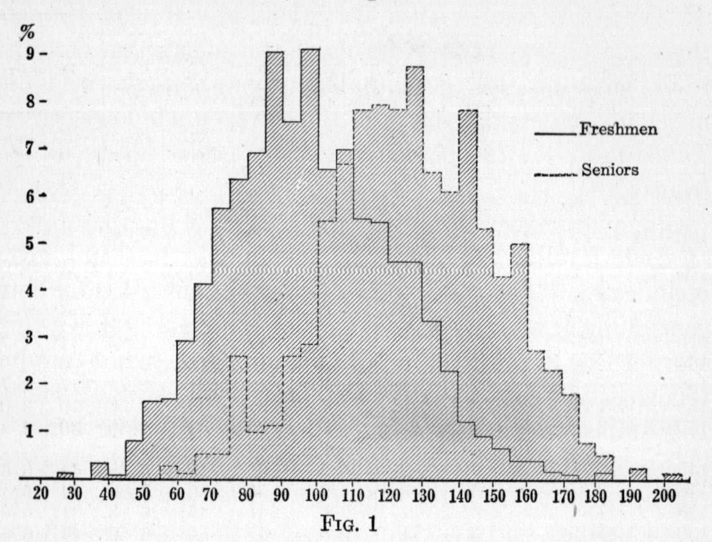

Fig. 1

It will be seen that in spite of the wide range of ability in pupils, of which every high-school teacher must be conscious, there are nevertheless few (less than 3 percent) who have not median intelligence (a score of 65) or better. This corresponds approximately to a mental age of thirteen years and two months on the Stanford Binet Scale. The pupils in academic high schools are, in fact, a limited group which covers just about half, the upper half, of the total distribution of American intelligence.

The facts of Tables 7 to 11 are corroborated by the scores made in the Army during the war by those who reported themselves as having left school after one year of high school, after two years of high school, and so on. The median scores for recruits so reporting were 98, 105, 111, and 115 for the successive high-school years.

Continuance in High School in Relation to Intelligence

The Army figures show very definitely that at the time these recruits were of high-school and college age, say five to ten years ago, the more intelligent youths remained longer in school, in general, than did those who made lower scores. Table 12 shows the percents at each Alpha level[1], derived from Table 281, page 748, of the Army report (National Academy of Sciences, Memoirs, 1921). The number of cases of each level always exceeds 1000, and the total group includes 48,102 recruits. Fig. 2 shows the situation for those continuing at least into the first year of high school. Of those scoring less than 35 on Alpha, for example, 4 in a hundred reported that they had entered high school; of those scoring 155 or better, 92 in a hundred, or twenty-three times as large a proportion, so reported. This comparison, based on Alpha scores, omits the illiterate group altogether. It is probable that, had they been included, the chance of entering high school would have proved to be at least thirty times as great for those over 155 as for those below 35.

Fig. 3 shows similarly the proportions who reported that they had become seniors and (in practically all cases) had graduated from high school. Of those scoring less than 35, less than 1 percent, and of those scoring 155 or over, 73 percent reached the senior year in high school. Thus the chance of reaching this level is over a hundred times as great for the highest as for the lowest group. Here again the contrast

[1] The reader will, of course, bear in mind that these Army tests were taken at ages from 21 to 30, not at the time of entrance to high school. There would be some variety in competent estimates of the extent to which the score means that the man was as a boy more intelligent than others and so continued longer in school, and the extent to which it means that by staying on in high school, he got information and power which enabled him to make a higher score. In view of the nature of the Alpha test, and the evidence gathered in connection with the Army testing, psychologists in general would estimate that relative scores in it are determined largely by innate qualities.

would have been intensified had we had a comparable measure of the illiterate group, and included them in the comparison.

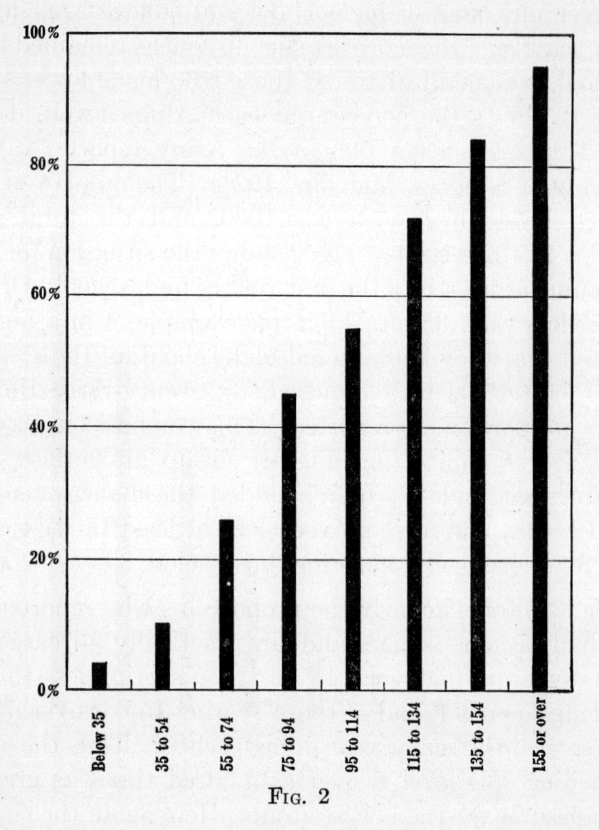

FIG. 2

Fig. 4 shows comparable figures for college entrance. A quarter of one percent of the lowest group, and 53 percent of the highest group, reported that they had entered college. The chance of college entrance at that time appears to have been almost two hundred times as good for those highly endowed intellectually as for the lowest fifth. Were illiterates included, the contrast would in this case also be strengthened.

For the average person, score 55 to 74, the chance of entering high school was at that time about 1 in 4; of entering

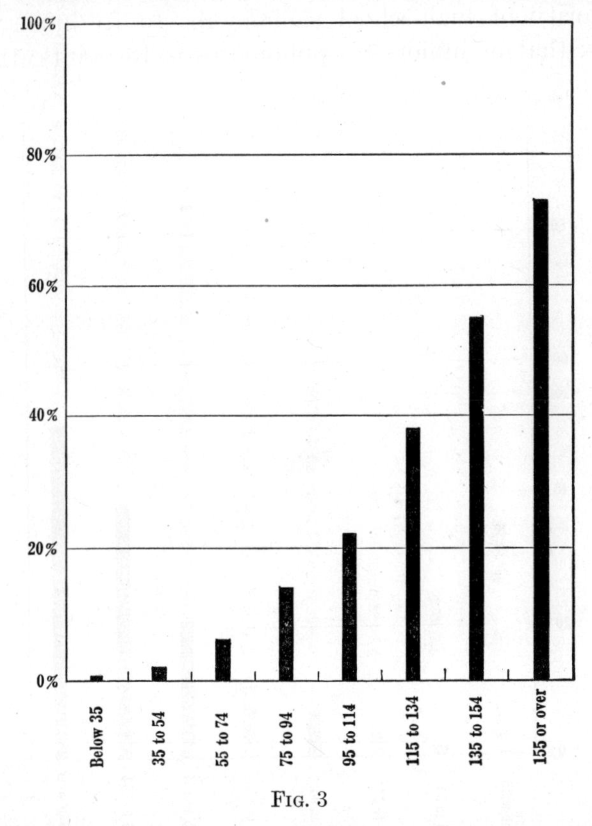

FIG. 3

college, about 1 in 30. It is at present somewhat greater. This educational selection of intelligence is evidenced also in the yearly increase in Alpha medians when the test is given throughout a school. Of course, not the whole amount of this increase is due to elimination of the less intelligent pupils. To determine the exact proportion which is due to this cause is at present impossible, but that part of it is so caused is easily proved.

Tables 13 and 14 show the available facts. The average superiority of high-school sophomores to freshmen is 15 points; that of juniors to sophomores is 10 points; that of

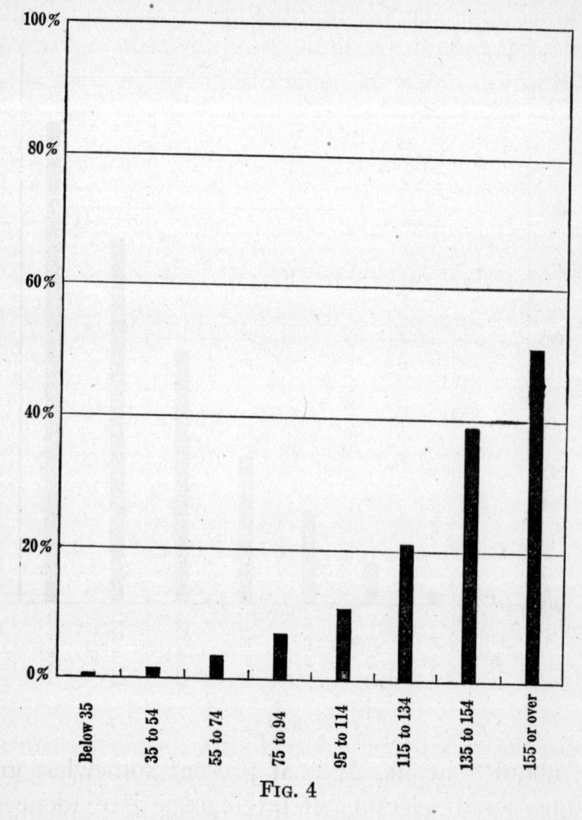

Fig. 4

seniors to juniors is 4 points. Probably a third or more of these differences is attributable to the greater elimination of the less intelligent pupils.

The influence of intelligence on continuance in school appears also when we look at the elimination which takes place among the freshmen of least intelligence. Careful consideration of Tables 13 and 14 has led us to conclude that a freshman who scores 77 will as a senior score about 90; on this basis we can get a rough notion of the extent of this elimination in the schools represented in our data without necessarily following a freshman group all the way through. The assumption must be made that successive entering

TABLE 11

MEDIAN ALPHA SCORES, HIGH SCHOOL AND COLLEGE GROUPS

	Freshmen	Sophomores	Juniors	Seniors	College Freshmen
Mt. Clemens	96	117	124	121	....
Milan	85	99	....	....	....
Mt. Pleasant	85	101	120	128	....
Alma	91	98	110	121	....
Michigan, Outside Det.	89	104	118	123	....
Ill., Ia., Wis.	99	112	124	127	....
Ill., Ia., Wis., Mich.	97	111	123	126	....
Detroit	....	....	....	123	....
N. Y. (Kansas Report)	92	104	118	132	....
Emporia, Kansas	80	105	101	111	....
Stanton, Va.	91	115	136	117	....
Kansas Report	....	....	....	....	129
University of Ill.	....	....	....	....	131
Ohio State Univ.	....	....	....	....	130
Oberlin College	....	....	....	....	148
Yale University	....	....	....	....	160
Recruits who had been Freshmen, Soph., etc.	98	105	111	115	119
Officers who had been Freshmen	141	141	142	143	143

classes are each 5 percent larger than that of the year before[1]; then we may compare the number in the freshman class who score below 77 with 115.8 percent of the number in the senior class who score below 90. Comparisons of this kind show that, in the Michigan schools from which we

[1] See the data given earlier in this chapter.

TABLE 12

THE PERCENT OF RECRUITS AT VARIOUS ALPHA LEVELS, REPORTING THEMSELVES AS HAVING CONTINUED IN SCHOOL TO THE FIRST YEAR OF HIGH SCHOOL, TO THE FOURTH YEAR OF HIGH SCHOOL AND TO THE FIRST YEAR OF COLLEGE

ALPHA SCORE	No. of Cases	School Continuance		
		Percent Entering H. S.	Percent H. S. Seniors	Percent Entering College
155 and above	1011	93	73	53
135—154	1680	84	55	33
115—134	3076	72	38	21
95—114	5002	55	22	11
75— 94	6282	45	14	7
55— 74	9750	23	6	3
35— 54	10709	10	2	1.2
Below 35	9592	4	0.7	0.3
Total	48102	36	22	20

TABLE 13

HIGH SCHOOL ALPHA SCORES

	Schools						Army	
	Mich.	Ill. Wis. Iowa	N. Y. (Kan. Rept.)	Emporia, Kan.	Stanton, Va.	Minn. Survey	Recruits	Officers
Freshmen	88	99	92	80	91	93	98	140.7
Sophomores	104	112	104	105	114.5	105	105	141.3
Juniors	118	124	118	101	136	111	111	142
Seniors	123	127	132	111	117*	120	115	142.6

*Recognized to have been a remarkably poor senior group for this school.

have data, 75 percent of the freshmen who score less than 77 on Alpha drop out of school before senior year. A score of 77 would be reached by not more than 35 percent of the population. In Madsen's group (Ill., Ia., and Wis.), about 73 percent of such freshmen drop out. In general, only about one in four of the group below 77 remains to graduate.

TABLE 14

YEARLY INCREMENTS

	Schools						Army	
	Mich.	Ill. Wis. Iowa	N. Y. (Kan. Rept.)	Emporia, Kan.	Stanton, Va.	Minn. Survey	Recruits	Officers
Freshman to Sophomore....	16	13	12	25	21.5	12	7	0.7
Sophomore to Junior........	14	12	14	–4	21.5	6	6	0.7
Junior to Senior.	5	3	14	10	–19	9	4	0.6
Freshman to Senior........	35	28	40	31	24	27	17	2

Success in Courses in Algebra in Relation to Intelligence

Pupils who choose algebra, or a course including algebra, are in general a more intelligent group than those who do not; pupils who pass in algebra are in general a more intelligent group than those who take it and fail. The groups overlap considerably, but the one is definitely better than the other. The graphs in Figs. 5 to 10 show the distribution

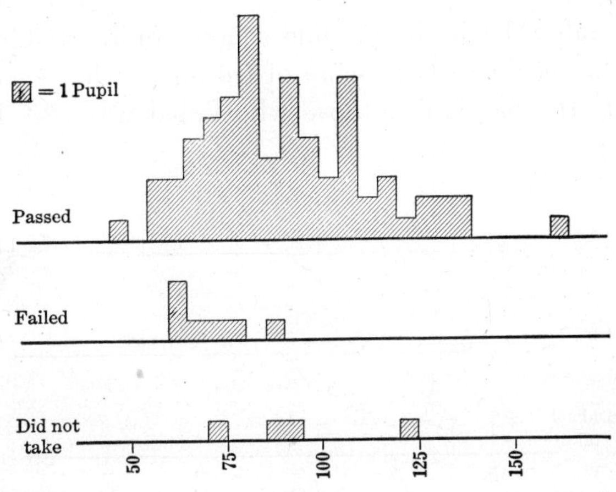

DISTRIBUTION OF ALPHA SCORES, MT. PLEASANT, MICHIGAN.

FIG. 5

of scores of pupils passing in algebra, of those who fail, and of those who do not take algebra. The contrast between the median scores of those who pass, those who fail, and those

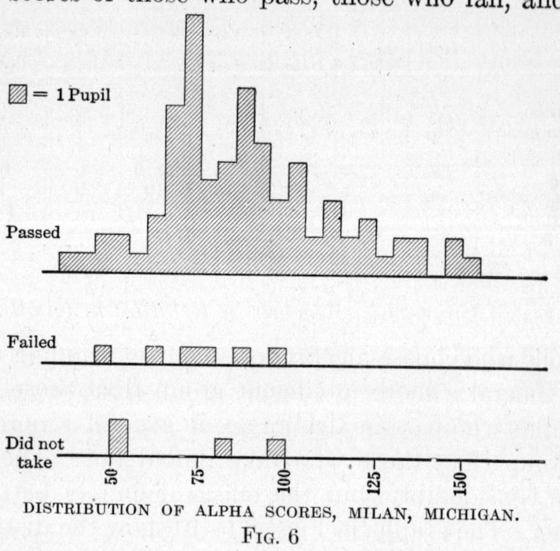

DISTRIBUTION OF ALPHA SCORES, MILAN, MICHIGAN.
FIG. 6

who do not take it, is in some schools quite striking. In Alma, the median Alpha score of freshmen who passed was 94, while the median of those who failed was 78. In Mt.

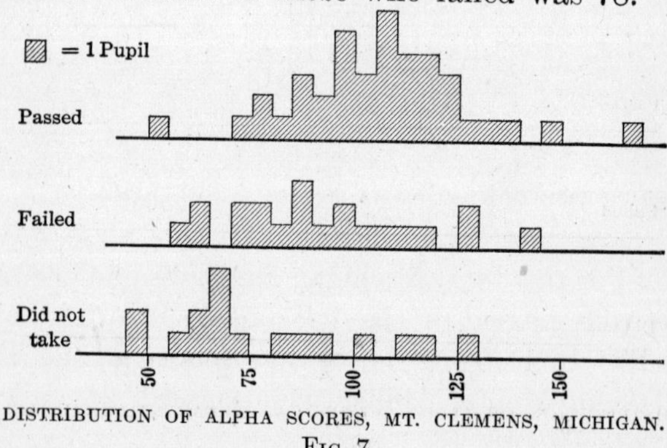

DISTRIBUTION OF ALPHA SCORES, MT. CLEMENS, MICHIGAN.
FIG. 7

Clemens the corresponding medians were, passed 107, failed 89, and, still more significant, those who did not take algebra 69. In Mt. Pleasant the median Alpha score of the freshmen who passed algebra was 89; of those who failed, 65; of those who did not take algebra. 88. In Milan, passed 86;

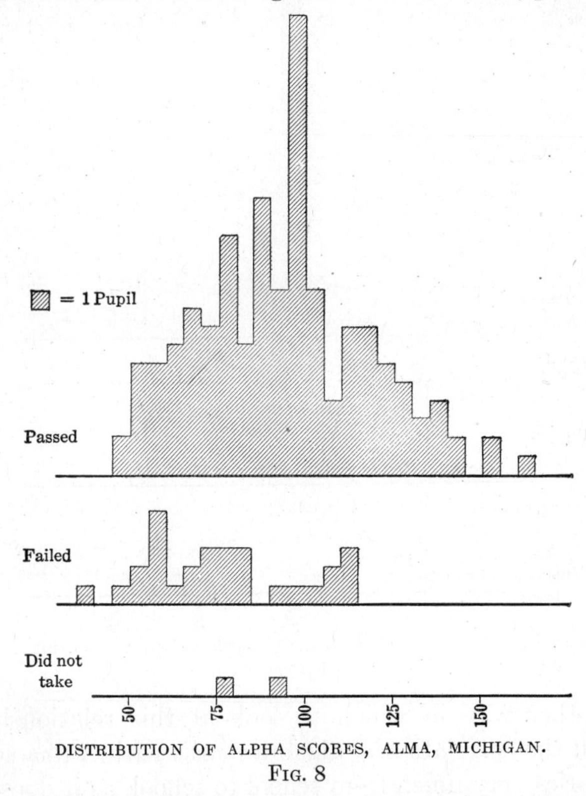

DISTRIBUTION OF ALPHA SCORES, ALMA, MICHIGAN.
FIG. 8

failed 75. For this total Michigan group we find the median of those who passed to be 92; failed, 80; did not take algebra, 75. In Detroit, the median scores on the Terman Group Test of Mental Ability were, passed, 94; failed, 85; did not take algebra, 73.

Our Michigan data have been analyzed to indicate also the expectation of failure when the Alpha score is below 55, 55 to 74, etc. Tables 15 and 16 show this, for the schools separately, and for the combined data.

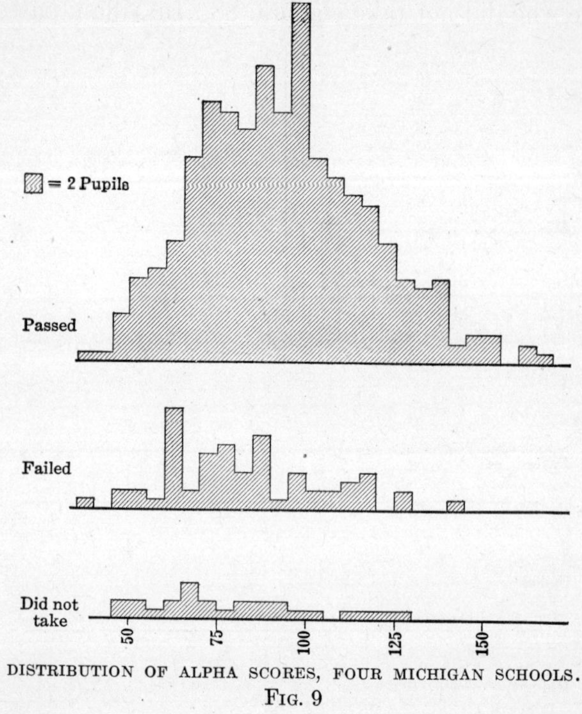

DISTRIBUTION OF ALPHA SCORES, FOUR MICHIGAN SCHOOLS.
FIG. 9

Another way in which to look at this relationship is through the correlation of algebra marks with Alpha scores. This varies very much from school to school, as it does with other school subjects, according to the content of the course, the skill of the teacher in motivating and in teaching both dull and bright pupils, in judging of their acquirements and their progress, and in assigning marks in keeping with these. Were these at their highest, and the Alpha test a perfect measure of general intelligence, the correlation would be very close, though never near 1.00, since algebra doubtless calls for a somewhat specialized type of intelligence. The

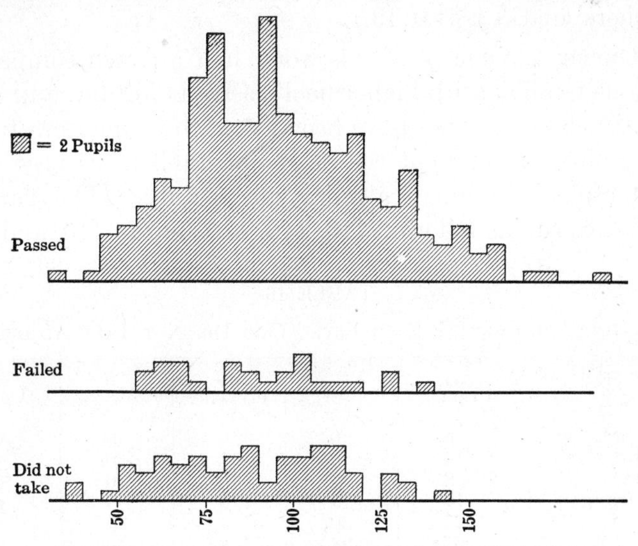

DISTRIBUTION OF TERMAN SCORES, DETROIT, MICHIGAN.
FIG. 10

TABLE 15

PERCENT FRESHMEN AT EACH LEVEL FAILING IN ALGEBRA[1]

	Mt. Clemens	Milan	Mt. Pleasant	Alma	Total
135+	25	0	0	0	5
115–134	25	0	0	12	11
95–114	23	6	0	9.6	10
75– 94	47	7	6	16.7	14
55– 74	67	7	23	21	20
Below 55	0	14	0	33	20
Median Alpha Score	89.2	75	67.5	79	79.6

[1] In Detroit, where the test used was the Terman Group Test of Mental Ability, the percents failing were: Below 50, no failures (8 cases); 50-69, 32 percent; 70-89, 26 percent; 90-109, 20 percent; 110-129, 12 percent; 130-149, 7 percent; 150 and over, 9 percent.

coefficients actually obtained from the Michigan data vary from +0.15 to +0.47, centering around +0.35. With the Terman Test, Detroit freshmen, the correlation with teachers' marks is +0.19.

During the course of this work, half a dozen competent persons familiar with high-school classes in algebra and with mental measurements have been asked to estimate in terms of intelligence quotient the degree of intelligence necessary to complete freshman algebra successfully. The various estimates run as follows: 110, 110, 110, 105 to 110, and 110

TABLE 16

PERCENT FRESHMEN AT EACH LEVEL WHO DID NOT TAKE ALGEBRA[1]

	Mt. Clemens	Milan	Mt. Pleasant	Alma	Total
135+	0	0	0	0	0
115–134	14	0	11	0	5
95–114	8	6	0	0	2
75– 94	17	3	6	3.6	6
55– 74	25	0	4	0	8
Below 55	67	22	0	0	16
Median Alpha Score	69.4	67.5	87.5	85	77.5

[1] In Detroit, where the test used was the Terman Group Test of Mental Ability, the percents not taking algebra were: Below 50, 27 percent; 50-69, 23 percent; 70-89, 9 percent; 90-109, 5 percent; 110-129, 6 percent; 130-149, 3 percent; 150 and over, 0 percent.

(this last was for a rapid advancement eighth grade class in algebra). An intelligence quotient of 110 on the Stanford Revision of the Binet Scale, at the age of fourteen years (that is to say, a mental age of 15 years, 5 months), corresponds to an Alpha score of almost 100 (98.5).

Doubtless interest and effort may compensate somewhat for ability, and doubtless special ability for mathematics may compensate for deficiencies in the abilities measured by Army Alpha. In general, however, a pupil whose first trial

Alpha ability is below 100[1] will be unable to understand the symbolism, generalizations, and proofs of algebra. He may pass the course, but he will not really have learned algebra. This would rule out more than half (56 percent) of the present first-year students. It would not be unreasonable to regard 105 or 110 as an approximate prerequisite, excluding five or ten percent more. Pupils excluded may be given a course in mathematics that is within their capacity, or may in some cases study the customary first-year course in algebra in their second or third year.

The reader will, of course, understand that the whole matter is one of steadily decreasing fitness of pupils with decrease in intelligence, not of any clear-cut distinction at 100 or 105 or 110.

[1]A first trial Alpha ability of 100 would not be accurately determined by a single trial with the Army Alpha. A number of different tests, such as the Army Alpha, the Terman Test for grades 7 to 12; the Otis Test, Part I of the Thorndike Examination for High-School Graduates, the Thorndike-McCall test in paragraph reading, or the Stanford Binet individual examination, should be used if a sufficiently accurate determination is to be made. A combined examination totalling two and a half hours may be expected to have a probable error of less than five points of the Alpha scale. Even this will mean that one pupil in three hundred will be given a score that is 20 or more Alpha points too low; so that any pupil who is thought not to have done himself justice should have the opportunity of a re-test.

Differences between communities must also be considered in the case of a general standard such as this. For example, in one state, not more than about 4 percent of all the school children—about 1 in 6 of those who enter the academic high school—are likely to profit by taking algebra, as now taught. In another state, about 24 percent, more than 1 in 3 of those who enter high school, may profit from the present algebra course. The proportion of pupils provided for in algebra courses and in academic high schools will need to be quite different in these two states.

Reproduction of Army Alpha, Form 5.

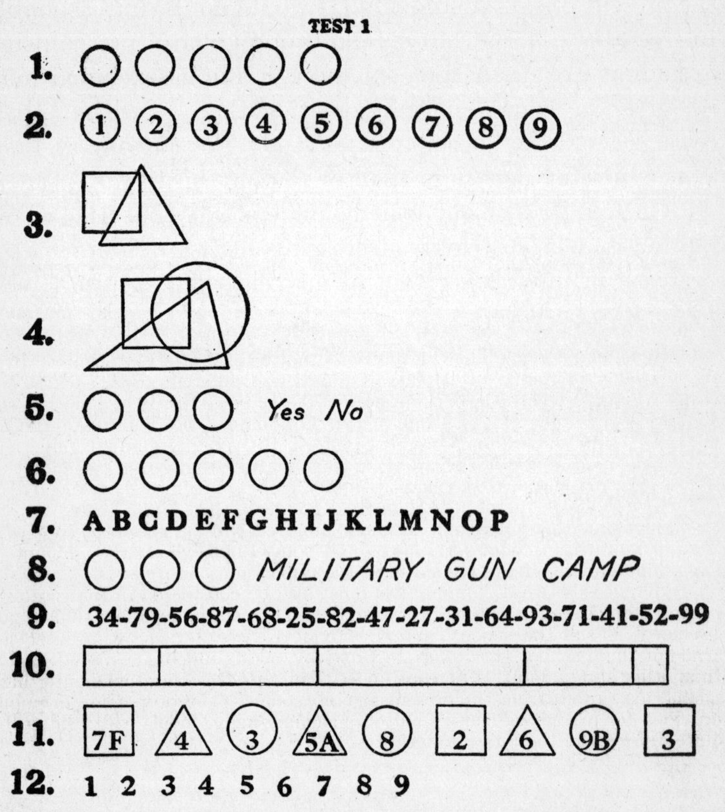

This is an oral directions test. The directions given are as follows:

TEST I, FORM V

1. "Attention! 'Attention' always means 'Pencils up.' Look at the circles at 1. When I say 'go,' but not before, make a cross in the first circle and also a figure 1 in the third circle.—Go!" (Allow not over 5 seconds.)

2. "Attention! Look at 2, where the circles have numbers in them. When I say 'go', draw a line from Circle 1 to Circle 4 that will pass *above* Circle 2 and *below* Circle 3.—Go!" (Allow not over 5 seconds.)

3. "Attention! Look at the square and triangle at 3. When I say 'go,' make a cross in the space which is in the triangle, but not in the square, and also make a figure 1 in the space which is in the triangle and in the square.—Go!" (Allow not over 10 seconds.)

4. "Attention! Look at 4. When I say 'go,' make a figure 1 in the space which is in the circle, but not in the triangle or square, and also make a figure 2 in the space which is in the triangle and circle, but not in the square.—Go!" (Allow not over 10 seconds.)

(N.B. Examiner.—In reading 5, don't pause at the word CIRCLE as if ending a sentence.)

5. "Attention! Look at 5. If a machine gun can shoot more bullets a minute than a rifle, then (when I say 'go') put a cross in the second circle; if not, draw a line *under* the word No.—Go!" (Allow not over 10 seconds.)

6. "Attention! Look at 6. When I say 'go,' put in the second circle the right answer to the question: 'How many months has a year?' In the third circle do nothing, but in the fourth circle put any number that is a wrong answer to the question that you have just answered correctly.—Go!" (Allow not over 10 seconds.)

7. "Attention! Look at 7. When I say 'go,' *cross out* the letter just before C and also draw a line *under* the second letter before H.—Go!" (Allow not over 10 seconds.)

8. "Attention! Look at 8. Notice the three circles and the three words. When I say 'go,' make in the *first* circle the first letter of the *first* word; in the *second* circle the *first* letter of the *second* word; and in the *third* circle the last letter of the third word.—Go!" (Allow not over 10 seconds.)

9. "Attention! Look at 9. When I say 'go,' *cross out* each number that is more than 20, but less than 30.—Go!" (Allow not over 15 seconds.)

10. "Attention! Look at 10. Notice that the drawing is divided into five parts. When I say 'go,' put a 3 or a 2 in each of the two largest parts and any number between 4 and 7 in the part next in size to the smallest part.—Go!" (Allow not over 15 seconds.)

11. "Attention! Look at 11. When I say 'go,' draw a line through every even number that is not in a square and also through every odd number that is in a square with a letter.—Go!" (Allow not over 25 seconds).

12. "Attention! Look at 12. If 7 is more than 5, then (when I say 'go') cross out the number 6 unless 6 is more than 8, in which case draw a line under the number 7.—Go!" (Allow not over 10 seconds.)

Test 2

Get the answers to these examples as quickly as you can.
Use the side of this page to figure on if you need to.

SAMPLES
1. How many are 5 men and 10 men?..............Answer (15
2. If you walk 4 miles an hour for 3 hours, how far do you walk?..............Answer (12

1. How many are 30 men and 7 men?..............Answer (
2. If you save $7 a month for 4 months, how much will you save? Answer (
3. If 24 men are divided into squads of 8, how many squads will there be?..............Answer (
4. Mike had 12 cigars. He bought 3 more, and then smoked 6. How many cigars did he have left?..............Answer (
5. A company advanced 5 miles and retreated 3 miles. How far was it then from its first position?..............Answer (
6. How many hours will it take a truck to go 66 miles at the rate of 6 miles an hour?..............Answer (
7. How many cigars can you buy for 50 cents at the rate of 2 for 5 cents? Answer (
8. A regiment marched 40 miles in five days. The first day they marched 9 miles, the second day 6 miles, the third 10 miles, the fourth 8 miles. How many miles did they march the last day?. Answer (
9. If you buy two packages of tobacco at 7 cents each and a pipe for 65 cents, how much change should you get from a two-dollar bill? Answer (
10. If it takes 6 men 3 days to dig a 180-foot drain, how many men are needed to dig it in half a day?..............Answer (
11. A dealer bought some mules for $800. He sold them for $1,000, making $40 on each mule. How many mules were there?.....Answer (
12. A rectangular bin holds 400 cubic feet of lime. If the bin is 10 feet long and 5 feet wide, how deep is it?..............Answer (
13. A recruit spent one-eighth of his spare change for post cards and four times as much for a box of letter paper, and then had 90 cents left. How much money did he have at first?..............Answer (
14. If $3\frac{1}{2}$ tons of coal cost $21, what will $5\frac{1}{2}$ tons cost?..........Answer (
15. A ship has provisions to last her crew of 500 men 6 months. How long would it last 1,200 men?..............Answer (
16. If a man runs a hundred yards in 10 seconds, how many feet does he run in a fifth of a second?..............Answer (
17. A U-boat makes 8 miles an hour under water and 15 miles on the surface. How long will it take to cross a 100-mile channel, if it has to go two-fifths of the way under water?..............Answer (
18. If 241 squads of men are to dig 4,097 yards of trench, how many yards must be dug by each squad?..............Answer (
19. A certain division contains 3,000 artillery, 15,000 infantry and 1,000 cavalry. If each branch is expanded proportionately until there are in all 20,900 men, how many will be added to the artillery?...Answer (
20. A commission house which had already supplied 1,897 barrels of apples to a cantonment delivered the remainder of its stock to 29 mess halls. Of this remainder each mess hall received 54 barrels. What was the total number of barrels supplied?..............Answer (

TEST 3

This is a test of common sense. Below are sixteen questions. Three answers are given to each question. You are to look at the answers carefully; then make a cross in the square before the best answer to each question, as in the sample:

SAMPLE
Why do we use stoves? Because
- ☐ they look well
- ☒ they keep us warm
- ☐ they are black

Here the second answer is the best one and is marked with a cross. Begin with No. 1 and keep on until time is called.

1. Cats are useful animals, because
- ☐ they catch mice
- ☐ they are gentle
- ☐ they are afraid of dogs

2. Why are pencils more commonly carried than fountain pens? Because
- ☐ they are brightly colored
- ☐ they are cheaper
- ☐ they are not so heavy

3. Why is leather used for shoes? Because
- ☐ it is produced in all countries
- ☐ it wears well
- ☐ it is an animal product

4. Why judge a man by what he does rather than by what he says? Because
- ☐ what a man does shows what he really is
- ☐ it is wrong to tell a lie
- ☐ a deaf man cannot hear what is said

5. If you were asked what you thought of a person whom you didn't know, what should you say?
- ☐ I will go and get acquainted
- ☐ I think he is all right
- ☐ I don't know him and can't say

6. Streets are sprinkled in summer
- ☐ to make the air cooler
- ☐ to keep automobiles from skidding
- ☐ to keep down dust

7. Why is wheat better for food than corn? Because
- ☐ it is more nutritious
- ☐ it is more expensive
- ☐ it can be ground finer

8. If a man made a million dollars, he ought to
- ☐ pay off the national debt
- ☐ contribute to various worthy charities
- ☐ give it all to some poor man

☛ Go to No. 9 above

9. Why do many persons prefer automobiles to street cars? Because
- ☐ an auto is made of higher grade materials
- ☐ an automobile is more convenient
- ☐ street cars are not as safe

10. The feathers on a bird's wings help him to fly because they
- ☐ make a wide, light surface
- ☐ keep the air off his body
- ☐ keep the wings from cooling off too fast

11. All traffic going one way keeps to the same side of the street because
- ☐ most people are right handed
- ☐ the traffic policeman insists on it
- ☐ it avoids confusion and collisions

12. Why do inventors patent their inventions? Because
- ☐ it gives them control of their inventions
- ☐ it creates a greater demand
- ☐ it is the custom to get patents

13. Freezing water bursts pipes because
- ☐ cold makes the pipes weaker
- ☐ water expands when it freezes
- ☐ the ice stops the flow of water

14. Why are high mountains covered with snow? Because
- ☐ they are near the clouds
- ☐ the sun seldom shines on them
- ☐ the air is cold there

15. If the earth were nearer the sun
- ☐ the stars would disappear
- ☐ our months would be longer
- ☐ the earth would be warmer

16. Why is it colder nearer the poles than near the equator? Because
- ☐ the poles are always farther from the sun
- ☐ the sunshine falls obliquely at the poles
- ☐ there is more ice at the poles

TEST 4

If the two words of a pair mean the same or nearly the same, draw a line under same. If they mean the opposite or nearly the opposite, draw a line under opposite. If you cannot be sure, guess. The two samples are already marked as they should be.

SAMPLES	good—bad	same—opposite	
	little—small	same—opposite	
1	wet—dry	same—opposite	1
2	in—out	same—opposite	2
3	hill—valley	same—opposite	3
4	allow—permit	same—opposite	4
5	expand—contract	same—opposite	5
6	class—group	same—opposite	6
7	former—latter	same—opposite	7
8	confess—admit	same—opposite	8
9	shy—timid	same—opposite	9
10	delicate—tender	same—opposite	10
11	extinguish—quench	same—opposite	11
12	cheerful—melancholy	same—opposite	12
13	accept—reject	same—opposite	13
14	concave—convex	same—opposite	14
15	lax—strict	same—opposite	15
16	assert—maintain	same—opposite	16
17	champion—advocate	same—opposite	17
18	adapt—conform	same—opposite	18
19	debase—exalt	same—opposite	19
20	dissension—harmony	same—opposite	20
21	repress—restrain	same—opposite	21
22	bestow—confer	same—opposite	22
23	amenable—tractable	same—opposite	23
24	avert—prevent	same—opposite	24
25	reverence—veneration	same—opposite	25
26	fallacy—verity	same—opposite	26
27	specific—general	same—opposite	27
28	pompous—ostentatious	same—opposite	28
29	accumulate—dissipate	same—opposite	29
30	apathy—indifference	same—opposite	30
31	effeminate—virile	same—opposite	31
32	peculation—embezzlement	same—opposite	32
33	benign—genial	same—opposite	33
34	acme—climax	same—opposite	34
35	largess—donation	same—opposite	35
36	innuendo—insinuation	same—opposite	36
37	vesper—matin	same—opposite	37
38	aphorism—maxim	same—opposite	38
39	abjure—renounce	same—opposite	39
40	encomium—eulogy	same—opposite	40

TEST 5

The words A EATS COW GRASS in that order are mixed up and don't make a sentence; but they would make a sentence if put in the right order: A COW EATS GRASS, and this statement is true.

Again, the words HORSES FEATHERS HAVE ALL would make a sentence if put in the order ALL HORSES HAVE FEATHERS, but this statement is false.

Below are twenty-four mixed-up sentences. Some of them are true and some are false. When I say "go," take these sentences one at a time. Think what each **would** say if the words were straightened out, but don't write them yourself. Then, if what it **would** say is true, draw a line under the word "true"; if what it **would** say is false, draw a line under the word "false." If you can not be sure, guess. The two samples are already marked as they should be. Begin with No. 1 and work right down the page until time is called.

SAMPLES	a eats cow grass	true . . false	
	horses feathers have all	true . . false	
1	lions strong are	true . . false	1
2	houses people in live	true . . false	2
3	days there in are week eight a	true . . false	3
4	leg flies one have only	true . . false	4
5	months coldest are summer the	true . . false	5
6	gotten sea water sugar is from	true . . false	6
7	honey bees flowers gather the from	true . . false	7
8	and eat good gold silver to are	true . . false	8
9	president Columbus first the was America of	true . . false	9
10	making is bread valuable wheat for	true . . false	10
11	water and made are butter from cheese	true . . false	11
12	sides every has four triangle	true . . false	12
13	every times makes mistakes person at	true . . false	13
14	many toes fingers as men as have	true . . false	14
15	not eat gunpowder to good is	true . . false	15
16	ninety canal ago built Panama years was the	true . . false	16
17	live dangerous is near a volcano to it	true . . false	17
18	clothing worthless are for and wool cotton	true . . false	18
19	as sheets are napkins used never	true . . false	19
20	people trusted intemperate be always can	true . . false	20
21	employ debaters irony never	true . . false	21
22	certain some death of mean kinds sickness	true . . false	22
23	envy bad malice traits are and	true . . false	23
24	repeated call human for courtesies associations	true . . false	24

TEST 6

SAMPLES								
	2	4	6	8	10	12	*14*	*16*
	9	8	7	6	5	4	*3*	*2*
	2	2	3	3	4	4	*5*	*5*
	1	7	2	7	3	7	*4*	*7*

Look at each row of numbers below, and on the two dotted lines write the two numbers that should come next.

3	4	5	6	7	8	....	....
10	15	20	25	30	35	....	....
8	7	6	5	4	3	....	....
3	6	9	12	15	18	....	....
5	9	13	17	21	25	....	....
8	1	6	1	4	1	....	....
27	27	23	23	19	19	....	....
1	2	4	8	16	32	....	....
8	9	12	13	16	17	....	....
9	9	7	7	5	5	....	....
19	16	14	11	9	6	....	....
2	3	5	8	12	17	....	....
11	13	12	14	13	15	....	....
29	28	26	23	19	14	....	....
18	14	17	13	16	12	....	....
81	27	9	3	1	⅓	....	....
20	17	15	14	11	9	....	....
16	17	15	18	14	19	....	....
1	4	9	16	25	36	....	....
3	6	8	16	18	36	....	....

TEST 7

SAMPLES

sky—blue : : grass— **table green warm big**
fish—swims : : man— **paper time walks girl**
day—night : : white— **red black clear pure**

In each of the lines below, the first two words are related to each other in some way. What you are to do in each line is to see what the relation is between the first two words, and underline the word in heavy type that is related in the same way to the third word. Begin with No. 1 and mark as many sets as you can before time is called.

1	gun—shoots : : knife— **run cuts hat bird**	**1**
2	ear—hear : : eye— **table hand see play**	**2**
3	dress—woman : : feathers— **bird neck feet bill**	**3**
4	handle—hammer : : knob— **key room shut door**	**4**
5	shoe—foot : : hat— **coat nose head collar**	**5**
6	water—drink : : bread— **cake coffee eat pie**	**6**
7	food—man : : gasoline— **gas oil automobile spark**	**7**
8	eat—fat : : starve— **thin food bread thirsty**	**8**
9	man—home : : bird— **fly insect worm nest**	**9**
10	go—come : : sell— **leave buy money papers**	**10**
11	peninsula—land : : bay— **boats pay ocean Massachusetts**	**11**
12	hour—minute : : minute— **man week second short**	**12**
13	abide—depart : : stay— **over home play leave**	**13**
14	January—February : : June— **July May month year**	**14**
15	bold—timid : : advance— **proceed retreat campaign soldier**	**15**
16	above—below : : top— **spin bottom surface side**	**16**
17	lion—animal : : rose— **smell leaf plant thorn**	**17**
18	tiger—carnivorous : : horse— **cow pony buggy herbivorous**	**18**
19	sailor—navy : : soldier— **gun cap hill army**	**19**
20	picture—see : : sound— **noise music hear bark**	**20**
21	success—joy : : failure— **sadness success fail work**	**21**
22	hope—despair : : happiness— **frolic fun joy sadness**	**22**
23	pretty—ugly : : attract— **fine repel nice draw**	**23**
24	pupil—teacher : : child— **parent doll youngster obey**	**24**
25	city—mayor : : army— **navy soldier general private**	**25**
26	establish—begin : : abolish— **slavery wrong abolition end**	**26**
27	December—January : : last— **least worst month first**	**27**
28	giant—dwarf : : large— **big monster queer small**	**28**
29	engine—caboose : : beginning— **commence cabin end train**	**29**
30	dismal—cheerful : : dark— **sad stars night bright**	**30**
31	quarrel—enemy : : agree— **friend disagree agreeable foe**	**31**
32	razor—sharp : : hoe— **bury dull cuts tree**	**32**
33	winter—summer : : cold— **freeze warm wet January**	**33**
34	rudder—ship : : tail— **sail bird dog cat**	**34**
35	granary—wheat : : library— **desk books paper librarian**	**35**
36	tolerate—pain : : welcome— **pleasure unwelcome friends give**	**36**
37	sand—glass : : clay— **stone hay bricks dirt**	**37**
38	moon—earth : : earth— **ground Mars sun sky**	**38**
39	tears—sorrow : : laughter— **joy smile girls grin**	**39**
40	cold—ice : : heat— **lightning warm steam coat**	**40**

TEST 8

Notice the sample sentence:

People **hear** with the **eyes** **ears** **nose** **mouth**

The correct word is **ears,** because it makes the truest sentence.

In each of the sentences below you have four choices for the last word. Only one of them is correct. In each sentence draw a line under the one of these four words which makes the truest sentence. If you can not be sure, guess. The two samples are already marked as they should be.

SAMPLES: People **hear** with the **eyes** **ears** **nose** **mouth**
France is in **Europe** **Asia** **Africa** **Australia**

1 America was discovered by Drake Hudson Columbus Balboa 1
2 Pinochle is played with rackets cards pins dice 2
3 The most prominent industry of Detroit is automobiles brewing flour packing 3
4 The Wyandotte is a kind of horse fowl cattle granite 4
5 The U. S. School for Army Officers is at Annapolis West Point New Haven Ithaca 5
6 Food products are made by Smith & Wesson Swift & Co. W. L. Douglas B. T. Babbitt 6
7 Bud Fisher is famous as an actor author baseball player comic artist 7
8 The Guernsey is a kind of horse goat sheep cow 8
9 Marguerite Clark is known as a suffragist singer movie actress writer 9
10 "Hasn't scratched yet" is used in advertising a duster flour brush cleanser 10
11 Salsify is a kind of snake fish lizard vegetable 11
12 Coral is obtained from mines elephants oysters reefs 12
13 Rosa Bonheur is famous as a poet painter composer sculptor 13
14 The tuna is a kind of fish bird reptile insect 14
15 Emeralds are usually red blue green yellow 15
16 Maize is a kind of corn hay oats rice 16
17 Nabisco is a patent medicine disinfectant food product tooth paste 17
18 Velvet Joe appears in advertisements of tooth powder dry goods tobacco soap 18
19 Cypress is a kind of machine food tree fabric 19
20 Bombay is a city in China Egypt India Japan 20
21 The dictaphone is a kind of typewriter multigraph phonograph adding machine 21
22 The pancreas is in the abdomen head shoulder neck 22
23 Cheviot is the name of a fabric drink dance food 23
24 Larceny is a term used in medicine theology law pedagogy 24
25 The Battle of Gettysburg was fought in 1863 1813 1778 1812 25
26 The bassoon is used in music stenography book-binding lithography 26
27 Turpentine comes from petroleum ore hides trees 27
28 The number of a Zulu's legs is two four six eight 28
29 The scimitar is a kind of musket cannon pistol sword 29
30 The Knight engine is used in the Packard Lozier Stearns Pierce Arrow 30
31 The author of "The Raven" is Stevenson Kipling Hawthorne Poe 31
32 Spare is a term used in bowling football tennis hockey 32
33 A six-sided figure is called a scholium parallelogram hexagon trapezium 33
34 Isaac Pitman was most famous in physics shorthand railroading electricity 34
35 The ampere is used in measuring wind power electricity water power rainfall 35
36 The Overland car is made in Buffalo Detroit Flint Toledo 36
37 Mauve is the name of a drink color fabric food 37
38 The stanchion is used in fishing hunting farming motoring 38
39 Mica is a vegetable mineral gas liquid 39
40 Scrooge appears in Vanity Fair The Christmas Carol Romola Henry IV 40

CHAPTER II

The Uses of Algebra

Few if any teachers of mathematics or students of education in general would value algebraic abilities in exact correspondence to the extent or intensity of their direct use in studies or productive labor. Their indirect or disciplinary value would probably always be given some weight. On the other hand all would attach some weight to utility, even if narrowly conceived.

To measure the utility of each ability acquired in elementary algebra to each of the persons who do or might study it would be a worthy enterprise, though its findings would become obsolete and misleading if uncorrected year by year (percents were once as rare in the shop and on the street as fractional exponents are now). It would have to be a very extended and intricate enterprise, and we have not undertaken it. We have limited ourselves to two aspects of it, the uses of algebra as a preparation for work in the sciences in college or professional school, and the uses of algebra in general reading and study. A consideration of these two sample topics will satisfy the main purposes of such a study fairly well. For algebra as a tool today is chiefly a tool for scientific work, for thinking about general relations. Only a few of its abilities are used by workers in agriculture, mining, manufacturing, transportation, trade, the ministry, teaching, medicine, and the law, except as they become students of the sciences.

It may be objected that the consideration of these two topics will give an exaggerated impression of the general need for algebra by first-year pupils, only a minority of whom will go to college or treat their problems by the methods of science; and also that, for similar reasons, it will give a distorted impression of the relative values of the different topics within algebra. There is some justice in these objections, and the reader should not transfer the absolute or relative amounts of utility which we have to report to his first-year class without corrections and allowances.

THE USES OF ALGEBRAIC ABILITIES AS A PREPARATION FOR WORK IN THE SCIENCES IN COLLEGES AND PROFESSIONAL SCHOOLS

We may begin by examining the results of a census of opinions made by the National Committee on Mathematical Requirements. "A number of college teachers, prominent in their respective fields, were asked to assign to each of the topics in the following table [Table 17] its value as preparation for the elementary courses in their respective subjects, according to the following scale: E essential, C of considerable value, S of some value, O of little or no value. Table 17[1] gives a summary of the results for the 'Physical Sciences' (astronomy, physics, chemistry) and for the 'Social Sciences' (history, economics, sociology, political science). The last column in each set gives the number of replies received; the numbers in the other column give the number (reduced to percentages for greater convenience of comparison) assigning

[1] I have added to these data three columns expressing roughly the general estimates revealed by the votes, counting each percent of E as 1, of C as $\frac{2}{3}$, of S as $\frac{1}{3}$, and of O as 0. Thus *Negative numbers* is rated as 86 $(79\times1)+(5\times\frac{2}{3})+(10\times\frac{1}{3})$, by college teachers of the physical sciences, as 64 by college teachers of the social sciences, and as 75 by both groups (equal weight being given to each).

E, C, S, O respectively to the various topics." [1921, p. 225.]

The first thing to be noted about these opinions is the great variation with respect to almost every topic. Some prominent college teachers can be found to rank equations of the third degree, imaginary numbers, conic sections, and polar coördinates as essential as preparation for the elementary courses in their subjects. Some can be found to rate as of little or no value negative numbers, simple formulas, ratio, linear equations in one unknown—in fact, any topic in the list.

Part of this variation is due probably to different conceptions of the algebraic topic in question; the teachers who regard knowledge of negative numbers as of little or no value may have thought of the more elaborate and recondite features of the topic such as that "a minus times a minus gives a plus," whereas those who regard it as essential may have been thinking of reading a thermometer. We shall see later that science uses very widely certain particular mathematical abilities within some of these topics but uses other abilities within the same topic very rarely. For example, the student of certain physical sciences may meet the standard form number very often, but very rarely have anything else to do with negative exponents.

Even if teachers of only one subject, say chemistry, voted and if the exact nature of each of the topics in the list were defined uniformly for all of them, much of the variation would still probably remain. An inspection of standard college textbooks for first-year courses in any science will show a variety in the algebraic abilities that are assumed and in the degree of each that is called for.

TABLE 17

Value of Topics as Preparation for Elementary College Courses

In the headings of the following table, E = essential, C = of considerable value, S = of some value, O = of little or no value, N = number of replies received. The figures in the first four columns of each group are percents of the number of replies received.

	Physical Sciences					Social Sciences					P. S.	S. S.	Both.
	E	C	S	O	N	E	C	S	O	N			
Negative numbers—their meaning and use	79	5	10	5	39	45	17	22	17	18	86	64	75
Imaginary numbers—their meaning and use	23	21	25	31	39	13	13	37	37	16	45	31	38
Simple formulas—their meaning and use	93	5	2		41	47	26	21	5	19	97	71	84
Graphic representation of statistical data	57	25	15	3	40	57	24	14	5	21	79	78	79
Graphs (mathematical and empirical):													
(*a*) as a method of representing dependence	62	16	22		37	15	54	15	15	13	80	56	68
(*b*) as a method of solving problems	46	20	28	6	25	18	18	46	18	11	69	45	57
The linear function, $y = mx + b$	78	14	8		37	29	29	14	29	14	90	53	72
The quadratic function, $y = ax^2 + bx + c$	59	21	17	3	34	8	8	33	50	12	79	24	52
Equations: Problems leading to—													
Linear equations in one unknown	98	2			41	40	7	20	33	15	99	45	72
Quadratic equations in one unknown	78	15	5	2	40	31	8	8	54	13	90	39	70
Simultaneous linear equations in two unknowns	71	24	3	3	38	33	8		58	12	89	38	64
Simultaneous linear equations in more than two unknowns	43	29	23	6	35	8	8	17	67	12	70	19	45
One quadratic and one linear equation in two unknowns	40	24	27	9	33		9	9	82	11	65	9	37
Two quadratic equations in two unknowns	31	19	28	22	32		9		91	11	53	6	30
Equations of higher degree than the second	10	32	32	26	31			9	91	11	42	3	23
Literal equations (other than formulas)	43	18	32	7	28		10	40	50	10	29	20	25
Ratio and proportion	84	8	3	5	39	37	26	32	5	19	90	65	78
Variation	50	13	20	17	30	17	33	25	25	12	65	47	56
Numerical computation:													
With approximate data—rational use of significant figures	61	36		3	39	40	27	20	13	12	85	65	75
Short-cut methods	27	38	24	10	37	29	35	23	12	17	60	57	59
Use of logarithms	62	29	7	2	42	12	29	29	29	17	84	41	63
Use of other tables to facilitate computation	24	45	26	5	38	18	29	41	12	17	63	47	55
Use of slide rule	24	39	26	12	38	11	39	28	22	18	63	46	55
Theory of exponents	36	31	25	8	36		21	21	57	14	65	25	43
Theory of logarithms	34	26	21	18	38	7	13	20	60	15	58	22	40
Arithmetic progression	16	32	38	13	37	23	29	12	35	17	50	46	48
Geometric progression	19	27	40	14	37	23	25	18	35	17	50	46	48
Binomial theorem	35	32	19	13	37	13	20	27	40	15	63	35	49
Probability	9	32	41	19	32	20	35	35	10	20	44	55	50
Statistics:													
Meaning and use of elementary concepts	23	28	31	17	29	55	36	5	5	22	53	81	67
Frequency distributions and frequency curves	15	19	35	32	26	47	33	10	10	21	59	72	66
Correlation	11	18	39	32	28	33	47	14	5	21	36	69	53
Numerical Trigonometry:													
Use of sine, cosine, and tangent in the solution of simple problems involving right triangles	68	21	3	8	38			25	75	12	83	8	46
Demonstrative geometry	68	15	12	6	34		21	43	36	14	82	28	55

TABLE 17—*Continued*

	Physical Sciences					Social Sciences					P. S.	S. S.	Both.
	E	C	S	O	N	E	C	S	O	N			
Plane trigonometry (usual course)	57	27	11	5	37	8	23	31	38	13	79	34	57
Analytic geometry:													
Fundamental conceptions and methods in the plane	32	45	19	3	31		15	38	46	13	68	23	46
Systematic treatment of—													
Straight line	34	37	20	9	35	9	9	18	64	11	65	21	43
Circle	29	43	20	9	35		18	9	73	11	64	15	40
Conic sections	18	41	26	15	34		9	18	73	11	54	12	33
Polar coördinates	18	26	41	15	34			18	82	11	49	6	28
Empirical curves and fitting curves to observations	12	38	38	12	34	8		25	67	12	50	16	33

The great variation amongst the ratings of each topic is consistent with a general placing of it as high or low for usefulness. These placements may be considered as absolute magnitudes meaning how much more than zero usefulness knowledge of the topic has; or as relative magnitudes, putting each topic in an order among the rest for its comparative usefulness. The National Committee wisely utilizes these opinions chiefly for the latter purpose and so may we.[1]

If we give a credit of 1 for each percent of "Essential" votes, a credit of $\frac{2}{3}$ for each percent of a "Considerable" vote, and a credit of $\frac{1}{3}$ for each percent of "Some" votes,

[1] Such ratings are the product partly of impartial observations and partly of certain loyalties and convictions. For example, the consensus of college teachers of the physical sciences has a three-fourth vote that problems leading to quadratic equations in one unknown are essential for the elementary courses. Yet to the best of his knowledge and belief the writer, who took such courses in college in physics, chemistry, geology, and astronomy, never solved a quadratic equation in any of these courses or met any problem for whose solution a quadratic equation was suggested as desirable. One out of three of these teachers votes that ability with problems leading to simultaneous quadratics is essential; but apart from determining from two of its points the constants of a curve known to be of quadratic form, which surely is not a task for an elementary course, there are hardly any such problems in science. The makers of textbooks have searched far and wide for problems on which to apply the technique of simultaneous quadratics and still have to remain content with such fabrications as:

a. Find the number of two digits in which the units' digit exceeds the tens' digit by 2, and such that the product of the number and its tens' digit is 105.

b. If the length of a rectangular field be increased by 2 rods and its width be diminished by 5 rods, its area becomes 24 square rods; if its length be diminished by 4 rods and its width be increased by 3 rods, its area becomes 60 square rods. Find its length and width.

we have 100 as a maximum credit and 0 as a minimum credit. These credits, for the physical sciences vote, for the social sciences vote, and for the average of the two are given in Table 18, the first three columns in order. In this table the topics are arranged in order by the average of the two votes. Some such method of weighting both votes seems desirable. Since it is probable that relative utility for the study of the physical sciences is more important in the case of mathematics than relative utility for the study of the social sciences, we have added a fourth column in which the weights are as 2 to 1.

The National Committee Report makes its list of "Topics in order of Value as Preparation for Elementary College Courses" in a different way from this; namely, according to the magnitude of the percentage of E votes in the case of the physical sciences (or in the case of the social sciences if the percent of E's is higher for them, which it is in only five cases). The sum of the "essentials" and "considerables" percentages is also given for the group where it is the higher (the physical sciences group, in all save seven cases). These ratings are repeated in columns 5 and 6 of our Table 18.

In examining Table 18, the reader may be helped by remembering that the median rating in column 1 is 65; in column 2 it is 40; in column 3 it is 51. The variability is somewhat greater in column 2 than in column 1, but no serious harm will be done by putting into comparison the deviations from 65, 40 and 51. For example, in the case of simple formulas the votes for the physical sciences, crediting 97, give +32; those for the social sciences, crediting 71, give +31; showing a close agreement. In the case of graphic representations the 79 and 78 signify +14 and +38, a considerable disagreement.

TABLE 18

COMPARISON OF TOPICS ACCORDING TO VARIOUS SYSTEMS OF CREDITS

I. Votes of teachers of physical sciences, E = 1, C = $\frac{2}{3}$, S = $\frac{1}{3}$, O = 0.
II. Votes of teachers of social sciences, E = 1, C = $\frac{2}{3}$, S = $\frac{1}{3}$, O = 0.
III. I and II combined with equal weight.
IV. I and II combined with I given double weight.
V. Highest percentage of E's, using either science.
VI. Highest percentage of E's+C's, using either science.
VII. Deviations of the entries under I from their median.
VIII. Deviations of the entries under II from their median.

	Physical Sciences. I	Social Sciences. II	(I+II)/2. III	(2 I+II)/3. IV	Highest E Vote. V	Highest (E+C). VI	Deviations of I. VII	Deviations of II. VIII
Simple formulas—their meaning and use	97	71	84	88	93	98	1.8	1.4
Graphic representations of statistical data	79	78	79	79	57	82	.8	1.7
Ratio and proportion	90	65	78	82	84	92	1.4	1.1
Negative numbers—their meaning and use	86	64	75	79	79	84	1.2	1.1
Numerical computation: With approximate data—rational use of significant figures	85	65	75	78	61	97	1.1	1.1
The linear function, $y=mx+b$	90	53	72	78	78	92	1.4	.6
Problems leading to linear equations in one unknown	99	45	72	81	98	100	2.0	.2
Problems leading to quadratic equations in one unknown	90	39	70	73	78	93	1.4	–.04
Graphs: as a method of representing dependence	80	56	68	72	62	78	.8	.7
Statistics: meaning and use of elementary concepts	53	81	67	62	55	91	–.7	1.8
Statistics: frequency distribution and frequency curves	59	72	66	63	47	80	–.4	1.4
Problems leading to simultaneous linear equations in two unknowns	89	38	64	72	71	95	1.4	–.1
Numerical computation: use of logarithms	84	41	63	70	62	91	1.1	.04
Numerical computation: short-cut methods	60	57	59	59	29	65	–.3	.7
Graphs: as a method of solving problems	69	45	57	61	46	66	.2	.2
Plane trigonometry	79	34	57	64	57	84	.8	–.3
Variation	65	47	56	59	50	63	0.0	.3

TABLE 18—*Continued*

	Physical Sciences.	Social Sciences.	(I+II)/2.	(2I+II)/3.	Highest E Vote.	Highest (E+C).	Deviations of I.	Deviations of II.
	I	II	III	IV	V	VI	VII	VIII
Use of slide rule	63	46	55	57	24	63	−.1	.3
Numerical computation—use of tables other than logarithms	63	47	55	58	24	69	−.1	.3
Demonstrative geometry	82	28	55	64	68	83	1.0	−.5
Statistics: correlation	36	69	53	47	33	80	−1.6	1.3
The quadratic function, $y=ax^2+bx+c$	79	24	52	61	59	80	.8	−.7
Probability	44	55	50	47	20	55	−1.2	.7
Binominal theorem	63	35	49	54	35	67	−.1	−.2
Arithmetic progression	50	46	48	4	23	52	−.8	.3
Geometric progression	50	46	48	49	23	48	−.8	.3
Numerical trigonometry	83	8	46	58	68	89	1.0	−1.4
Analytic Geometry: fundamental conceptions and methods in the plane	68	23	46	53	32	77	.2	−.7
Problems leading to simultaneous linear equations in more than two unknowns	70	19	45	53	43	72	.3	−.9
Theory of exponents	65	21	43	50	36	67	0.0	−.8
Analytic geometry: systematic treatment of straight lines	65	21	43	50	34	71	0.0	−.8
Analytic geometry: systematic treatment of circle	64	15	40	48	29	72	−.1	−1.1
Theory of logarithms	58	22	40	46	34	60	−.4	−.8
Imaginary numbers—their meaning and use	45	34	40	41	23	44	−1.1	−.3
Problems leading to one quadratic and one linear equation in two unknowns	65	9	37	46	40	64	0.0	−1.4
Analytic geometry: systematic treatment of conic sections	54	12	33	40	18	59	−.6	−1.2
Empirical curves and fitting curves to observations	50	16	33	39	12	50	−.8	−1.0
Problems leading to two quadratic equations in two unknowns	53	6	30	37	31	50	−.7	−1.4
Polar coördinates	49	6	28	35	18	44	−.9	−1.4
Literal equations other than formulas	29	20	25	26	43	61	−2.0	−.9
Problems leading to equations of higher degree than the second	42	3	23	15	10	42	−1.3	−1.6

For the sake of the reader who cares for more adequate commensurability, columns 7 and 8 give the credit numbers of columns 1 and 2, each as a deviation from the median of its column divided by the variability (mean square deviation) of the column.

The most notable features of Table 18 are the much higher relative positions awarded to the algebra of statistics, to graphic presentations generally, and to numerical computation with many-place numbers, than have been given to them in the teaching of the past (or of the present, save in a few schools). Almost equally notable is the very low opinion of the utility of literal equations other than formulas. The more detailed matters of interest in the table may be left to the reader's examination of it.

So far we have paid no heed to an important feature of the National Committee's data — the individuals who failed to reply to the circular at all, and those who omitted certain topics from their ratings. The forty-two[1] who replied for the physical sciences were from a total of not over fifty teachers of the physical sciences to whom the circular was sent; the twenty-two [1] who replied for the social sciences were from a total of not over twenty-five teachers of the social sciences to whom the circular was sent.[2] Of the forty-two, as few as twenty-six replied to one of the questions; of the twenty-two, as few as ten. This is exclusive of "Graphic Representations — construction and interpretation," with only fifteen and eight ratings, the small number of which seems to be due to misunderstanding.[3] It is too hazardous to estimate what those who did not reply at all would have

[1]Assuming that the maximum number of ratings for any topic equals the total number replying.

[2]I am indebted for these facts to a letter from Professor Young of the National Committee.

[3]This item was not reported in the Committee's printed report or in our tables and may be dropped from consideration.

said, had they replied, beyond the probability that they would in general have given lower ratings. Failure to reply to such circulars is due in part to lack of interest, which would be correlated in this case with the attachment of less importance to algebra and mathematics in general. It is due in part to general unreadiness to spend time on work for others, and to accidental circumstances.

Since the Committee succeeded in getting so high a percentage of replies it is not necessary to make any more specific allowance for the non-answerers. Any reasonable allowance would affect the *relative* ranking very little.

The failure to rate certain particular topics is also due to the operation of many forces, but we can learn something about the magnitude of these from internal evidence. Mere unreadiness to spend time is probably not in the present instance a large cause, since the items late in the list are sometimes very well reported on. "Use of logarithms," over half way down the list, is reported on by forty-two for the physical sciences; and "plane trigonometry," almost at the end, by thirty-seven. "Statistics," three-quarters down the list, is reported on by a maximum of teachers of the social sciences.

Lack of interest is a cause; twenty-one or twenty-two of the teachers of social sciences report on the four topics about statistics, which are among the most neglected by the teachers of the physical sciences. Only eleven to fifteen teachers of the social sciences rate the various topics under "Problems leading to Equations." Lack of knowledge of the meanings of the words is probably a cause. Assuming forty-two as the number of teachers of the physical sciences replying, nine topics out of ten are rated by nearly nine-tenths of them. Only one topic out of ten is rated by nine-tenths of the teachers of the social sciences. This difference seems too great to be explained by lack of interest alone.

Relative lack of interest in a topic means that the person in question would, if he had rated it, rate it relatively low. Lack of knowledge of the meaning of a topic means that the person in question ought to have rated it as of little or no value. If, for example, he did not even know what polar coördinates were, they surely could not be essential for learning the elementary course in his subject!

We have evidence that for these reasons or others, failure to rate a topic means that on the average low estimates of the topic would have been made, if the persons had been induced to rate. This evidence is the correspondence between "number failing to rate" and "lowness of rating by those who did rate." This correspondence is in terms of coefficients of correspondence or correlation, +.51 for the physical sciences and +.84 for the social sciences.

We might roughly compensate for this error by some such rule as: "Count half of the non-raters of a topic as rating it *little or no*. Count the other half as assigning ratings of E, C, S, and O in the same proportions as those who did rate it."

The effect of applying any such rule would be to lower the absolute values of topics in proportion to the percentage of non-raters, and so to increase the variation amongst topics. The relative values would not be much disturbed, because in the case of the physical sciences the proportion of non-raters is not large, while in the case of the social sciences the correlation between "proportion rating" and the present *order* of values is so close. The effect upon the values for physical and social sciences combined would be to give more weight to the latter, since the increase in variation would be greater than in the values for the physical sciences.

Since we do not know just what rule to use in treating the non-raters of certain topics, much less what rule to use in treating those who rated none, it seems best not to try to estimate what a true complete vote would give. It is, however, worth while to note that we gave as high weight as we did to the votes of the teachers of the social sciences, taken at their face value, because a true complete vote would surely give them a greater relative weight than the vote as taken gave them.

The study by the National Committee has been repeated in substance by Koch and Schlauch [1] [as yet unpublished] and they have kindly allowed us to study their report.

Their ratings were those made by 77 out of 500 to whom the task was proposed. These are unspecified as to the subjects which they teach, but were, we judge, predominantly teachers of the physical sciences. The correlation of their ratings with our weighted total for the National Committee ratings is .60; if our weighted total for teachers of the physical sciences is taken, the correlation with the Koch and Schlauch results is .83. In view of these facts, it seems best not to try to combine their results with those of the National Committee, but to wait for them to present and discuss such matters as they think suitable. If we did combine their results according to our own judgment, it would be with the ratings of teachers of physical sciences; and such a combination would produce only very slight alterations in the general account given here.

[1] "A College Professor's Questionnaire was sent to 500 professors or their assistants. There were about 125 returns of which 77 were retained for tabulation." It included the following: "Indicate a numeric order of the following topics by assigning number (1) to the most important topic, (2) to the second topic in importance, and so on through the list. Place an (e) against those topics which you consider essential and an (x) against those topics which you would like to see omitted from the secondary school syllabus.

A. Linear equations in one unknown.
B. Simple formulas — their meaning and use.
C. Ratio and proportion.
D. Negative numbers — their meaning and use.
E. Geometric reasoning — ability to give logical demonstrations," and 40 more topics.

Attention was called in the discussion of the National Committee ratings to the fact that the rating of a topic as "essential," "of much use," "of some use," and "of little or no use," depends upon some sort of combination of judgment of the different abilities included in the topic. A rating might be "of some use" for the topic as a whole and be "essential" for some features of it, but "of little or no use" for other features of it.

In many cases of uncertainty about what to include in the elementary course in algebra, the uncertainty is precisely about how far a certain topic shall be carried, or how elaborate tasks of a certain sort shall be undertaken, or whether this or that application shall be made. A census of opinions concerning detailed features within a topic is therefore desirable as a supplement to such a census as we have been studying. It seems desirable also to have these judgments in terms of actual samples of abilities or tasks.

To this end the fifty-six tasks of Table 19 were sent to certain college teachers of science, with the following statement:

> In connection with an investigation of the psychology and teaching of algebra, I need to know how complex and difficult tasks a student entering college should have mastery of in order to make such use of algebra as is desirable in the elementary course in .... in college. I hope you can be so good as to take a few minutes to inspect the enclosed lists and check each task as follows:
>
> If the ability to handle such a task is essential for your elementary course in ...., write *E* before it.
>
> If though not essential, it is of considerable value, write *C* before it.
>
> If it is of less value, but still of some value, write *S* before it.
>
> If it is of no value, write *O* before it.

TABLE 19

Average Ratings of Fifty-Six Computational tasks by College Teachers of (A) Physics, (B) Biology, Botany and Agriculture, (C) Economics and Statistics, and (D) Anthropology, Psychology and Sociology. *Essential* is counted as 3.0; *of Considerable Value*, as 2.0; *of Less but of Some Value*, as 1.0; *of No Value*, as 0.

LINEAR EQUATIONS	A 17 PHYSICS	B 6 BIOLOGY, BOTANY, AGRICULTURE	C 6 ECONOMICS STATISTICS	D 8 SOCIOLOGY, PSYCHOLOGY, ANTHROPOLOGY	AVERAGE OF B, C AND D
1. $3\,n=12$	3.0	1.5	2.5	1.0	1.7
2. $2n=\frac{k}{5}$ when $k=30$	3.0	1.8	2.5	0.9	1.7
3. $3\,n+2=5\,n+12$	3.0	1.2	2.5	0.6	1.4
4. $8\,n+(n+3)=2\,n+5$	3.0	1.2	2.3	0.6	1.0
5. $\frac{2}{3}n-10+3n=-5+\frac{7}{8}n$	3.0	1.2	2.2	0.6	1.3
6. $\frac{2}{3}n+45=\frac{1}{3}n+55$	3.0	1.0	2.2	0.6	1.3
7. $5-3n=6-(n+15)$	2.9	1.0	2.2	0.6	1.3
8. $\frac{n-1}{n-2}-2-\frac{n}{1-n}=0$	2.9	0.7	1.5	0.6	0.9
9. $3.4n-.17\,(n-2)=51\left(\frac{n}{5}-3\right)$	2.9	0.7	1.7	0.5	1.0
10. $\frac{4\,(n+3)}{9}=\frac{8n+37}{18}-\frac{7n-29}{5n-12}$	2.5	0.7	1.3	0.5	0.8
11. $\frac{1}{3}\,(n-\frac{1}{3}\,(n+1)\,)+\frac{1}{4}\,(4n-5)=\frac{27}{4}-\frac{1}{3}\,(n+1)$	2.5	0.5	1.0	0.5	0.7
12. $\frac{5n-.4}{.3}+\frac{1.5n-.05}{2}=\frac{4.15-8n}{1.2}$	2.5	0.5	1.0	0.5	0.7
SIMULTANEOUS LINEAR EQUATIONS					
13. $y=mx+b$. Find m and b if $y=13$ when $x=5$ and $y=27$ when $x=12$	2.9	1.5	2.2	0.6	1.4

TABLE 19—Continued

SIMULTANEOUS LINEAR EQUATIONS Continued	A 17 Physics	B 6 Biology, Botany, Agriculture	C 6 Economics Statistics	D 8 Sociology, Psychology, Anthropology	Average of B, C and D
14. $y=mx+b$. Write the equation if $y=10.5$ when $x=3.5$ and $y=20.1$ when $x=6.7$	2.5	1.0	2.2	0.6	1.3
15. $y=mx+b$. Write the equation if $y=1.6$ when $x=0.7$ and $y=12.0$ when $x=8.7$	2.4	0.8	2.2	0.6	1.2
16. $x+y=8$ $x-y=2$. Solve for x and y	3.0	1.0	1.8	0.6	1.1
17. $2x+5y=15$ $3x-4y=11$ Solve for x and y.	3.0	1.0	1.8	0.6	1.1
18. $\frac{1}{x+y}+\frac{1}{x-y}=0$ $\frac{x-3y}{y-2x}=2$ Solve for x and y	2.6	0.7	1.5	0.5	0.9
19. $1.6x-2.05y=.39$ $5.2x+4.1y=3.42$ Solve for x and y	2.8	0.7	1.3	0.5	0.8
20. $3x+2y=12\frac{1}{4}$ $\frac{5}{x+3}:\frac{4}{3y+\frac{1}{2}}=3:\frac{2}{5}$ Solve for x and y	1.9	0.7	0.8	0.4	0.6
21. $\frac{5}{3x}+\frac{2}{5y}=7$ $\frac{7}{6x}-\frac{1}{10y}=3$ Solve for x and y	2.1	0.7	0.8	0.4	0.6

TABLE 19—Continued

QUADRATIC EQUATIONS	A 17 Physics	B 6 Biology, Botany, Agriculture	C 6 Economics Statistics	D 8 Sociology, Psychology, Anthropology	Average of B, C and D
22. $\frac{7x^2}{3}=m$. Find x when $m=21$	3.0	0.8	2.2	1.0	1.3
23. $\frac{4x^2}{5}=k$. Find x when $k=17$	2.9	0.8	2.2	0.8	1.3
24. $x^2+10x=24$	2.9	0.7	2.2	0.6	1.2
25. $x^2+7x+11=0$	2.9	0.7	2.2	0.6	1.2
26. $8x^2+2x+13=0$	2.9	0.7	2.2	0.6	1.2
27. $3x^2+2x=11$	2.9	0.7	1.8	0.6	1.0
28. $5x^2+16x-12k+2=0$. Find x when k is 3	2.6	0.5	1.3	0.5	0.8
29. $9x^2+(1+k)x+4=0$. For what values of k are the values of x equal?	2.2	0.5	1.3	0.5	0.8
30. $\frac{2}{3x-3}+1+\frac{4}{2x-3}=0$	2.1	0.7	1.0	0.4	0.7
31. $5x^2+1.5x+.88=0$	2.6	0.7	1.3	0.4	0.8
32. $\sqrt[5]{x^2}+8\sqrt[5]{x}=9$	0.9	0.5	0.8	0.1	0.5
33. $\frac{x+1}{2x^2+3x-2}+\frac{2-x}{1-x-2x^2}+\frac{x-2}{x^2+3x+2}=0$	0.7	0.5	0.8	0.1	0.5
34. $(x^2+1)^2-11(x^2+1)+24=0$	1.4	0.5	1.2	0.1	0.6
SIMULTANEOUS LINEAR AND QUADRATIC					
35. $x+y=-4$; $xy=-21$	2.4	0.8	1.2	0.25	0.8
36. $x-y=-25$; $4x-4y=xy$	2.1	0.8	1.2	0.25	0.8
37. $y+2x^2-3x-9=0$; $y+x-3=0$	1.6	0.7	1.2	0.25	0.7
38. $x-4y=4$; $x^2+y^2=16$	1.6	0.7	1.2	0.25	0.7

TABLE 19—Continued

SIMULTANEOUS LINEAR AND QUADRATIC Continued	A 17 Physics	B 6 Biology, Botany, Agriculture	C 6 Economics Statistics	D 8 Sociology, Psychology, Anthropology	Average of B, C and D
39. $2x+y-2=0$ $4x^2+6xy+2x-6y+1=0$	1.2	0.7	1.2	0.25	0.7
40. $x-y=1$ $\frac{x}{y}-\frac{y}{x}=\frac{5}{6}$	1.1	0.7	1.2	0.25	0.7
41. $y=x^2-4x+3$ $4x+4y-7=0$	1.4	0.7	1.2	0.25	0.7
EXPONENTS					
42. Write 86,290,000 as a standard number.	2.6	0.8	1.2	0.4	0.8
43. Express 9.76×10^{-7} as an ordinary number.	2.6	0.8	1.2	0.25	0.8
44. $\sqrt{10000}$ equals what?	2.9	1.2	1.7	0.5	0.8
45. $\sqrt{a^2b^3}=$what?	2.8	1.0	1.5	0.25	0.9
46. $100^{1/2}=$what?	2.9	1.2	2.2	0.25	1.2
47. $125^{1/3}=$what?	3.0	1.2	2.2	0.25	1.2
48. $7^{-2}=$what?	2.8	1.0	1.3	0.1	0.8
49. Which is larger 3^{-2} or 2^{-3}?	2.5	1.2	1.3	0.1	0.9
50. Write without radicals $\sqrt{2\sqrt[3]{4}}\div8\sqrt{4^{-2}\times3^3}$	1.6	1.2	0.9	0.1	0.7
51. Express $.00034792\times10^4$ as an ordinary number	2.5	1.2	0.9	0.1	0.7
52. What is the value of $8^{-2/3}\times16^{3/4}\times2^0$?	1.6	0.8	0.9	0.0	0.6
53. $\frac{1}{a^{-1}-b^{-1}}+\frac{\sqrt{a^5b^{-1}}(a^{1/2}b^{1/2})^3}{a-b}=$what?	1.3	0.8	0.6	0.0	0.5
54. What is the value of $\frac{1}{2^{-1}+3^{-1}}$?	1.8	0.8	0.6	0.0	0.5
55. Divide $2a-a^{1/2}b^{-1/2}-3b^{-1}$ by $2a^{1/2}-3b^{-1/2}$	0.6	0.8	0.6	0.0	0.5
56. Simplify $\left[\sqrt[5]{\frac{a^{1/2}x^{-2}}{x^{1/2}a^{-2}}}\div\sqrt{\frac{x^{-1}\sqrt{a}}{a\sqrt{x}}}\right]^{-4}$	0.6	0.8	0.6	0.0	0.5

> The National Committee on Mathematical Requirements have published valuable ratings of this nature for certain entire topics, but ratings of tasks within these topics are almost necessary to interpret their results. They will be useful in many other ways also.

Table 19 presents the ratings of computational tasks by four groups of college teachers—seventeen teachers of physics, six teachers of biology, botany and agriculture, six teachers of economics (4) and statistics (2), and eight teachers of sociology (4), psychology (2), and anthropology (2)[1].

The ratings, and still more the letters which accompanied them, show certain tendencies which should be kept in mind in connection with a study of the table.

There was a tendency to rate by the large groups, in spite of the obvious range of difference in complexity and importance of the tasks within each group.

There was a tendency to confuse the ability to do one of these tasks with the ability to learn to do it or with the general ability of which the ability to do one of these tasks is a symptom. Teachers of a college course naturally prefer to have in their classes the sort of pupils who have the capacity to learn to handle such work as 32, 33, 34, 52, 53, 54, 55, and 56, even though no such tasks ever occur in the course, and they tend to rate these computations not for their actual intrinsic usefulness, but for the kind and degree of intelligence which they indicate.

As an offset to these two tendencies we have the natural tendency to be influenced somewhat by the general arrangement from simple and often used computations to complex and rare ones within each topic.

Our original intention was to have these same fifty-six tasks rated when presented in a random order and also to

[1] These replies were 85, 75, 75, and 100 percent, respectively, of the number of inquiries sent.

have some of them rated when presented along with many non-mathematical tasks, so that there would be little or no suggestions about relative simplicity and commonness of use. This would, however, require about an hour and a half of time from each man of science; and it seemed, on the whole, unfair to ask them to give so much as that.

In spite of a certain blurring of distinctions by these tendencies, the general results of Table 19 tell a fairly clear story. The utility of an algebraic technique diminishes in proportion as its application is complicated by elaborate and rare data, or by the need for ingenious rearrangements, factorizations, and acute perceptions of relations. On the contrary, complications of a sort actually found in scientific work (as by decimals, or by the use of fractional exponents instead of the radical form) are voted to be useful. College teachers of physics wish a thorough mastery of algebraic computation. Their average vote falls below "Some" only for such obvious monstrosities as 32, 33, 55, and 56. College teachers of the biological and social sciences (other than economics and statistics) make little use of even the simplest algebraic technique in their first-year courses. Their average vote rises above "Some" only for $3n=12$, $2n=\frac{k}{5}$ *when* $k=30$, and *finding the constants in* $y=mx+b$, *two points on the curve being given.* The ratings of two tasks which are in the list but are not in the usual teaching of algebra, and almost certainly were not parts of algebra as it was taught to these men twenty-five years ago, are of special interest. One is the use of simultaneous equations to solve for the constants; the other is the use of exponents in understanding and using the standard form number.[1] In

[1] The ratings here would have been still higher probably except for certain objections to the use of the terms "standard" and "ordinary."

both cases the ratings are so high[1] as to suggest that these tasks be at once made a part of the course in algebra. The high ratings for $\frac{1}{2}$ and $\frac{1}{3}$ as exponents suggests that these usages be introduced earlier in the course than at present.

Five teachers of physics or chemistry (three of the former) and five teachers of economics, psychology, and sociology (two, two, and one respectively) rated the nineteen problems of Table 20. Using 3 for "essential," 2 for "Of considerable value," 1 for "Of some value," and 0 for "Of no value," we show their average ratings.

The general voice of the teachers of physics and chemistry is that any one of these problems is useful for their work, though they put the commercial problems (6, 14, and 17) relatively very low, and put the genuine problems somewhat below the fantastic ones. The general voice of the teachers of the social sciences is that in no case is the ability of any considerable use for their elementary course. Three (one economist, one psychologist and the sociologist) rate all the nineteen "zero"; one psychologist assigns value to five of the simpler and genuine tasks; the bulk of the credits are due to one economist.

TABLE 20

AVERAGE RATING OF NINETEEN VERBAL PROBLEMS BY COLLEGE TEACHERS OF THE PHYSICAL AND SOCIAL SCIENCES

	Average Rating	
	Physical	Social
1. How much water should be added to 10 gal. of a 20% solution to make an 8% solution?	2.6	0.8
2. Listerine contains 25% alcohol. When used as throat spray it must be diluted. How much water to 100 parts listerine to make a mixture containing 15% alcohol?	2.6	0.8

[1] Even if a generous allowance is made for the suggestion from their position in the list.

TABLE 20—*Continued*

		Average Rating Physical	Average Rating Social
3.	Two grades of spice worth 25c. and 45c. a lb. are to be mixed so that the mixture can be sold at 50c. a lb. with a profit of 25%. How much of each grade should be taken to make 8 lbs. of the mixture?	2.2	0.4
4.	Find where to cut a board 50 in. long into 2 parts, one being 26 in. longer than the other.	2.4	0.2
5.	A bar 60 in. long is to be so cut that one piece is to be $\frac{2}{5}$ as long as the other. Find where to cut it.	2.8	0.2
6.	A clothing merchant puts on sale boys' suits costing $8.00 each and advertises them at a 25% reduction from the marked price. What is the price if the actual profit after the reduction is 20% of the cost price?	1.4	0.6
7.	A belt runs over a pulley 48 in. in diameter, making 180 revolutions a minute. It is desired to reduce the size of the pulley, keeping the belt at the same speed by increasing the number of revolutions to 216 a minute. How much must the diameter be decreased?	2.0	0.6
8.	2.6 seconds after a gun fired the sound of its hitting the target 440 yd. distant is heard. If sound travels 1,100 ft. per second, what is the average velocity of the bullet?	2.6	0.8
9.	Two boys are to mow a lawn 60 ft. by 32 ft. The first boy is to mow half by		

TABLE 20—*Continued*

		Average Rating Physical	Social
	cutting a strip of uniform width around the plot. How wide is the strip?.....	2.6	0.0
10.	The specifications for a cardboard box are: "4 in. deep, twice as long as wide, contents 512 cu. in., to be made by cutting a square from each corner of a rectangular piece, turning up the sides and pasting along the edges." How large must the rectangle from which it is cut be?........................	2.0	0.0
11.	A board 42 in. long is to have a hole bored so that its distances from the ends are in ratio $\frac{3}{4}$. Where must the hole be bored?........................	2.8	0.4
12.	In a factory there are 3 large machines and 5 small ones making the same product. An order comes in which would require one large machine 60 days or 1 small machine 90 days to make goods to fill. How long will it take all 8 machines?................	1.8	0.4
13.	A book page is said to be most satisfying to the eye if its length is a mean proportional between its width and the sum of its length and width. If it is to be 5 in. wide, how long should it be made?............................	2.4	0.8
14.	By getting a 5% discount for cash in buying a merchant can increase his net profit 8% without increasing the selling price. What is his gain when he receives no discount?.................	1.0	0.4

TABLE 20—Continued.

		Average Rating Physical	Average Rating Social
15.	A car slipping down hill moves 2 in. the first second, 6 in. the 2d second, 10 in. the 3d second, etc. How far will it move in 20 seconds?	2.0	0.2
16.	At two stations A and B, 6 miles apart, the price of coal is $10 and $12 per ton, respectively. The cartage rate is $1.00 and $1.50 per ton, respectively. Find where on the road from A to B a customer lives if the price of coal is the same delivered for him from either A or B	1.6	0.6
17.	A dealer bought grapefruit for $1.04. After throwing away 4 spoiled ones, he sold the rest at 6c. apiece more than he paid for them, making a profit of 22c. How many did he buy?	1.7	0.2
18.	A can do $\frac{1}{2}$ as much work as B and B $\frac{1}{2}$ as much as C, and together they can complete the work in 24 days. How long would it take each to do it alone?	1.7	0.0
19.	Given 3 metals of the following composition by weight: (1) 5 parts gold, 2 silver, 1 lead; (2) 2 parts gold, 5 silver, 1 lead; (3) 3 parts gold, 1 silver, 4 lead. To obtain 9 oz. of a metal of equal quantities of gold, silver, lead, how many oz. of each of the 3 given metals must be taken and melted together?	2.2	0.0

THE USES OF ALGEBRA AS SHOWN BY AN INVENTORY OF HIGH-SCHOOL TEXTBOOKS[1]

As a crude but impartial method of estimating which algebraic abilities are useful and how great the need for them is, we have examined from two to four high-school textbooks under each of the titles of Table 21, noting the nature and extent of each demand or opportunity for the use of algebraic processes.

The books examined were as follows:

Robinson and Breasted: History of Europe, 1914.
Hazen: Modern European History, 1919.
Beard and Bagley: History of the American People, 1920.
Muzzey: American History, Revised Edition, 1921.
Ely and Wicker: Elementary Principles of Economics (Revised), 1917.
Marshall and Lyon: Our Economic Organization, 1921.
Giles: Vocational Civics, 1919.
Hart: Community Organization, 1920.
Hughes: Community Civics, 1917.
Hodgdon: Elementary General Science, 1918.
Barber, Fuller, Pricer, Adams: First Course in General Science, 1916.
Snyder: Every Day Science with Projects, 1919.
Salisbury: Physiography; Briefer Course, 1919.
Wright: Manual in Physical Geography, 1906.
Peabody and Hunt: Elementary Biology, 1912.
Gruenberg: Elementary Biology, 1919.
Hodge and Dawson: Civic Biology, 1918.
Hegner: Practical Zoology, 1915.
Packard: Zoology, 1904.
Allen and Gilbert: Botany, 1917.
Bergen and Caldwell: Introduction to Botany, 1914.
Jackson and Daugherty: Agriculture, 1908.
Warren: Elements of Agriculture, 1909.
Conn: Elementary Physiology and Hygiene, 1910.
Fitz: Physiology and Hygiene, 1908.
Calkins: First Book in Psychology, 1910.
Pillsbury: Essentials of Psychology, 1920.
Kalenberg and Hart: Chemistry, 1913.
Brownlee, Fuller, Hancock, Sohon, Whitsett: Elementary Principles of Chemistry, 1921.
McPherson and Henderson: First Course in Chemistry, 1915.
Henderson and Woodhull: Elements of Physics, 1900.
Carhart and Chute: Physics, 1920.
Millikan and Gale: Practical Physics (Pyle's Revision), 1920.
Matteson and Newlands: Foods and Cookery, 1916.
Greer: Textbook of Cooking, 1915.
Greer: School and Home Cooking, 1920.
Kinne and Cooley: Foods and Household Management, 1914.

[1]This section is reprinted from *School Science and Mathematics* for May and June, 1922.

Kinne and Cooley: Shelter and Clothing, 1913.
Baldt: Clothing for Women, 1916.
Matthews: Sewing and Textiles, 1921.
Cooley and Spohr: Household Arts for Home and School, 1920.
Stillwell: School Print Shop, 1919.
Matthewson: Notes for Mechanical Drawing, 1904.
Griffith: The Essentials of Woodworking, 1908.

TABLE 21

Subject	Total Number Linear Inches Examined	Total Number Linear Inches Using Algebra	Percent Linear Inches Using Algebra	Number Separate Times Algebra Was Used	Average Number Uses of Algebra Per Hundred in.[4]
Social Sciences					
American History........	7,070	210	3.0	56	0.8
Community Civics.......	4,010	6	0.1	4	0.1
Economics..............	5,220	146	2.8	38	0.7
European History........	8,200	48	0.6	11	0.1
Total: Social Sciences..	24,500	410	1.7	109	0.4
Physical and Biological Sciences					
Agriculture..............	4,690	30	0.6	10	0.2
Biology.................	5,200	154	3.0	44	0.8
Botany.................	3,990	6	0.2	1	0.03
Chemistry[1].............	7,540	161	2.1	424	5.6
		332	4.4	6,079	80.5
Total Chemistry......	.....	499	6.5	6,503	86.1
General Science..........	10,440	304	2.9	111	1.1
Physics[2]...............	7,550	307	4.0	837	11.0
		4	.1	132	2.8
Total Physics.........	.....	311	4.1	969	12.8
Physiography...........	4,290	513	12.0	127	3.0
Physiology..............	3,730	10	0.3	5	0.1
Psychology.............	4,410	23	0.5	8	0.2
Zoology................	4,330	6	0.1	2	0.05
Total: Physical and Biological Sciences.....	56,170	1,856	3.3	7,780	13.8
	.....	1,520	2.7	1,559	2.8[3]
Practical Arts					
Cookery................	8,920	89	1.0	47	0.5
Mechanical Drawing.....	340	8	2.4	141	41.5
Sewing.................	7,680	1	.01	1	0.01
Woodworking...........	1,029	0	0.0	0	0.0
Total: Practical Arts...	17,960	98	.5	189	1.1
Total for all subjects...	98,630	2,364	2.4	8,078	8.2
	.....	2,028	2.1	1,857	1.9[3]

[1] In Chemistry 332 inches, 6,079 separate uses, are chemical formulas and reaction equations, each writing of either being counted.

[2] In Physics 4 inches, 132 separate uses, are chemical formulas and reaction equations, each writing of either being counted.

[3] This line gives the data with the chemical formulas and reaction equations discarded.

[4] In the ordinary high school text 100 inches is approximately eighteen pages.

The amount of text examined in each subject, the frequency of the uses of algebra, and the total use of algebra are reported in Table 21.

The units of measure—"an inch examined," "an inch of algebra used," and "a use"—are, of course, extremely crude, but entirely impartial.

A detailed study was made of the kind of algebra called for under the five following heads:

(1) Manipulation of complicated polynominals;

(2) Formation and solution of identities and equations;

(3) Formation and evaluation of formulas;

(4) Development and use of the notion of function;

(5) Construction and interpretation of graphs for
  (a) statistics
  (b) functions.

Table No. 22 presents the facts as found.

The following conclusions from a survey of these tables would appear to be justifiable:

(1) Omitting from consideration courses in mathematics itself, there is no need in high-school studies for facility in complex manipulation of polynomials.

(2) In present textbooks there is no use made of the mathematical concept of function.

(3) Except in chemistry and physics and agriculture the study of equations has at present no utilization in high-school work.

(4) The making of formulas is practically not required of the high-school student.

(5) The comprehension and evaluation of formulas is required only in physics, chemistry, and in the physical and chemical parts of general science.

(6) The mathematical graph, either as illustration of the scope of the formula, or as vivifying the concept of function, practically does not occur in any high-school work.[1]

[1] Its only utilization in high-school physics lies in the reading and construction of graphs illustrating the resolution of forces.

(7) The statistical graph is used to a greater or less degree in all high-school subjects investigated, the number of linear inches given to graphs (in cuts and explanations) being twice as great as for all other types of the utilization of algebra combined.

It is important to note to what extent the algebra of which use is made is necessary to a comprehension of the texts considered.

TABLE 22

	Manipulation of Symbols		Equations				Formulas				Function		Graphs			
			Identities		Conditions		Formation		Evaluation				Statistical		Mathematical	
	Inches	Occurrences	Inches	Occurrences	Inches	Occurrences	Inches	Occurrences	Inches	Occurrences	Inches	Occurrences	Inches	Occurrences	Inches	Occurrences
European History	...	...	...	...	...	...	...	...	...	...	...	...	48	11	...	...
American History	...	...	...	...	...	...	...	...	...	...	...	...	210	56	...	...
Economics	...	...	...	...	...	...	...	...	...	...	...	...	146	38	...	...
Civics	...	...	...	...	...	...	...	...	...	...	...	...	6	4	...	...
General Science	...	...	...	...	...	...	...	...	13	4	...	...	287	107	...	...
Physiography	...	...	...	...	...	...	...	...	...	...	...	...	513	127	...	...
Biology	...	...	...	...	...	...	...	...	...	...	...	...	144	44	...	...
Zoology	...	...	...	...	...	...	...	...	...	...	...	...	6	2	...	...
Botany	...	...	...	...	...	...	...	...	...	...	...	...	6	1	...	...
Agriculture	...	...	...	...	6	1	...	...	...	...	...	...	24	9	...	...
Physiology	...	...	...	...	...	...	...	...	...	...	...	...	10	5	...	...
Psychology	...	...	...	...	...	...	...	...	...	...	...	...	23	8	...	...
Chemistry	...	...	235	906	124	276	1	38	132	5281	...	...	7	2	...	...
Physics	...	...	...	...	240	570	...	...	63	254	...	...	...	...	4	13
Cookery	...	...	...	...	3	5	...	...	...	...	...	...	27	6	...	...
Sewing	...	...	...	...	...	...	...	...	...	...	...	...	1	1	...	...
Wood Work	...	...	...	...	...	...	...	...	...	...	...	...	...	...	...	...
Mechanical Drawing	...	...	...	...	...	...	...	...	8	141	...	...	...	...	...	...

To read a statistical graph is within the power of the high-school student untrained in algebra. Pictograms and cartograms have become part of the common language of newspapers and magazines. But to make them offers such difficulty that only the exceptionally gifted child might be expected to master it without specific instruction. Still less would one anticipate that a critical attitude concerning the veracity of a graph and ability to detect falseness and graphic misstatement would develop without drill in that precise function.

In chemistry the formula is not the equivalent of the formula in algebra. To quote from a recent text (Brownlee, Fuller, Hancock, Sohon, Whitsett, Elementary Principles of Chemistry, pp. 114, 115, 123), "The formula of a molecule is formed by grouping together the symbols of the atoms composing it. . . . When a molecule contains more than one atom of the same kind the symbol is not usually repeated, but the number of atoms is written as a subscript to the symbol. . . . . The only thing that decides the formula of a compound is a chemical analysis. The symbol of an element and the formula of a compound represent more than the name. The symbol of an element stands for one atom of that element . . . and for a definite quantity of the substance" (by weight).

Further, in the chemistry texts reviewed the chemical reaction identities in the two newer books are written with arrows, while in the older one the sign of equality is used. For example, the identity written $NaCl+H_2SO_4=Na\ H\ SO_4+HCl$ in the older books, in the newer is written $NaCl+H_2SO_4 \rightarrow Na\ H\ SO_4+HCl$. The student is instructed to read such an equation "NaCL and H_2SO_4 will yield NaH SO_4 and HCl." If the reaction is reversible two arrows are used thus: $2NH_3 \rightleftarrows N_2+3H_2$.

To quote again from the same source (p. 133): "These equations are not like mathematical equations. They are brief statements of experimental facts and should not be written unless it is known that the reactions actually take place. If we know by experiment:

(1) That the substances do react;
(2) The composition of each of these substances;
(3) All the products formed;
(4) The composition of each product;

we can represent the reaction by an equation, and calculate the relative quantities involved."

The treatment of formulas and equations in algebra may then do as much harm as good to the student's learning of chemistry, especially if "emphasis" is put on the equation of condition rather than on the equation of identity.

It is, therefore, only in the "problems" of the chemistry text that the student would find his elementary algebra of advantage. The equations which will be made are, in the large fraction of cases, of the form of a proportion in which three terms are known numerical quantities. A child thoroughly grounded in proportion in arithmetic would be able to make and solve most of these without recourse to algebra, though the student who has had equation drill in algebra might fairly be expected to form and to solve such proportions with increased readiness and smaller percentage of error.

A similar conclusion regarding the equation problems of agriculture is legitimate. The problems deal chiefly with balanced rations and are similar to the algebra problems of alloy or mixture. A knowledge of the proportion equation suffices for the solution.

In high-school physics the significant demand for algebra lies in the reading and evaluation of formulas. Typical formulas are:

$P_1/P_2=D_1/D_2$; $S=1/2\,gt^2$; $I=E/(R_e+R_i/n)$; $V=\sqrt{2aS}$; $E=R/n$; $(P_1V_1)/(P_2V_2)=T_1/T_2$.

These formulas are selected from those stated in Millikan and Gale, *Practical Physics* (Pyle's Revision); 1920; pp. 36, 76, 77, 109, 137, 280.

To read these and such other formulas as the student will probably encounter in high-school physics requires knowledge of the signs +, −, =; the fraction line, the algebraic expression of multiplication, the use of subscripts and exponents,[1] the radical sign and the parenthesis, in addition to the fundamental notion of the expression of quantity by letters as well as by digits.

The student well prepared in arithmetic probably knows all of these except the expression of multiplication, and the subscript and literal notations, before he studies algebra.

It would be only just, however, to expect that one who had studied algebra would bring to these formulas a richness of content and a clarity and certainty of meaning that should make his comprehension of what he reads in them more vivid.

The evaluation of such formulas from an algebraic point of view means the substitution (in an algebraic expression) of numerical values for the various literal quantities and the performing of the indicated operations. The answer sought will require no more than addition, subtraction, multiplication, division, squaring, and square root, in various combinations. These do not demand mastery of algebra, but it is likely that the student of algebra would be spared the labor and ignominy of certain errors in the manipulation of his quantities.

For instance, in $S = 1/2\, gt^2$ the student unacquainted with algebra is in danger of multiplying the values of g and t, dividing by 2, and squaring the result, or of multiplying,

[1] It is useful for the student to understand both positive and negative exponents for the expression of standard numbers, though rarely beyond 6, and the fractional exponent 1/2, and, rarely, 1/3 and 1/4 as an alternative for the use of the radical sign.

squaring, and then dividing. Any teacher made familiar with like errors by their continual recurrence could supply equivalent illustrations. A very small amount of properly learned algebra would free the student from such practices.

Further, the student unacquainted with algebra would find himself quite at a loss in the derivation of new formulas from those given.

To illustrate, by the algebra student $S = 1/2\ gt^2$ is transformed into $g = 2s/t^2$ or $t = \sqrt{2s/g}$ with a small amount of labor, but to the uninitiated each new form is a totally new experience. He can neither make it for himself nor readily understand how it was obtained by the other student. The foundation formula acquired through reasoning or experimentation is thus deprived of the richness of application legitimately to be expected from it.

In high-school physics as in chemistry a large number of problems is given in the texts examined. These problems present almost exclusively data to be dealt with by means of formulas found in the body of the text. The student must select the formula appropriate to the problem, make the needed substitutions, and solve the resulting equation.

The complexity of the operations required in such solutions rarely goes beyond the clearing of fractions in a proportion, transposing, combining terms, and extracting a numerical square-root.

The following formulas present from the algebraic point of view the maximum of difficulty:[1]

$1/p + 1/q = 1/f$ with numerical quantities given for p and q, to solve for f;

$K = (l_2 - l_1)/l_1\ (t_2 - t_1)$, with K, l_1, l_2, t_1, known, to solve for t_2;

[1] From Carhart & Chute: *Practical Physics;* 1920, pp. 254, 290, 383, 384, 134, 139.

$I = E/(r+r')$; I, E, r, being known, to solve for r';
$I = nE/(r+nr')$; I, E, r, r' given, to solve for n;
$f = mv^2/r$; f, m, r, being known, to solve for v;
$t = \pi\sqrt{l/g}$; t and l being given, to solve for g.

The algebra required in these is certainly much less than that ordinarily taught in the first year course.

To recapitulate: I. As a preparation for high-school subjects outside the mathematics courses as they are at present taught, so far as may be judged from textbooks, the most extensive need for algebra is in the reading of statistical graphs, and to a far lesser degree, in the making of them.

II. As a preparation for physics the chief value lies in the mastery of the formula, particularly in handling the transformation known as "changing the subject of the formula." The student unacquainted with algebra is entirely at a loss in proceeding with this type of work.[1]

III. The solution of physics problems demands the ability to choose the appropriate formula, to make substitutions, and to solve the resulting equations. This involves skill in clearing of fractions (chiefly in proportions), transposing, combining, and taking numerical square root.

IV. For the solution of chemistry problems, the ability to form a correct proportion is of high value. The algebra problems which can be applied with advantage in this field are those dealing with proportions, mixtures, and alloys.

The next question that presents itself is: Can algebra be utilized by other high-school subjects to greater degree than at present with profit to these subjects, and with reinforced vitality in the algebra itself?

[1] See Nunn: *The Teaching of Algebra*, 1914, p. 78, also pp. 14, 15.

Let it be assumed for convenience of classification that possible utilization of algebra may occur under the same five heads already noted:

(1) Manipulation of complicated polynomials;
(2) Formation and solution of equations;
(3) Formation and evaluation of formulas;
(4) Development and use of the mathematical concept of function;
(5) Construction, interpretation, and criticism of graphs—both statistical and functional.

In no subject of the high-school curriculum outside the mathematics courses is any need for the manipulation of complicated polynomials discovered, nor does any such need seem probable.

By the formation and solution of equations certain science courses might be much enriched. However, the need would seem to be for extreme facility in the handling of easy equations rather than the formation and solution of complicated ones. Especially is this true of the ordinary proportion equation. Charles' Law, Boyle's Law, the laws of the five simple machines (usually given in first-year general science as well as in physics and chemistry); changing recipes in cooking; adjusting patterns in sewing; obtaining lengths for use in mechanical drawing, or in scientific free-hand drawing; plan-making and model-making in wood and metal work and other shop work; all these demand ready use of the proportion.

A pseudo-use of the identity occurs in such dictionary and word-analysis forms as "biology = bios, life + logos, study of."[1] It is probable that such a form more readily conveys its meaning to the child who has studied algebra than "Biology is derived from two words, 'bios' and 'logos' meaning 'life' and 'study of,'" while if fifty such derived words are to be taught the advantage of symbolism from the

[1] Certain texts use this form "biology, bios = life + logos = study of," a form distinctly reprehensible from the algebra teacher's viewpoint.

mere consideration of space required for printing is patent. The identity in chemistry already has been exploited. Reaction away from the algebraic identity is the present trend.

This form of work readily passes over into the formation and evaluation of formulas. It seems certain that added usefulness could be found in certain subjects were the child equipped to make a formula, or to use one when given. Differing minds, to grasp and retain with ease, require differing forms. In cookery it might be profitable to the student if he were required to transform rules into equivalent formulas. For instance, the length of time necessary to cook a roast of beef—"a quarter of an hour to the pound and twenty minutes extra"—or the amount of coffee to be used—"a spoonful for each cup and a spoonful for the pot"—are doubtless more easily remembered if presented in formula as well as in rule. Formulas for balanced rations, for infant feeding, for the use of leavening agents, might profitably occur. A move in this direction is the present custom of writing recipes by presenting first a table of the ingredients and their amounts, then a paragraph describing the method of combining them. The advantage here is the advantage of all algebraic symbolism, such compactness as to make possible a complete survey of the "elements of the problem" with a minimum expenditure of time and thought.

In sewing, formulas on the allowance of goods for the making of ruffles, for accordion and other kinds of plaiting, length of the bias strip for goods of given width, are devices for storing needed facts in the memory in an economical form. The allowance of embroidery threads to cover certain spaces is usually made nowadays by the experienced intuition of the worker or the sales-woman, with consequent over or under supply. Perhaps in time formulas for them will be developed that will prove of advantage.

In physiography it would be of interest and profit to present such formulas as $d = 1.22\sqrt{h}$, where h = the height of the observer above sea level and d the greatest distance at which an object on the sea is visible to him.

The utilization of the statistical or pictorial graph has already become general in all high-school subjects. What is needed is the development of the critical faculty in reading and in evaluating such graphs. Especially is this true of the pictograms which utilize natural three-dimension forms. In determining lengths for proportional facts due regard should be paid to formulas for surface, volume, and perspective. The student needs to be taught to avoid the false inferences which are natural when a pictogram naturally interpreted in units of volume is presented in linear units.

Teachers of many subjects would find it advantageous if they could utilize the mathematical graph—that is, the graphic presentation of a law or a function. It is the aim of science to formulate its conclusions not as mere statistics, but as mathematical laws. But at present such laws are rarely expressed for the student in a mathematical form. Whether this is because the student has not been qualified by his study of algebra to grasp such statement, so that the vernacular must be used instead, or because the writers of the texts are unable to put their conclusions in such form, is not pertinent to the present discussion.

Practically the only use of this highly refined tool of exposition discovered in our textbook survey is to be found in Ely and Wicker's *Elementary Principles of Economics* (revised edition, 1917, pp. 183-201).

Certainly, many of the laws of economic dependency; certain basic notions of continuity and sequence and causation in history; facts of growth versus soil conditions and the composition of soils in agriculture; laws of growth increases in botany; interdependence of animal and plant life in biology; time and rate in interest, and cost-price relations in commercial work; all these might be made clearer to the student by correct and appropriate use of the mathematical graph.

The advantage of such graphic representation to supplement data by interpolation and extrapolation probably needs no emphasis with high-school students; rather a caution against reading unfounded conclusions into graphs is needed. The fallacy of coördinates, the dangers inherent in optical delusions, the misunderstanding occasioned by the omission of zero reference points, should be pointed out. Also, warning is needed against over-use. It would be easy in such work to go beyond the point of diminishing returns for high-school students who are little equipped for scientific generalization, but a small amount of such work might be made a source of illumination and progress.

For most high-school students any larger use of the function concept than that suggested in the mathematical graph is of doubtful value. It is one of those fundamentally powerful conceptions whose elaboration has been one of the half dozen significant achievements of the race, but to the high-school student it is vague and tantalizing and stimulating rather than clarifying. To sum up:

(1) Involved manipulation of polynomial expressions is not a justifiable way of using the high-school student's time.[1]

(2) Since the application of equations in other high-school subjects is chiefly in the proportion form, mastery of that form and other easy equation forms should be secured.

[1] This statement is meant to cover addition and subtraction of long expressions; multiplication and division of polynomials; manipulation of fractions with polynomial terms; squares and cube root of polynomials resulting in answers of more than two terms; involved factorization of polynomials; reduction of radicals (or fractional exponents) whose index is greater than three; operations with polynomials containing radicals; the factor and remainder theorems; and the binomial theorem, when the exponent is larger than five.

(3) It would be profitable to extend the field of application of the construction of formulas as well as their evaluation.

(4) There is need for the careful development of the art of criticism as applied to graphs.

(5) The presentation of laws by means of mathematical graphs should be encouraged.

(6) The function concept should be used when advantageous, but with economy.

THE USE OF ALGEBRA AND GEOMETRY SHOWN BY AN INVENTORY OF THE ENCYCLOPEDIA BRITANNICA

To arrive at a notion of how far algebra has penetrated into the intellectual and practical life of men of intelligence and achievement comparable to that of high-school graduates, an examination of the Encyclopedia Britannica has been made, noting all references to mathematics beyond arithmetic. The first two hundred pages of each volume from I to XXVIII have been read and the mathematical references collected.

Table 23 gives the facts as found:

TABLE 23

Encyclopedia Counts.

	Number of Articles Concerned	Number Linear In. Space Utilized
Mathematical Definitions	12	22
Long Articles with Slight Mathematics	7	3,714
Requiring Vocabulary of Geometric Shapes:		
Description of Crystals	39	874
Shapes of Buildings, Land, etc.	60	2,301
Other use of Terms Indicating Geometric Shapes	68	4,606
Biographies of Mathematicians	32	663
Requiring Algebra Only:		
Graph	5	375
Formula	2	123
Requiring more than Elementary Algebra	44	7,063
Total Mathematics Usage	269	19,741
Total Examined	7,551	106,400

A further analysis may be illuminating.

The twelve articles listed under "Mathematical Definitions" are: Abscissa, Angle, Angulate, Anthelion, Axiom, Axis, Azimuth, Correspondence, Diagonal, Diameter, Gnomon, Ordinate. General intelligence, rather than knowledge of mathematics, is required to understand these definitions.

The seven articles which are listed under the head of "Long Articles with Slight Mathematics" are: Bacteriology, Egypt, Glass, Harmony, Medical Education, Mendelism, Respiratory System. The person who knew no high-school mathematics would be handicapped in reading not more than twenty-five linear inches of the 3,714 inches used in these articles. He would find the words conical, plane, parallel, angle, cylindrical, spherical, ellipsoid, rectangular, abscissae, ordinates, of whose meanings he would be ignorant or doubtful, and he would be at a loss to interpret the following five formulas:

$$V=(nd-1)/(c-F)$$
$$v/(V+v)=p/100$$
$$V=v(100-p)/p$$
$$ab^2=AB^2+(Bb-Aa)^2+2AB(Bb-Aa)\cos x$$
$$(c'd')^2=AB^2+(Ac'-Ba')^2-2AB(Ac'-Bd')\cos x$$

In addition he would find one statistical graph of meager meaning and would be doubtful concerning the exact idea conveyed by this sentence: "And hence the enharmonic circle of fifths is a conception of musical harmony by which infinity is rationalized and avoided, just as some modern mathematicians are trying to rationalize the infinity of space by a non-Euclidean space so curved in the fourth dimension as to return upon itself." Otherwise, he probably could read all the articles with as much profit as if he were skilled in mathematical lore.

Again the one hundred sixty-seven articles listed under "Requiring Vocabulary of Geometric Shapes" make small demand upon knowledge of mathematics. The thirty-nine of these listed under "Description of Crystals" contain usages of the technical terms describing the shapes in which crystals of various substances occur and are illustrated and explained in texts dealing with these subjects. This nomenclature originally was taken from solid geometry but it has become a part of the special equipment of the student of those branches of geology which utilize it, and may reasonably be expected to be acquired there by one who needs it. The second division, "Shapes of Buildings, Land, etc.," includes such terms as rectangular, rhombic, quadrilateral, conical, cylindrical, triangular, trapezoidal, spherical, square, octagonal, hexagonal, quadrangle, as these are applied to grounds, towers, tunnels, and the like. The third division, "Other Use of Terms, Indicating Geometric Shapes" is a similar usage of terms as applied to plants, water systems, tools, cameras, explosives, glaciers, harps, ivory, labyrinths, pencils, perfumery, and the like. The person who had not studied geometry might find his comprehension of the articles less easy and accurate than his more favored friend who had studied geometry in school, but he would probably not suffer greatly.

The thirty-two biographies of mathematicians do not use any mathematics in the articles, but they list the achievements and books of their subjects and therefore probably would have slight meaning for the person untrained in mathematics.

The seven articles appearing under the head "Requiring Algebra Only," are: Animal Heat, Dew, Electricity Supply, Hydrozoa, Sunshine (which are illustrated by graphs using the terms abscissae and ordinates in their explanation), and Anemometer and Hearing, in the latter of which occur these formulas:

$$P = .0005v^2$$
$$ut_2 = 256$$
$$ut_3 = 512$$
$$ut_6 = 2,048$$
$$re_6 = 2,304$$
$$sol_6 = 3,072$$
$$si_6 = 3,840$$

The reader ignorant of algebra would be handicapped in these articles to the extent to which these are an essential part of the explanation.

There remain, from the 7,551 articles occupying 106,400 linear inches of printed matter, forty-four articles for the understanding of which more than high-school mathematics is required.

Table 24 gives a detailed analysis of these articles. A check mark in any column indicates a need of the subject at the head of that column for the comprehension of the article. For instance, "calorimetry" requires algebra beyond elementary and the calculus; "Diagram" needs geometry, analytics and projective geometry for its complete mastery.

Assuming that the importance of articles is roughly proportioned to their length the following percents are interesting. The articles using mathematics beyond arithmetic in any form are 3.57 percent of the total number of articles, but they occupy 18.55 percent of the total space. If from these we discard the long articles in which very slight reference to mathematics occurs the articles remaining are 3.47 percent of the total number of articles, but use 15.06 percent of the space. The articles which call for a geometric vocabulary, though not necessarily geometric knowledge, comprise 2.21 percent of the total number and use 7.31 percent of the total space. If only those articles using elementary algebra and advanced mathematics be considered, the percent of the total number of articles is

.68 and of the space 7.11. Those articles which utilize advanced mathematics are .58 percent of the total number of articles, but require 6.64 percent of the total space.

These facts suggest the following conclusions:

TABLE 24

Mathematics Required to Read	Adv. Alg.	Geom.	Trig.	Anal.	Calc.	Other Mathematics
Aberration	√	√	√	√	√	Method of Least Squares
Absorption of light	√					
Accumulator	√					
Anomaly						Astronomy
Calibration			√			Method of Least Squares
Calorimetry	√				√	
Continued fractions	√		√		√	
Density	√				√	
Determinant	√			√	√	
Diagram		√		√		Projective Geometry
Dial		√	√			Astronomy
Elasticity	√	√	√	√	√	
Electrical machine	√	√				
Electricity	√	√	√		√	
Heat	√	√			√	
Hydraulics	√	√	√	√	√	Method of Least Squares
Hydromechanics	√	√	√	√	√	
Hydrometer		√				
Hyperbola		√		√		
Lapidary		√				
Lubrication	√				√	
Mensuration	√	√	√	√	√	
Ohmmeter		√	√			
Orbit	√	√		√		Astronomy
Order				√	√	Differential Equations
Ordnance	√	√	√			
Perpetual motion		√				
Polygon	√	√	√			
Polygonal numbers	√	√				
Polyhedral numbers	√	√				
Polyhedron	√	√	√			
Porism		√				
Reflection of light	√	√				
Refraction of light	√	√	√			
Sights		√	√			
Sun	√	√	√	√		Astronomy
Surface	√	√	√	√	√	
Surveying	√	√	√	√	√	
Traction		√		√		
Transformers	√	√	√	√	√	Standard Deviation
Transit circle	√	√	√			Astronomy
Vision	√	√	√	√		

(1) The parts of elementary algebra that have made a place for themselves in the reading matter of the man who seeks information upon topics of widespread interest are the graph—particularly the statistical graph—and the formula.

(2) Elementary algebra is not sufficient to enable one to read encyclopedia articles dealing with technical topics in physics, engineering, astronomy, chemistry, mathematics and allied subjects; but rather, speaking in general, one must have acquaintance with the mathematics ordinarily presented to the student in courses to be pursued from three to five or more years subsequent to the year of elementary algebra.

(3) The importance of mathematics to an understanding of subjects of general interest is much greater than the frequency of its use since the ratio of about 1 to 11 is maintained between the percent of the number of articles and the percent of space they occupy.

There are certain inferences from the foregoing that seem worthy of consideration in connection with the curriculum in first year algebra.

(1) For utilitarian values in the first year of algebra one must pay much attention to the formula and to the statistical graph.

(2) The student whose aptitudes and capacities will probably lead him into any one of a large number of fields of advanced study such as physics, engineering, psychology, chemistry, education, electricity, economics, aeronautics, social sciences, ballistics, navigation, etc., should have preparatory work not merely in algebra but in many other branches of mathematics to which rigorous algebra is the necessary antecedent and tool.

In comparison with careless statements implying that a knowledge of all that is taught in the usual course in elementary algebra is of frequent use in the study of science, the facts which have been presented here may seem to reduce the utility of algebra to a small matter. Similar analyses of textbooks and of the Britannica with reference to other high-school subjects would, however, show that algebra was at or near the top in this respect. Since, further, science in general is steadily becoming more quantitative, the utility of ability to read formulas and graphs is increasing. Consider for example the prominence of frequency curves and correlations in biology, psychology, sociology, and education today, remembering that thirty years ago they were almost unknown there.

Algebra is a useful subject, but its utility varies enormously.

Very few, if any, cases appeared in all the textbooks examined or in the 5000 pages of Britannica where ability to factor a^3-b^3 or a^3+b^3 was demanded. There were few, if any, cases where the ability to factor a^2-b^2 was demanded. Algebraic abilities form a series of diminishing utility, beginning with reading formulas and graphs of the relation of one variable to another, and ending with short-cut methods of division by knowledge of special identities.

The mere knowledge of the language of algebra has more utility than educators have thought, while skill in computing has less. Educators are prone to think it folly to learn algebraic symbols at all if you don't compute with them. They forget that we have to read them, and perhaps twice for every once that we compute with them. The knowledge of and ability to use formulas in solving quadratics, answering problems about progressions, and determining coefficients and exponents in the binomial expression which figure so largely in school examinations are toward the low end for utility.

In general, we may divide this continuous series into five groups or sections.

Section 1 includes:

The notion of symbolism, ability to "let . . . = the number of . . ."

Ability to read simple formulas, including knowledge of positive integral exponents, of $\sqrt[2]{a}$ or $a^{\frac{1}{2}}$, $\sqrt[3]{a}$ or $a^{\frac{1}{3}}$, of single parenthesis, and of the omission of the $\times$ sign and of 1 as coefficient.

Ability to evaluate and solve such formulas, including the interpretation of negative numbers as answers, when the unknown forms one member of the equation.

Ability to read graphs of the relation of one variable to another in the $++$ quadrant, when the nature of each variable is understood.

Section 1 seems worth requiring of all students in Grade 9 if anything is worth requiring of them at all, and if they have not learned its content previously in arithmetic.

Section 2 includes:

The ability to read formulas involving a parenthesis within a parenthesis and complex fractions. This involves learning algebraic addition and subtraction, algebraic multiplication and division by monomials, and certain elementary facts about radicals and surds.

The ability to transform formulas (after evaluation) where the desired number has a coefficient or is found in a denominator, or appears twice, or is only one term of one member of the equation into formulas where it appears alone as one member. The ability to do so before evaluation, provided no division of a polynomial by a polynomial is required.

Section 3 includes:

The elementary algebra of the relation of one variable to another, including graphic presentations.

The solution by linear and quadratic equations of genuine problems for which such solutions are desirable.

Simultaneous linear equations.

The elementary facts concerning surfaces of frequency, central tendencies, and variability or dispersion around a central tendency.

The use of logarithms, tables of powers, roots and reciprocals, the slide rule, and other means of facilitating computation.

Approximations and significant figures.

Section 4 includes:

Negative and fractional exponents.

The general treatement of radicals (probably without imaginary numbers).

Simultaneous equations. One linear, one quadratic.

Elementary theorems in probability.

The exponential curve and the theory of logarithms.

Correlation.

Training with general formulas, especially the binomial theorem and the progressions.

Computations rarely used, such as multiplication and division by polynomials and factorization of polynomials.

Section 5 includes:

Simultaneous equations, both quadratic.

Literal equations other than useful formulas.

Imaginary numbers.

Equations higher than the second degree.

Fitting curves to observations.

It is, of course, the case that these sections are not clean-cut, but overlap somewhat, certain features of a topic being more useful and others less useful than the general place assigned to the topics would indicate.

The abilities in Section 1, which can be acquired in a couple of weeks of the course, are probably many times as useful[1] to the general run of high-school pupils as all the abilities of Section 4, on which twelve weeks or more are often spent. If we had sufficient knowledge to measure the utility per hour of time spent we should, in fact, probably find it diminishing at least as rapidly as is shown in Fig. 11.

The fact of diminishing returns in the way of direct practical utility probably characterizes many of the high-school subjects. The awareness that nature is regular, lawful, and predictable, and a few chosen facts about mechanics, heat hydrostatics, and electricity, teachable in a month, would probably outweigh for utility to the general run of high-school pupils as they live today, what they might learn in the next two months, or in the five thereafter, or in an entire second year. The acquisition of a vocabulary of five hundred French words in fifty hours and such common facts and laws of French grammar as could be learned in another fifty hours would outweigh, for utility to the general run of high-school pupils as they live today, what they would

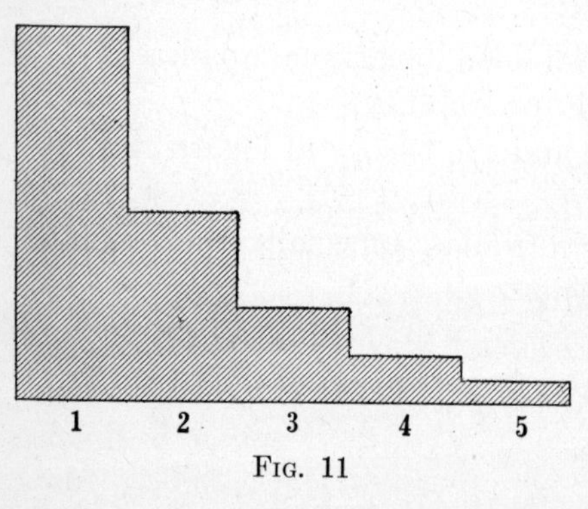

Fig. 11

[1] "Useful" and "utility" are used here as hitherto in this article of the direct practical utilities in contrast to utility as training, but including utility as preparation for other studies.

acquire in the 500 hours spent in extending the vocabulary from 10,000 to 12,500, and in acquiring the mastery of subtle matters of form and syntax.

With certain individuals, the case may be very different. The last thousand hours of study which makes him the world's authority on some field of mathematics may be for a certain person the most useful of all. It may for example quadruple his service to the world and double his salary. It is useful for the world not only to have many persons competent in certain often used abilities but also to have a few persons competent in specialties. If a high-school pupil is cut out by nature to be a scholar and discoverer in mathematics, the chief utilities of the course in algebra for that pupil are not those which we have been discussing but the amount of force with which the course in algebra helps him on toward being a mathematician and advancing the world's knowledge.

Actual, Possible, and Desirable Uses of Algebra

We find in the case of arithmetic that people often do not use it when it would be to their advantage to do so, because they have not mastered it. A woman seeing "$7\frac{5}{8}$lb. roast beef, $2.67" may spend five minutes in telephoning the butcher to ask how much the roast was per pound, because she cannot readily divide by a fraction. Many people waste much time in dividing by long numbers when it is much easier to multiply by their reciprocals. The same may be true of algebra. Writers may use long and clumsy verbal descriptions because they lack the ability and habit of framing a formula to describe a state of affairs or relation. Very little use of algebra is made by accountants, but this may be due in part to traditions from the bookkeeping of the small shop whence scientific accounting developed, rather than to essential needs and desirabilities.

It is even possible that the writers of the textbooks which we surveyed occasionally failed to use algebra in ways that would have been desirable, because they lacked the required knowledge or skill.

The Values of Algebraic Abilities for Direct Usefulness and for Culture

There are notable discrepancies between the serial order in the five sections above and the serial order in which the abilities should probably be put in respect to their value for what we vaguely call culture.

Negative and fractional exponents, which represent a system which is in mathematics what a master poem or painting is in art, or what the law of the conservation of energy is in empirical science, rank very low for direct usefulness. The use of logarithms, tables, and slide rule rank in the utility series along with the conception that a relation between two variables may be expressed as the line made by a moving point, or by the equation telling the location of this point with respect to a defined coördinate system. The former would usually be regarded in respect to culture, as like learning to use a crowbar, or take the elevator; the latter is a conception comparable by its unity, range, and power to any epic poem or symphony.

In many important features, however, there is agreement. The general lesson of algebraic symbolism ranks high in both series. Elaborate manipulations and simplifications by canny use of certain combinations and factorizations rank low in both. The study of a variable quantity, expressed by a surface of frequency, and measured by describing the form of that surface and reporting its central value and variability would be left by raters for cultural value not far from its place for usefulness. The progressions as ordinarily taught are about as devoid of culture as of utility.[1]

[1]If they are used as introductions to and parts of more general truths, as by Nunn, they gain notably in cultural value.

CHAPTER III

The Nature of Algebraic Abilities[1]

THE PRESENT CONCEPTION OF ALGEBRA AS A SCHOOL SUBJECT

During the generation from 1880 to 1910 which witnessed the popularization of high schools in America, algebra[2] became fixed as a required first year study, and with a content which I shall call for convenience the "older" content, or the "older" algebra. The "older" algebra sought to create and improve the following abilities: to read, write, add, subtract, multiply, divide, and to handle ratios, proportions, powers and roots with negative numbers and literal expressions, to "solve" equations and sets of equations, linear and quadratic, and to use these techniques in finding the answers to problems. These abilities were interpreted very broadly in certain respects and very narrowly in others. If anybody had asked Wentworth, for example, what negative numbers and literal expressions the pupil should be able to add, he would probably have answered, "Any"; and the pupils did indeed add an enormous variety, including many which were never experienced anywhere in the world outside of the school course in algebra[3]. On the other hand,

[1]This chapter appeared originally in the *Mathematics Teacher*, Jan. and Feb., 1922.

[2]Throughout these articles "Algebra" will be used for what is commonly called in this country "Elementary Algebra."

[3]For example:

1. Add: $4x^4y^5z^6-3x^3y^4z^5+17x^2y^3z^4-8xy^2z^3$; $14x^2y^3z^4+4xy^2z^3+5x^3y^4z^5-3x^4y^5z^6$; $-4x^4y^5z^6-2x^3y^4z^5+4xy^2z^3+19x^2y^3z^4$; $2x^3y^4z^5+5xy^2z^3-7x^4y^5z^6+9x^2y^3z^4$; $-12xy^2z^3+4x^4y^5z^6-15x^2y^3z^4-x^3y^4z^5$; $3x^4y^5z^6+41x^2y^3z^4-x^3y^4z^5+7xy^2z^3$.

Pg. 51, ex. 11; Wentworth: *Elementary Algebra*. Edition 1906 (reprinted from older editions).

decimals were very rarely used, and angles were almost never added, in spite of the definite need for that ability in the geometry of the following year.

The actual content with which these abilities were trained was determined largely by two forces. The first was faith in indiscriminate thought and practice—the resulting tendency being to have the pupil add, subtract, multiply, and divide, anything that could be added, subtracted, multiplied, or divided; and to have him solve any problem that the teacher could devise. The second was the inertia of custom, the resulting tendencies being, among others, to make algebra parallel arithmetic, to continue puzzle problems, to use applications conceived before or apart from the growth of quantitative work in the physical sciences, and to be unappreciative of graphic methods of presenting facts and relations.

The faith in indiscriminate reasoning and drill was one aspect of the faith in general mental discipline, the value of mathematical thought for thought's sake and computation for computation's sake being itself so great that what you thought about and what you computed with were relatively unimportant.

The paralleling of arithmetic was perhaps most noticeable in the order of topics, and in the almost monomaniac devotion to problems with one particular set of quantities and conditions so that there was some one number as the "answer." There was no reason why $a \times a = a^2$ and $a \times a^2 = a^3$ should not have been taught before $a + a = 2a$, but to do so probably never even occurred to the generation of teachers in question. That a general relation as an answer was a much more important matter than the number of miles a particular boat went, or the number of dollars a particular boy had, and more suitable as a test of algebraic achievement—this again hardly entered their minds.

This older algebra survives in whole or in part in some courses of study, instruments of instruction, and examination procedures. As the accepted view of leaders in the teaching of mathematics and in general educational theory, it is, however, now a thing of the past. I shall use the word "algebra" from now on to refer to the algebra which these leaders recommend as content for teaching in Grade 9 (sometimes Grades 9 and 10), or as a part of the mathematics of Grades 8, 9, and 10.

These leaders are not, of course, in exact agreement concerning details of content and degrees of emphasis, but, approximately, they would subtract from and add to the "older" algebra as follows:

They would omit such computations as occur never or very seldom outside of the older algebra. Addition, subtraction, multiplication, and division with very long polynomials, special products except $(a+b)^2$, $(a-b)^2$, $(a+b)(a-b)$, $(ax+b)(cx+d)$, the corresponding factorizations, fractions with polynomials in the denominator more intricate than $a(b+c)$, elaborate simplifications involving nests of brackets, compound and complex fractions, and rationalizations other than of $\sqrt{a}$, $\sqrt{a}+\sqrt{b}$, $\sqrt{a}-\sqrt{b}$, L. C. M.'s and H. C. F.'s except such as are obtainable by inspection—all these are taboo except in so far as some emphatic need of the other sciences or of mathematics itself requires the technique in question. Clumsy traditions in ratio and proportion (such as the use of "means," "extremes," "antecedent" and "consequent") are eliminated. Bogus and fantastic problems are forbidden wherever a genuine and real problem is available that illustrates or applies the principles as well. The actual uses of algebra in mathematics, science, business, and industry are canvassed and merit is attached to those abilities which are of service there. The mere fact that an operation, *e. g.*,

$$(2a^4+3a^3b^2c-7bc^2d-8d^2)(a-3b^2)$$

can be performed is not a sufficient reason for asking school pupils to perform it. The mere fact that a problem can be framed is not a proof that pupils will profit from solving it.

Thus from one-fourth to one-half of the time spent on the older algebra is saved. This is used to establish and improve the following abilities:

To understand formulas, to "evaluate" a formula by substituting numbers and quantities for some of its symbols, to rearrange a formula to express a different relation[1], to compute with line segments, angles, important ratios, and decimal coefficients, to understand simple graphs, to construct such graphs from tables of related values, and to understand the Cartesian coördinates so as to use them in showing simple relations of y to x graphically.

The discussions of Nunn [1914] and Rugg and Clark [1918], and the reports of the Central Association of Science and Mathematics Teachers [1919], of the N. E. A. Commission on the Reorganization of Secondary Education [1920], of the Commission appointed by the Committee of Review of the College Entrance Examination Board [1921], and of the National Committee on Mathematical Requirements [announced for early publication], would, if combined into an average consensus, tally rather closely with the foregoing statement.

Whereas the older algebra, giving in the main an indiscriminate acquaintance with negative and literal numbers and their uses, expected an undefined improvement of the

[1] I. e., "Changing the subject" of the formula, or "solving" for one of the variables without substituting particular values.

mind, this algebra is selective and expects to improve the mind by extending and refining its powers of analysis, generalization, symbolism, seeing and using relations, and organizing data to fit some purpose or question. It expects to improve these greatly for algebraic analyses, generalizations, symbolisms, and relations and for the organization of a set of quantitative facts and relations as an equation or set of equations, and hopes for a profitable amount of transfer to analyses, generalizations, symbolisms, relational thinking and organizations outside of algebra. It expects further to give better special preparation to see the more direct needs for algebra in life at large and to use it to meet them effectively.

This program for algebra is fairly clear and comprehensible as educational programs go. Nevertheless, a hundred teachers and a hundred psychologists and a hundred mathematicians who should try to act on it as stated, would probably do three hundred things, no two of which would be identical.

We need fuller and more exact statements of the nature of algebraic abilities and of the uses of algebra in mathematics, science, business, and industry. In particular we need clearer knowledge of what is, and what should be, meant by "ability to understand formulas," "ability with equations," "ability to solve problems," and "ability to understand, make, and use graphs." Still more do we need clearer knowledge of what "analysis," "generalization," "symbolism," "thinking with relations," and "organization" mean.

ABILITY TO UNDERSTAND AND FRAME FORMULAS

The ability to understand formulas may mean simply the ability to understand the face value of the symbols involved.

Such is the case when a pupil understands that $A=p+prt$ means the "amount is equal to the principal plus the product of the principal, rate, and time;" or, being given the formula and also,

Let the case be one of simple interest, and let the interest accrue without fixed reinvestment,

Let A = the amount in dollars,
Let p = the principal in dollars,
Let r = the percent paid per year for the use of the money, and
Let t = the time in years,

he understands that $A=p+prt$ means "Fill in p, r and t and A will be the correct amount."

The ability to understand formulas may, however, mean the ability to understand the face value of the symbols and also to supply such units and make such interpretation of the situation and the result of using the formula as fits the case and insures the right answer. Thus if, in the case above, the pupil was given only $A=p+prt$, and knew when and how to use it he would really understand much more than the formula. Many pupils, for example, who could translate $A=p+prt$ and use it as they had been taught to do habitually would fail with "What would be the amount of 74 pounds at 1 percent per month after eight years, the interest being paid every 2 years but left uninvested?" They would not know how to use the formula or even perhaps whether to use it.

The extent to which pupils shall be expected to read between the lines of a formula, knowing when it applies and when it does not, and choosing such units that the result will be correct, is a matter of dispute in theory and practice.

On the one hand it is argued that such interpretations are a matter of physics or geometry or business practice or the like, not algebra, and also that the mixture of such interpretations with rigorous mathematical thinking lessens the instructiveness of the latter. Algebraically, for example, it is correct if $A=p+prt$ and $I=\frac{E}{R}$ to conclude that $AI=\frac{pE+prtE}{R}$. That it happens to be nonsense to say that the amount of money times the current equals the principal times the voltage, etc., is not for the learner of algebra to know or care. So the extremists might argue.

On the other hand, it is argued, first, that algebraic technique divorced from its applications to lengths and weights and dollars and years and amperes and volts is a barren game; second, that absolute clearness and rigor in the statement of formulas so that nothing needs to be read between the lines spoils the best feature of a formula, its brevity. Only two principles are needed, the extremists on this side would say. First, "Use formulas only in ways such as common sense and the facts of the case tell you are reasonable." Second, "Use such units that the answer will be right."

From the point of view of the psychology of the learner either extreme seems tolerable, provided it is operated with consistency and frankness, and provided, in the case of the second plan, too much sacrifice of comprehensibility to brevity is not made. The learner may be taught to insist that every symbol in a formula be defined as a quantity, expressed as a number of such and such units, and to separate sharply his operations with a formula from his choice of which formula. He would then simply refuse to try to operate with most formulas as commonly given. $I=\frac{E}{R}$

would have to be defined as, if $I=$ the current in amperes; E$=$ the potential in volts and $R=$ the resistance in ohms,— then $I=\frac{E}{R}$. $A=\frac{1}{2}BH$ for a triangle would have to be extended to:—"Let $B=$ the number of inches in the base of the triangle. Let H$=$ the number of inches in the altitude of the triangle. Let $A=$ the number of square inches in the area of the triangle. Then $A=\frac{1}{2}BH$. If B and H are numbers of feet, A will equal the number of square feet in the area of the triangle," etc., etc. If he chooses the right formula the result of correct computation is *ipso facto* the right answer.

He may, on the contrary, be taught that most formulas, such as $I=\frac{E}{R}$ or $S=at+\frac{1}{2}at^2$, are simply hints to guide memory and thought in framing the right choice and arrangement of symbols and numbers, and that he is responsible for that arrangement, and for the interpretation of any results of evaluating or solving it.

The former plan secures abilities easier to learn and requires less skill in the teaching; the latter secures abilities which are perhaps more educative and a better preparation for dealing with formulas as they actually occur in books on science and technology. If so, however, it is because time and thought are spent in the algebra course in learning science and technology, or in solving ambiguities of statement by reasoning out what probably is or should be meant.

The greatest danger in the second plan is in the pupil's framing of formulas. Suppose, for example, that he is told to express in a formula the fact that Profit equals Sales less the Number of Articles Produced times the Production Cost per Article, less Selling Costs plus Overhead, and writes $P=F-NC_p-C_s+O$. Is he to be blamed? His algebra and symbolism are correct. It is only his knowledge of business facts and terms that is at fault. If he writes $P=F-NC_p-C_s-O$, is he to be praised? C_p in actual business may well be a number of cents, not dollars, so that his formula may produce a preposterous answer. Or suppose that he is asked to frame a formula for the number of acres in a rectangular plot, the length and width in feet being given, and writes $A=\frac{lw}{a}$. This is true enough if a is correctly defined "between the lines" as the number of square feet in one acre, but of what use is it? In *framing* formulas it seems best to teach the pupil to demand such rigor and adequacy in the conditions given to him that his task is simply translation into an arrangement of numbers and symbols and to demand the same rigor and adequacy of him.

In *reading* formulas it seems reasonable to train the pupil to a certain extent to read between the lines, to be judicious and consistent in his selection of units, and in other respects to use formulas as suggestions and clues rather than as adequate, unambiguous rules. It will be convenient and probably sometimes necessary for him to do so in his actual contacts with formulas in books and elsewhere.

In either case the pupil may profitably understand that from the moment that he begins to operate with the formula until he completes the operations by reaching the desired result or "answer" all the symbols are simply numbers. Nothing needs to be labeled as inches, feet, dollars, years, volts, ohms, foot-pounds, or the like during the operations. What the quantities are must be considered before operating in choosing or framing the formula, and after operations are done in order to put the right label or interpretation on the "answer." But for the purposes of operation Amperes $=\frac{\text{volts}}{\text{ohms}}$ is just like Ans. $=\frac{\text{Number } a}{\text{Number } b}$ or $x=\frac{a}{b}$.

Other things being equal, genuine formulas useful in mathematics, science, industry and business are to be preferred for training in understanding, evaluating, transforming, and framing formulas. Other things, especially convenience, are not always equal. The genuine formulas that are of significance to pupils may be too simple, or too much burdened with long numbers, and there may not be enough of them to give the practice considered necessary. So teachers and textbooks tend to make up formulas of just the desired complexity, involving just the relations with which practice is needed, and with just as little or much numerical difficulty as the occasion demands.

The use of these artificial formulas is not essentially more vicious than the use of multiplications like 465×9817. It is probable that not one pupil in a hundred will ever have to multiply 9817 by 465. But we do not object to such work in moderation in arithmetic because the elementary abilities practiced are all useful; and this is a good way to give them practice. In the same way practice with a formula like $P = M^2N + \frac{M(O-N)}{N}$ may be defensible although the formula has only a very slight probability of occurrence outside of school.

ABILITY WITH EQUATIONS

Ability with equations includes two groups of abilities which are, at least psychologically, very different. The first is to manipulate the equation so as to obtain a numerical value for the literal element, or so as to obtain a value for one of the literal elements in terms of the others. The equation is "solved." The second is to understand the equation as the expression of a certain relation whereby we can correctly prophecy what value a certain element will have, according to the values which one or more other elements have.

Thus $\frac{Q}{2}=KR+4$ is "solved" for Q by finding that $Q=2KR+8$. $\frac{Q}{2}=KR+4$ is "understood with respect to the relation between Q and R" by understanding that if K is a constant, Q is in direct proportion to R, that if Q is expressed as ordinate to fit R abscissa value, we have a set of points on a straight line cutting the y axis at $+8$, with a slope depending on what K is, and that every increment added to R produces, other things being equal, an increment of $2K$ in Q. The older algebra neglected the second ability almost entirely, and even yet the first ability is given far more time and attention in most textbooks, courses, and examinations. Yet the second ability seems of equal or greater importance.

There are three cases of "solving." First, the pupil is taught to organize all the data needed to secure the answer to his problem in the form of an equation with x or n or Ans. or ? or an empty space to be filled, to represent his desired result. "Solving" then means the computation needed to get the x or n or Ans. or ? or empty space on one side of = and to get the other side free from the x or n and, where desirable, in simplest form. Sometimes two or more equations with x and y or n_1 and n_2 or Part I of Ans. and Part II of Ans. are used. The ability to manage this organization and manipulation of data is useful. The problems of life, when of this sort, almost never lead to quadratic equations. The computations are rarely literal.

Second, the pupil is taught to "solve" a formula or equation already organized, as when he derives a formula for finding the radius of a circle from its circumference, from $C=2\pi r$.

Third, the pupil is taught to solve equations of the type $y=ax+b$ or $y=x^2+ax+b$ for a and b, being given related pairs of values of x and y, and to discover the two values of x corresponding to a given value of y in equations of the type, $y=ax^2+bx+c$. This third sort of "solving" is valuable if a certain mastery of the "understanding of the relation" has been attained. Otherwise it is dangerously near to being an aimless mental gymnastic. The older but still common practice of solving quadratics only for the case where $y=0$, out of all relation to the general problem, seems indefensible. The only argument in its favor seems to be that x in $ax^2+bx+c=0$ is an unknown quantity and that you should therefore find its value regardless of whether knowledge of its value is of any consequence.

The understanding of equations as the expression of relations, goes straight to the heart of all applied mathematics, showing the formula and equation as the story of a rule or law which certain events in nature follow or approximate; it introduces the most important idea of mathematics, that of quantitative dependence or functionality; it is a vital and potent review of the principle that algebra tells what will happen to *any* number under certain conditions; it furnishes a principle of organization for graphics; it furnishes the treble parallelism between certain important relations, certain graphs, and certain equations which will arouse respect for algebra.

It may be retorted that the understanding of an equation as a story of a relation or law is too hard and varied an ability for pupils in the ninth or tenth grade to acquire, in comparison with the more mechanical and uniform "solving." We shall see in a later chapter that it has been made needlessly difficult by unfortunate usage of terms and unwise building up of certain mental habits which get in each other's way and trip each other up; and we shall there show how to reduce these difficulties greatly.

ABILITY WITH PROBLEMS

One of the most revered features of algebra is training in organizing a set of facts given in a problem described in words, into an equation or set of equations such that solving will produce the desired answer. This, so far as the problems are genuine ones whose answers a sane person in the real world might seek, is admirable. The genuine problems appropriate to a reasonable life do not, however, often lead to such fractional expressions, or quadratic equations, or denominators that need to be rationalized, or sequences wherein x appears three or four times, as are being studied and await "application to problems." The genuine problems are mostly of a type where x appears once, the other elements being numbers, and require only straightforward arithmetic plus certain conventions with respect to parentheses. Consequently, problems have been made up to give the pupil training in applying his more subtle algebraic techniques. These pervade the textbooks, courses of study, and examinations. The following are samples:

The earth and seven other planets revolve around the sun. Twice the number of planets which are nearer to the sun than the earth is plus one equals the number of planets which are farther from the sun than the earth is. Find the number of planets nearer to and the number more remote from the sun than the earth is.

If a railroad train is made up of five sleeping cars, one parlor car, and a certain sort of engine, its cost is $129,200. The cost of each sleeping car is $300 more than the cost of the engine. The cost of the parlor car is five-sixths of the remainder when the value of the engine is diminished by $100. Find the cost of the engine, the parlor car, and of a sleeping car.

The front wheel of a cart makes 16 revolutions more than the rear wheel in going 360 feet. If, however, the circumference of the front wheel were increased by a third, and that of the rear wheel by a fifth, the front wheel would make only 10 more revolutions than the rear wheel in going the same distance. Find the circumference of each wheel.

The hours required by Mr. A to travel a certain distance equals the number of miles he travels per hour. Mr. B goes the same distance in 2 hours less time by going three miles more per hour. Find the rate of travel and time taken by Mr. A.

There are two angles, one of which is 5° less than the other. If the number of degrees in each is multiplied by the number in its supplement, the product obtained from the larger of the given angles exceeds the other product by the square of the number of degrees in the smaller of the given angles. Find the angles.

Shall the ability to solve problems mean the ability to solve such as these, or shall it mean the ability to solve genuine problems of a sane life?

It has been customary to select and arrange the problem materials almost wholly from the point of view of the algebraic technique to be applied. The teacher or textbook maker, having taught the pupil how to operate with algebraic fractions, for example, looks about for problems which will lead to fractional equations. Other characteristics are treated as of minor importance. From the functional point of view, emphasizing ability to use algebra in solving problems which life will offer, it seems desirable to consider the lives of boys and girls and men and women as students, citizens, fathers and mothers, lawyers, doctors, business men or nurses, and select problems which they may usefully solve and which are properly solved by algebraic methods. These may then be arranged according to the technique involved if this is desirable; but from the functional point of view much is to be said in favor of arranging them with consideration also for their natural connections in the world of fact and their logical connections in the mind. Problems about public health, for example, may well be on the same page even though one should involve a simple equation without fractions, one a fractional equation, one a radical, and one a quadratic.

The arrangement is of much less consequence than the choice. Most important of all is the general choice of attitude — are problems to be an exercise ground for algebraic insight, or is algebra to be a tool for life's needs?

The quantitative problems of life usually come in connection with real things, events and relations. There are real fields and floors, or at least maps and house plans; real ships to be navigated, guns to be pointed, alloys to be compounded, medicines to be diluted, electric cells to be connected. The abilities eventually desired are preferably abilities to deal with real situations.

Since it is inconvenient to provide these real situations in schools, we have recourse to verbal descriptions of them. We cannot, however, take it for granted that the ability to manage a certain quantitative problem as it is described in words is identical with the ability to manage the same problem when it actually arises in a real situation. The difficulties of the described problem may be largely linguistic; to take an extreme case, a person obviously could not solve a problem no matter how easy it was in reality, if it were put to him in an unknown language. On the other hand, the verbal description may be far more suggestive of the procedure to follow than the real situation would be. If a boy should think "In how many years shall I be half as old as my father?" and proceed to solve the problem he would have to know enough to ask, "How old is he now?" "How old am I now?"; and he might puzzle about exact birthdays, even the time of day of birth or allowance for leap years. In the described problem, "A boy is 14. His father is 40. In how many years will he be half as old as his father?" he is given all the data needed, and encouraged not to bother about anything more than getting a certain number which, if added to 40 and 14, makes one result twice the other.

In proportion as we retain the older view that the main value of problem-solving is its formal disciplinary value as a mental gymnastic, the distinction between ability with a problem as it occurs in reality and ability with a similar problem described in words approaches zero importance. We may even deliberately make the verbal description much easier or much harder to understand than the corresponding real situation would be.

This view is, however, hardly tenable with respect to problems that assume to have anything to do with real situations of business, science, technology, or the home. If problems have only formal disciplinary value as mental gymnastic, we may as well use unreal problems about the square of somebody's age being equal to the cube of half of his age less $2\frac{1}{2}$ times his age, or about consecutive numbers, or about fractions whose numerators and denominators are related in divers ways. If we have problems about realities at all, we probably have them to train the pupil to manage realities themselves rather than to manage words about them.

One special difference between problems arising in connection with situations actually present to sense and the customary verbal problems of the algebra class is that in the former needed data may be missing and irrelevant data may be present in large numbers, whereas in the latter we have practically accepted it as a rule that every problem should be solvable from the data given without any further additional data and that all the data given in the problem must be used in order to attain the solution.

There may be in use in some schools problems where the pupil has to decide whether the data are adequate for its solution and problems where he searches elsewhere for the additional data needed, but they are surely very rare. Problems where the data to be used in framing the equations have to be chosen from amongst many data irrelevant for that particular problem are almost equally rare.

Our custom in schools of making each problem a little paragraph of statements all to be used without adding or subtracting one jot or tittle, like the pieces in a picture puzzle, is so fixed that we find one of our standard textbooks announcing the following:

1. Every problem gives a relation between some unknown numbers.
2. There are as many distinct statements as there are unknown numbers.
3. Represent one of the unknown numbers by a letter; then, using all but one of the statements, represent the other unknowns in terms of that same letter.
4. Using the remaining statement, form an equation.

[Wells and Hart, 1912, p. 100.]

Is problem solving in algebra to be only a puzzle game of fitting translated phrases into a proper equation?

There is, then, a wide range of possible opinion about what problem solving does mean and about what it should mean. We should surely try to make it mean something more educative than the solution of more or less attractive puzzles made up to exercise algebraic technique or to give indiscriminate practice in "thinking." We shall return to this subject later when the attempt will be made to clear up the psychology and pedagogy of problem solving in algebra.

The psychological demands and the psychological effects of organizing the facts of a problem into an equation or equations such as will, when solved, give the desired answer varies greatly according as the problem is, so to speak, an "original" which the pupil thinks out, or follows other similar problems which he has learned to handle by special training *ad hoc*.

The approved theory has been, and still is, that the former should be the process in the main, but skillful teachers seem to think or fear that actual proficiency in solving the problems that are met in life (or at least those that are met in examinations) is best secured by special training in certain routines such as "let x equal the smaller number, when there is a choice," or "be sure to use all the numbers that are given," and by still more specialized training with rate problems, mixture problems, tank-and-pipe problems, and the like.

Rugg and Clark argue that directed practice with many different kinds of problems will make the pupil "able to use the method in solving any kind that you may happen to meet later" [1918, p. 208]; and provide this directed practice with (I) Problems relating to age, (II) Problems in which a number is divided into two or more parts, (III) Problems based on coins, (IV) Problems based on relations between time, rate, and distance, (V) Problems involving percents, (VI) Problems concerning perimeters and areas, and (VII) Problems based on levers. Their treatment of the first group is as follows:

"Section 97. *Need for tabulating the data of word problems.*— Many problems involve so many different statements that it is practically *necessary to arrange the steps in the translation in very systematic tabular form.* Take an example like this:

"'John's age exceeds James' by 20 years. In 15 years he will be twice as old as James. Find the age of each now.'

"Before we can write this statement in the form of an equation we must express in algebraic form *four* different things: (1) John's age *now;* (2) James' age *now;* (3) John's age in 15 years; and (4) James' age in 15 years. These four facts can best be stated in a table like this:

"(First step) Let n represent James' age *now*.

"(Second step) Tabulate the data.

TABLE

	Age Now	Age in 15 Years
John's age	n+20	n+20+15
James' age	n	n+15

"With all the facts expressed in letters we can now state the equation which *tells the same thing as the original word* statement; namely:

"(Third step) $n+20+15=2(n+15)$

"We are now ready for the

"(Fourth step) the solution of the equation; the steps are as follows:"

Then follow explanations of solving and checking and sixteen carefully graded tasks leading up to five of the customary age problems such as, "A man is now 45 years old and his son is 15. In how many years will he be twice as old as his son?" [1918, p. 208.]

Shall we treat the problem material as a series of originals, or as a collection of typical groups of problems, each group of which the pupil learns how to solve much as he learns how to subtract a negative number or multiply x^a by x^b by adding exponents? Or shall we treat part of the problem material in the one way and the rest in the other? This last is just what we do in arithmetic. Certain groups of problems (as about areas, perimeters, discounts, insurance, compound interest, taxes and commissions) are prepared for by special training with each. Certain other problems are left to the undirected ingenuity of the pupils.

ABILITY WITH GRAPHS

The ability to understand, construct, and use graphs needs definition in at least two respects. (A) Shall school algebra present elementary facts concerning all graphs that are simple and important, or shall it limit itself to the graphic presentation of a relation between two variables through the Cartesian coördinate system, and to such introductory matter as facilitates that? (B) Assuming the second answer to A, shall it deal with graphs of irregular relations not presentable in any equations which the pupil can be expected to manage, or shall it restrict itself to the straight line, parabola, hyperbola, and the like? Question A may be made clear and vivid in this form: "Which of the types of graphs on pages 115 to 117 shall the pupil be taught to understand, construct, and use?"

The graphs which are simple and important may be classified: (I) descriptions of the way in which a certain

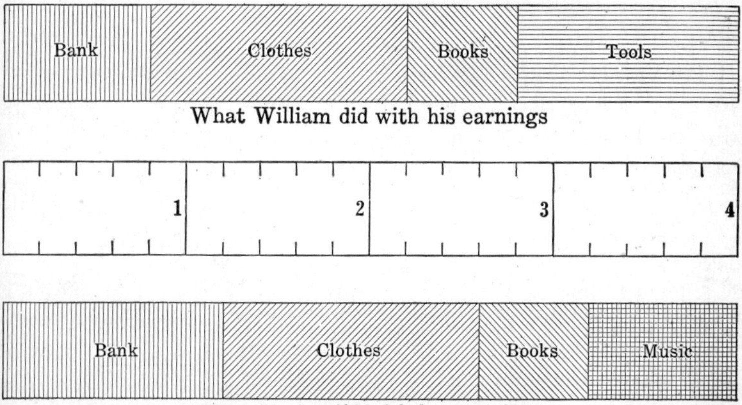

Bank Clothes Books Tools

What William did with his earnings

1 2 3 4

Bank Clothes Books Music

What Louise did with her earnings

Fig. 12

quantity is divided; (II) general comparisons of two or more magnitudes; (III) comparisons of two or more magnitudes which are put in order by their relation to some characteristic; and (III-A) comparisons of two or more magnitudes which are themselves frequencies of occurrences and are put in order by their relation to the magnitude of some characteristic.

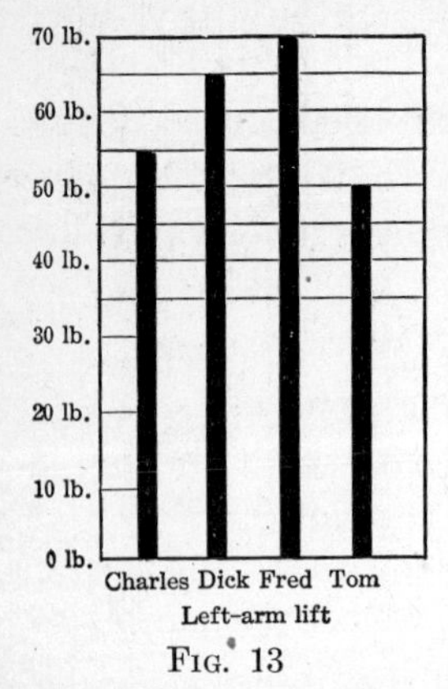

FIG. 13

TYPE I. DIVISION OF A QUANTITY

1. These diagrams show what William and Louise did with the money which they earned last summer. Make a table showing what each of them did with the money. What percent of his money did William put in the bank?

2. What percent did he spend for clothes? For books? For tools?

3. What percent did Louise spend for clothes? For books? For music?

TYPE II. COMPARISON OF MAGNITUDES

1. According to the diagram how many pounds can Charles lift? How many can Dick lift? Fred? Tom?

2. Draw a diagram to show how many of the exercises in the arithmetic practice each of these children did correctly, in 5 min.:

Alice, who had 30 correct
Anna, who had 19 correct
Nell, who had 24 correct
Sarah, who had 36 correct

Let $\frac{1}{8}$ in. of distance up and down equal 1 exercise correct, 1 in. equal 8 exercises correct, etc.

TYPE III

COMPARISON OF ORDERED MAGNITUDES

3. Using thin paper trace and complete this diagram or graph, which tells how Dick Allen improved in the broad jump. His records were 8 ft. in 1911, 9 ft. in 1912, 9 ft. 6 in. in 1913, 10 ft. 6 in. in 1914, 13 ft. in 1915, and 13 ft. 6 in. in 1916.

4. Draw a diagram or graph to show how Elsie improved in repeated trials with the practice test.

Her scores in the 10 successive weeks were 17, 17, 19, 23, 22, 23, 24, 25, 24, 26.

5. Do the practice of page —— twice a day for five days. Draw a graph showing how well you did in each trial and how much you improved.

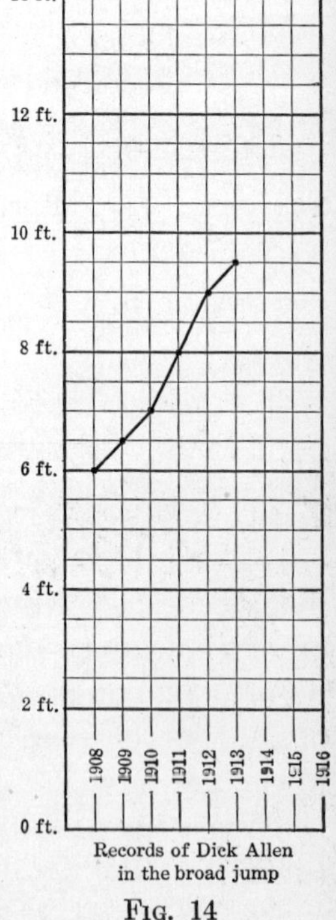

Records of Dick Allen in the broad jump

FIG. 14

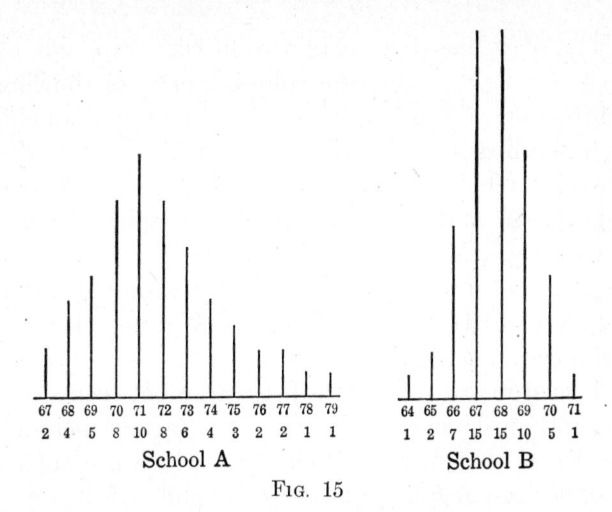

FIG. 15

TYPE III-A. GRAPHS OF RELATIVE FREQUENCIES

5. Examine the diagrams. Read the diagram for School A, saying, "On 2 days, the temperature was 67°; on 4 days it was 68°; on 5 days it was 69°"; etc. Find what percent of the days were "satisfactory" as to temperature in School A. In School B. (A temperature of 66, 67, 68, 69 or 70 is called "satisfactory.")

Figure 12 is an illustration of graphic descriptions of the way in which a certain quantity is divided, as when we wish to show how a family spends its income, or how the population of a country is divided with respect to race or how a pupil spends the day.

Figure 13 is an illustration of general comparisons of magnitudes, as of the number of boys and of girls in a school, or of the values of certain products of a farm, or of the number of voters of each political affiliation, or of the size of a school ten years ago and now, or the per cent of illiteracy in each of forty states.

When the magnitudes to be compared form a series easy to put in order by their relation to some characteristic, this is usually done. If, for example, we have the size of a school now, five years ago, ten years ago, fifteen years ago, etc., it is obviously best to put the columns or bars showing size in a chronological order. If we have the number of children at each age in a school, it is obviously best to put the columns or bars in the order: aged 5, aged 6, aged 7, etc.

Figures 14 and 15 illustrate such graphs of class III. The typical graph of the relation of one variable to another is the most notable case of class III. Graphs of surfaces of frequency or distribution (III-A) are a group of increasing importance in the social sciences.

It is customary to draw a distinction between statistical graphs and mathematical graphs. The former include all of classes I and II and those of class III which are not readily analyzable into regular relations conveniently expressed in the relation of y to x. This distinction is not sharp or rigid, some weight being also given to custom and convenience.

Return now to our question A, "Shall school algebra deal with graphs of classes I and II, or shall it limit itself to class III?" The future will probably save us the trouble of answering, because acquaintance with graphs of classes I and II, and with simple cases in class III, will probably be given in Grades 4 to 8. Figures 12, 13, 14, and 15 are, in fact, all from a recent textbook in arithmetic. When it has not been given already, such work may deserve a place in ninth grade mathematics because of its intrinsic worth, and because of the interest it lends to the graphic presentation of the relations between two variables.

Graphs of class I and class II are, however, psychologically useful in algebra only as possible introductions to those of class III. For pupils at the high-school level, the erection of a series of columns and the formation of the curve joining the midpoints of their tops, seems adequate. Also it seems to the writer that the development of a serial graph from the graph of mere comparison will introduce more interferences than aids to comprehension of the former. Logically, the graph of a systematic, ordered relation may be thought of as a sub-class of, and development from, the graph comparing any quantities — the population of states, or the scores of ball players, or the heights of species of trees. But it seems sounder psychology to teach the systematic graph, ordered in relation to time, or number of articles, or size of the object, or force exerted, etc., by itself.[1]

The elementary facts of surfaces of frequency deserve a place in the curriculum somewhere in Grades 6 to 9 (preferably in Grades 7 or 8), the pupil being taught to understand such graphs and to construct them from the tabular data. Their relation to the curve $y=e^{-x^2}$, and to the coefficients of the binomial expansion may perhaps be shown toward the end of a course in algebra.[2] Their consideration along with the graphs of simpler relations of one variable to another will be confusing and should be avoided.

Our second question was: Assuming that algebra in Grades 9 and 10 limits itself to "mathematical" graphs of the relation of one variable to another and to such introductory matter as is of value therewith, shall it deal with graphs not presentable in any equations that the pupil can master, or shall it restrict itself to the straight line, rectangular hyperbola, parabola, circle, and the like? It may be answered provisionally as follows: Enough work with irregular relations should be given so that the pupil will appreciate the regularity of regular ones by contrast, and also the place of these regular relation lines amongst relation lines in general. Except for that, it seems best to spend

[1] That is, to make the principle of organization the facts and laws to be expressed, rather than the general nature of graphic presentations.

[2] More suitably, probably, in an advanced course.

the time on standard forms whose mathematical significance can be realized.[1] The curves for $y=ax$, $y=ax+b$, $y=x^2$, $y=x^2+a$, $y=x^3$, $y=x^3+a$, $y=\frac{a}{x}$, $y=x^2+ax+b$, $y^2+x^2=a^2$, and $y=a^x$ will probably provide sufficient variety to make the principles general.

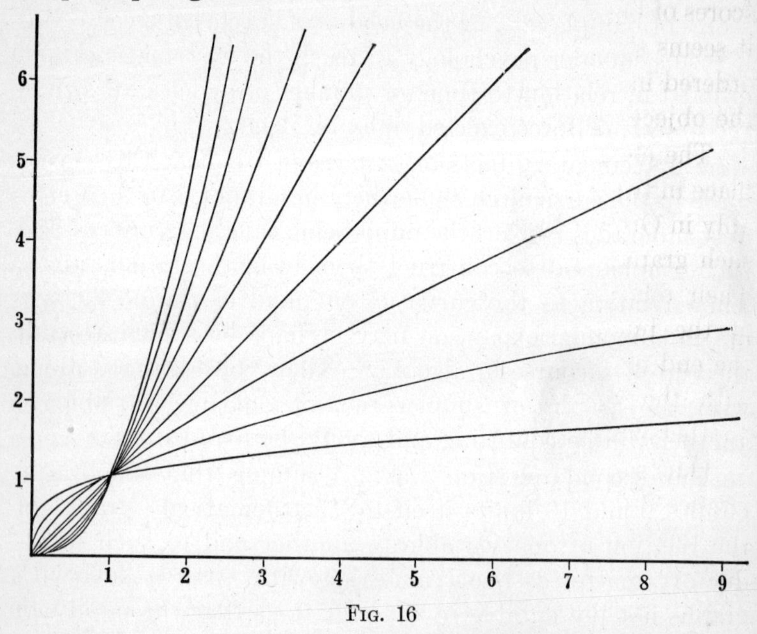

Fig. 16

When the topic of fractional exponents is taken up, a review and extension of the graphic treatment of regular relations may be had by presenting the curves for $y=x^{\frac{1}{2}}$, $y=x^{\frac{2}{3}}$, $y=x^{\frac{3}{4}}$, $y=x^{\frac{3}{2}}$, $y=x^{\frac{4}{3}}$, and the like, as in Fig. 16. If logarithms are taught, the construction of such curves provides excellent exercises in computation and approximations, and a useful application of knowledge of fractional exponents. Simple facts about $y=k^x$ may be taught also.

[1] Including, perhaps, standard forms obscured by chance errors which disappear in a coarser grouping of the data.

CHAPTER IV

The Psychology of the Equation[1]

The equation has two different uses. First, it is an organization of data in such a way as to indicate the operations required to obtain a certain numerical result, this result being the answer to a question which led the worker to frame the equation. So $60-x=x-45$ is a good way to organize data to answer the question, "What number is as much less than 60 as it is greater than 45?" The equation is here a thing to be *solved.* Second, the equation is the expression of a relation between a variable and one or more other variables. The important thing in this case is to understand the relation or law.

So in $y=kx$, or $y=\frac{k}{x}$, or $y=x^2$, or $x^2+y^2=k^2$.

In the first case the equation may, of course, represent a special instance of some important relation or law to be understood, and in the second case the equation, then or later, may be solved for some special values of the variables. But in the great majority of cases one or the other purpose is primary, as stated above. The difference is recognized to some extent in the early distinction between (*A*) organizing numerical data into an equation with x or two equations with x and y, and (*B*) framing a general formula or equation.

[1] This chapter appeared originally in the *Mathematics Teacher*, March, 1922.

Consider these samples of *A:*

1. Ten times a certain number is diminished by 6, the result being 36 more than four times the number. What is the number?

2. How much water must be added to three pints of a 20 per cent solution of carbolic acid to make it a 5 per cent solution?

3. A man walks at the rate of 3 miles an hour. Two hours after he started another man sets out to overtake him in an automobile going 25 miles an hour. How far will the first man have gone before the second man overtakes him?

No teacher probably expects the pupil to do anything with the resulting equations except solve them. He has to understand certain particular conditions to frame the equation. He does not have to understand the equation as an expression of a general relation or law in order to solve it. Usually he is not expected to.

Consider these samples of *B:*

1. Make a formula or an equation which tells the cost of any number of pounds of starch at 11 cents a pound.

2. Using m, s, and d for minuend, subtrahend, and difference, respectively, what equations can you make from them?

3. Using l, w, and h for the inside dimensions of any rectangular tank in inches, write a formula or equation for the cubic capacity of the tank in gallons, counting one gallon as equal to 231 cubic inches.

4. If x is any even number, what is any odd number?

In these cases the pupil is expected only to frame the formula, not to solve it. To frame it he must understand the general relation or law and, to at least a large extent, the formula or equation as its shorthand expression.

The pupil probably realizes a difference between work like that of *A* and work like that of *B*. Also he may be influenced by being given the name *equation* for things like $10x-6=36-4x$ or $.05(3+x)= .20(3)$, and the name *formula* for things like $C=\pi D$, $m-s=d$, $m-d=s$, or $J=\frac{lwh}{231}$.

From this point on, however, almost everything that he

is taught about equations blurs the distinction. He is given many literal equations like

$$(a-1)x=a^2-1,$$

$$\frac{x}{a+b}=a-b,$$

$$a(a-x)=b(b-x),$$

$$\frac{\dfrac{m+n}{x}}{\dfrac{1}{m}}=\frac{a}{b}.$$

Presumably these represent some important relations between x or y or z and the variables a, b, and c; otherwise they would not be set out as generalized rules. But, in fact, they do not, and all that he is told to do with them is to solve them.

He is taught the coördinate system and set to study $y=x+4$, $y=x$, $y=x-4$, and the like. He is much perplexed because hitherto x has always been unknown but only one number when you finally got it known; whereas now you know what it is, but it is 1 or 2 or 3 or 4, or whatever you like. Also, he has been painstakingly learning in simultaneous equations that you can do nothing useful with x and y unless you have two equations, whereas now you cannot have one equation of the new sort without both of them in it.

To the mathematician or logician these may seem to be childish and trivial perplexities. Childish they may be, but since we are teaching children, childish perplexities are precisely the ones we need to prevent. Trivial they certainly are not, at least to the psychologist. For the most uniform and stable connections or bonds that x or y has formed are with "not known," "to be found," "one number and only one right when you find it." The most frequent and emphatic connection or bond that "x and y both to be dealt with" has formed is with "you must make two equations." The pupil's strongest habits of thought with respect to x and y (with ordinary teaching, *all* his habits of thought with respect to x and y) make the $y=x+4$ a monstrous perversity. The coördinate system and the facts of the linear equation would in fact be more easily taught to the pupil who has had the customary training with equations to be solved, if NS and EW, were used instead of YY and XX, and if V_1 and V_2 were used instead of Y and X. Almost everything in the usual previous study of equations interferes with the understanding of $y=ax+b$.

Conversely, under the customary methods of teaching, the habit of regarding y as a variable whose value depends on the value attached to x, the habit of shifting x and seeing what happens to y, and the habit of thinking how y depends on x rather than hastening to solve for something, are likely to interfere with the old solving habits. The pupil who was wont to proceed readily, and even automatically, to solve for any posterior segment of the alphabet that came into view, now hesitates, wonders whether he is to solve it, or graph it, or evaluate it, or perhaps even consider what it and its context mean!

Partly because of a more or less explicit sense of this interference, the majority of teachers and textbooks retain the disturbing y only as a necessary evil to help explain the coördinate system and the graph of an equation, banishing it soon and replacing "the equation $y=2x+3$" by "the expression $2x+3$," and then quietly shifting to "$2x+3=0$," which can be "solved" in peace. But this shift is destructive to the understanding of $y=2x+3$ as the expression of a straight-line relation or law whereby one variable always equals 3 more than twice another. This treatment also tends to change the coördinate system from an easy and beautiful organization of what is known about equations into a puzzle to be reconciled with what you do with equations. If this peculiar y always becomes 0 before you do anything to the equation, why bother to learn about y? So works the potent unconscious argument of mental habit.

With quadratics this mysterious appearance and disappearance of y is repeated. All his experiences with equations except the brief interlude with $y=ax+b$ almost forbid the pupil to do aught with $y=ax^2+bx+c$ save regard the y as a misprint for 4 or 7. After a renewed exposition of the coördinate system has given him a dawning insight into what such equations mean, the y is spirited away again, and he has the equations of the form $ax^2+bx+c=0$ to solve. Why he should "solve" them he probably has not the faintest notion. An additional degree of mystery is added by now calling the values of x the *roots*.

A final element of confusion is introduced by simultaneous quadratics. The y comes back, but now it is (in many modern courses) not a mere second unknown to be solved for, as it was in $3y-4x=2$, $y-x=1$, but is the y of of the coördinate system. The two equations are not mere *corpora vilia* for solving, but two real relation-lines, the question being, "Do they cross; if so, at what point or points?"

The algebra of a generation ago was free from this confusion because it did not attempt to teach the equation as the expression of a general relation between a variable and one or more other variables, and did not introduce graphs and the Cartesian coördinates. The equation was a thing to be solved and nothing more.

When teachers of mathematics began to introduce the formula, the concept of a general relation or function, and its graphic treatment, two courses were open to them. They could try to assimilate the new aspect to the old, insisting on the new treatment as if it were only an extension and enrichment of the old. Or they could make a clear distinction, almost a contrast, between the equation as an organization of facts to find some unknown or hidden fact and the equation as an expression of a relation between variables.

The force of the teacher's mental habits and a superficial pedagogy favored the former course, and it was taken. For example, graphs were used as an aid in solving, or in checking solutions of, equations which gave specific numerical values of an unknown. Solving simultaneous quadratics was (and is) taught not as a means of determining the constants in an equation expressing a relation between the variables y and x, but as a means of answering such specific questions as:

1. The sum of the squares of two numbers is 130, and the product of the numbers is 63. Find the numbers.

2. A number is formed by three digits, the third digit being the sum of the other two. The product of the first and third digit exceeds the square of the second by 5. If 396 is added to the number the order of the digits is reversed. Find the number.

In almost every way the new aspect was made to seem as far as possible an outgrowth and extension of the old, or at least a peaceable ally of the old.

A superficial pedagogy might defend this as a case of "apperception," of basing the new idea on familiar ideas, of gradually extending the concept of the equation. A deeper psychology shows that the other course is the one that should have been taken and should be taken now. It appears, in fact, that the two aspects of the equation should be kept distinct from the start and to a large extent throughout; that they should, other things being equal, be given different names, taught at different times and in different ways and with different applications.

Teaching Particular Equations

The equation to be "solved" in order to find some particular quantity should appear early in arithmetic, say in Grade 3, in the form

$$5 + \ldots = 9$$
$$3 \times 9 = \ldots$$
$$24 = \ldots \text{ 3's}$$
$$7 \times \ldots = 21$$
$$\ldots \text{ 5's} = 30$$

It should be used freely thereafter in all computations where it is the most serviceable form for thought. For example,

$$\frac{3}{4} = \frac{\ldots}{8} \qquad \frac{3}{4} = \ldots \times \frac{1}{2} \qquad \$24 = \ldots \% \text{ of } \$400$$

It should be used as a way of organizing data in the solution of problems where it is a desirable way. Here *Ans.* or ? or "the number of dollars" may replace the empty space to be filled. This use may be continued in algebra, with such intricacies as are there desirable, but the name for the missing number should under no circumstances be x or y or z. *Ans.* or *Num.* or A or N seem to be the best names ($Ans._1$ and $Ans._2$, etc., being used where two or more equations are framed to state the given facts). Small n (for number) would be a good name, except for the later interference with n for "any number." Q (for question) would be almost as good as Ans. or Num. or N, possibly better. Whenever the equation was used to indicate an exercise in numerical computation or a search for some number, its earmarks would be an equality sign, and Ans. or N. or Q. In so far as problems requiring quadratic equations are perpetuated, we shall have $Ans.^2$ or A^2 or N^2 or Q^2, in their solutions.

Such equations are an organization for convenient computation. To frame them rightly requires sagacious handling of the particular facts and relations of the problem situation. To solve them when they are framed requires competent computation. They may be called equations or equalities or arrangements for solving, or even be given no name at all, so far as learning algebra is concerned. Since the number who study arithmetic is enormously greater then the number who study algebra, this type of arrangement should be called an equation.

This work, so far as done during the study of algebra, should be organized under the principle that "Any real question having a discoverable number as its answer can be answered by putting the data together in a suitable equation and solving, providing the data are sufficient to give the answer." Probably it should be completed before the systematic study of the coördinate system, and of linear equations as such, is begun. The questions answered should be in the main genuine ones. Only a few resulting in elaborate fractional equations are necessary, simply to show that, no matter how intricate the relations, they can be handled by the equation technique.

Teaching General Equations

The equation as the expression of a general relation between variables is prepared for in arithmetic in two ways. The idea of such a general relation has its first stage in the implicit use of rules like "Length in feet = (length in inches) ÷ 12"; "Cost in cents = (cost in dollars) × 100." In Grades 7 and 8 such rules or formulas are more explicitly got in mind, in the case of:—

$C = 2\pi R$, for Number of inches in circumference $= 2\pi$(No. of in. in radius),

$I = PRT$, for Number of dollars interest = (Number of dollars in principal) times (the rate in hundredths) times (the time in years),

$H^2 = S_1^2 + S_2^2$, for the hypotenuse rule, and the like.

The idea of space representation of a relation between two variables becomes familiar in a modern course in arithmetic by reading and making and using as problem matter such graphs as of the growth of a plant, height in relation to age, score in successive practices, population change, rise in costs or wages, and the like.

In algebra the work with formulas will be extended to give training in reading and understanding any formula which expresses correctly any useful relation which could be understood by the pupil in words, in expressing such relations in formulas, and in finding the value of any variable (whose value is worth finding), when the values of the others are given. As a rule, not one such "solving" should be set for any one variable, but many. For example, in $\text{Amp.} = \frac{\text{volts}}{\text{ohms}}$ the task will be, How many Amperes

(a) when volts = 110 and ohms = 22
(b) when volts = 110 and ohms = 25
(c) when volts = 220 and ohms = 20
(d) when volts = 12 and ohms = 2 ?

The use of x and y and z will be avoided in such formulas, as Nunn has advised. The work with graphs will be extended to the comprehension of the Cartesian coördinate system, habituation to y and x as names for the two distances and the understanding of $y = x$, $y = 2x$, $y = \frac{1}{2}x$, $y = x + 2$, $y = x - 2$, $y = \frac{1}{x}$, $y = \frac{4}{x}$, and other instructive relations. Then comes the systematic algebra of systematic selected types of equations. These may be called "equations of variables," or "equations of relation lines." This will include the study of $y=cx$ and $y=\frac{c}{x}$ in connection with "varies directly as" and "varies inversely as," of $y=x^2$ and $y=\sqrt{x}$ in connection with tables of squares and square roots and with interpolating in such tables, and of the quadratic equation in general. It seems preferable to use c_1, c_2, c_3, or k_1, k_2, k_3, rather than a, b, and c to express constants in such equations.

The case of solving for x when $y=0$ will be treated as simply one special case of all the possible solvings. First, y will be found for various assigned values of x, then x will be found for various assigned values of y, including 0, which are specially instructive. Probably graphical solutions, if taught, should come before solutions by computation. Simultaneous equations with x and y will be taught chiefly as a means of answering the question, "Do these relation lines cross? If so, where?"—and as a means of answering the question, "If in a linear equation $y=c_1x+c_2$ one point of the curve is (7,4) and another is (13,6), what do c_1 and c_2 equal?"[1]

In connection with the study of exponents, curves such as $y=x^{\frac{1}{5}}$, $y=x^{\frac{1}{4}}$, $y=x^{\frac{1}{3}}$, $y=x^{\frac{1}{2}}$, $y=x^{\frac{3}{5}}$, $y=x^{\frac{2}{3}}$, $y=x^{\frac{3}{4}}$, $y=x^{\frac{6}{5}}$, $y=x^{\frac{5}{4}}$, $y=x^{\frac{3}{2}}$, $y=x^7$, may be briefly inspected and compared. Some of the practice with logarithms may well be given up to the computations required for plotting a few such curves. The equations $y=a^x$, $(x+c_1)^2+(y+c_2)^2=c_3^2$, and others of notable interest may be studied, if time permits.

Finally the "function" $ax+b$ or the "function" ax^2+bx+c may be studied as a function without any "$y=$ " to introduce it, but it will then no longer be or bear the name of equation.

[1]Similarly, of course, with $y=c_1x^2+c_2x+c_3$.

As a consequence of this reorganization, the indiscriminate practice with what are now called literal equations would be replaced by two distinct lines of work. First there would be given, in connection with real formulas, practice in expressing any one of the variables in terms of the others, that is, in solving for that variable. Second, there would be given, in connection with typical forms of relation lines, practice in understanding the meaning of the constants concerned as well as the meaning of the two variables.

CHAPTER V

The Psychology of Problem Solving[1]

In a previous chapter we called attention to some of the differences of opinion concerning the application of algebraic technique to the solution of problems. In the present chapter we shall take up systematically the questions there suggested, together with others, and present the results of certain investigations which we have made, to aid teachers in deciding what to do with problem solving, when to do it, and how to do it.

At the outset we need to distinguish certain types of work all of which might with some justification be called applications of algebraic technique to the solution of problems, but which differ notably in the psychological demands they make of a pupil, in the psychological effects they have upon him, and in their uses in the algebra course. All are different from mere computation, evaluation, reading or making graphs, or the solution of equations already framed. The more important types are shown on page 133.

The psychology of the I-1 type is clear. Such problems, carefully chosen and graded, may be used very helpfully to teach meanings and to test, strengthen, extend and refine knowledge of meanings. Teachers of algebra should study the problem material of this sort devised by Nunn [1914] and by Rugg and Clark [1918a].

[1]This chapter appeared originally in the *Mathematics Teacher*, April and May, 1922.

Type I. Problems to answer which no explicit equation or formula is needed or supposed to be used.

1. Concerned with knowledge of meanings, *e. g.*, of literal numbers, negative numbers, exponents.

   If pencils cost c cents each what will n pencils cost?

   Express 248 ft. below sea level if sea level is called 0.

   If $a \times a \times a \times a$ is a^4, how will you express the product of a row of n a's?

2. Concerned with knowledge of operations.

   What was your average score in a game in which you made these separate scores: $-8, +4, -6, -2, +7$?

3. Concerned with combinations of meanings and operations or with other aspects of algebra.

   Under what conditions will $a = \frac{1}{a}$?

   In $a = p + \frac{q}{r}$ what will be the effect upon a of an increase in r?

   The value of v_2 for $v_1 = 4$ was omitted from this table by the printer. What do you think it was, approximately?

   If v_1 is 0 v_2 is 4
   If v_1 is 1 v_2 is 5.02
   If v_1 is 2 v_2 is 7.96
   If v_1 is 3 v_2 is 12.98
   If v_1 is 4 v_2 is ....
   If v_1 is 5 v_2 is 29.03
   If v_1 is 6 v_2 is 40.05

Type II. Problems to answer which an equation or formula is supposed to be used.

A. The equation or formula is one of a group of known formulas.

(1) Which of these formulas fits the problem is also known.

$F = 32° + \frac{9}{5} C.$

What does 86° F equal on the Centigrade scale?

(2) Which of them fits the problem is not known. Pupil must select the formula as well as fit the facts properly to it.

The meanings of these are known,

$V = at \quad v = gt$

$S = \frac{1}{2} at^2 \quad S = \frac{1}{2} gt^2$

$v = \sqrt{2aS} \quad v = \sqrt{2gS}$

$g = 32.$

Neglecting the resistance of the air, how far will a bullet dropped from a height of 5000 ft. fall during the sixth second?

B. The equation or formula is not known, but must be constructed by the pupil.

(a) The equation or formula is primarily a rule for all cases of a certain relation.

Frame equations for finding the dimensions of a rectangle twice as long as it is wide to be of a sq. in. area. Find the dimensions when a is 10, when a is 20, when a is 100.

(b) The equation (not usually called a formula) is primarily an organization of data to secure the result in one special case.

A girl wishes to have a rectangular card twice as long as it is wide, and 10 sq. in. in area. How long shall she make it?

Problems of the I-2 type, which apply algebraic computations in useful ways, are very rarely used. This may be because teachers think there is no need of applying computation; or it may be because genuine uses for algebraic computations are not to be found in such matters as a first-year high-school pupil can understand.

The first reason is almost certainly a bad one. Pupils, save some of the very intellectual, are stimulated by seeing what a computing procedure is for, and by associating it with the world outside of mathematics. In arithmetic it is found serviceable to introduce each new item of computational method by some genuine and interesting problem whose solution is facilitated by the computation in question. Students of algebra can doubtless get along without such stimulants better than students of arithmetic, who are younger and duller; and may gain less from them. But they, too, will gain much from such introductory problems showing the service which the computation performs. Again the teacher should examine the problem material of this sort in Nunn and in Rugg and Clark.

The second reason is in part valid. Much of the computational work often done in courses in algebra cannot well be introduced by or related to problems from the world the pupil knows. The problems that one might invent would not make the computation clearer or easier or more esteemed or longer remembered. The value of such computation is questionable.

The miscellaneous group listed as I-3 represents a borderland between what we ordinarily call problems and tasks calling for mathematical inference and conclusions of all sorts. Too little attention has been paid to this group by teachers of algebra. There has been, in fact, so little of such work that we can hardly judge of its value; but it would at least add variety, give more scope for "original" thinking, and assist in integrating a pupil's algebraic abilities into something which for lack of a better term we may call an "algebraic sense"—a somewhat general readiness to see algebraic facts and to think about them with all his algebraic equipment.

The work of II-A represents what is regarded by teachers of the physical and social sciences as the most essential contribution of algebra in preparation for their study. It has the merit that the facts operated with have some chance of being themselves worth thinking about—are not mere valueless items about A's age, or B's time in rowing up a stream, or C's buying and selling of sheep; and that the relations among these facts are likely to be important relations in nature; and that the results obtained are such as a sane man might obtain in that way.

It has also the merits that it emphasizes the fact that a letter can be used to mean any one of the class of numbers that fulfil certain conditions, and that it leads up to the general treatment of the relation of one variable to another.

Two cautions are useful in connection with such work. The first is to be careful not to burden pupils unduly with learning physics or astronomy or engineering for the sake of having genuine formulas. Formulas whose meanings are obvious from a careful reading should be preferred. They should not be complicated by unfamiliar terms, or by the need of difficult inferences to secure consistency in units. Tabular and graphic work may be usefully coördinated with the problem work. The same formulas may often be used in exercises in formula reading and formula framing, evaluation, and the understanding of relations, before they are used in connection with verbal problems.

The problem material from science and engineering that

can be adapted so as to satisfy wise teachers of algebra that it is as good for their purposes as the material in customary use about familiar facts like the ages of boys, the hands of clocks, boats on streams, or tanks and pipes, may be rather scanty. The question then arises of using made-up formulas, such, for example, as:

A boy's father gives him each month half as much as the boy earns.

1. Let e = what the boy earns in any month. What will equal what his father gives him for that month?

2. Let T = what the boy gets in all that month. Make a formula for finding T.

3. What will the boy receive in all in each of these months: January, when he earns $10? February, when he earns $12? March, when he earns $8.60?

4. How much must he earn in a month to get (in all) $20 that month?

It is for such cases that we need the second caution, namely, that we avoid in such made-up formulas the unrealities and trivialities that have characterized so much of the problem material of the past.

We have left the II-B type, where the equation or formula is not known but must be constructed by the pupil. This is problem material par excellence, and is, in fact, all that certain teachers would consider worthy of the name. Within it, the II-B-b type is in far wider use than the II-B-a type. In the rest of this chapter, consequently, unless the contrary is specified, we shall mean by a problem one of the type of II-B-b where an equation is built up, organizing the data given in the problem about some one state of affairs, so as to secure the answer to one or more quantitative questions about that particular state of affairs.

We shall deal with the following matters, in the order given:

The genuineness of the problem.

The importance of the problem.

Shall every technique be applied to problems?

How far shall the problems be worked as originals, and how far shall routine procedures for solving a certain kind of problem be taught?

The overestimation of the educative value of the verbal problem.

The use of problems at the beginning of a topic to show the need for certain technique and to facilitate the mastery of the technique, as well as at the end to test the ability to apply the technique.

Criteria in selecting problems.

Problems as tests.

Real versus described situations.

Isolated and grouped problems.

Problems requiring the selection of data, as well as their organization.

Problems requiring the discovery of data as well as their selection and organization.

Problems requiring general solutions.

Problems of puzzle and mystery.

The election of problems by students.

Genuineness

Relatively few of the problems now in use are genuine. First of all, over half of them are problems where in the ordinary course of events the data given to secure the answer would themselves be secured from the knowledge of the answer. For example, "In ten years John will be half as old as his father. In twenty years he will be three-fifths as old as his father. How old is John now? How old is his father?" In reality such a problem would only occur in the remote contingency that someone knowing that John was 10 and his father 30, figured out these future age ratios, then forgot the original 10 and 30, but remembered what the future ratios were!

We have made the count for three representative text-books of excellent repute with the results shown in Table 25.

TABLE 25

PERCENTAGES OF "ANSWER KNOWN" PROBLEMS[1]

	Up to the Beginning of Fractions	The Beginning of Fractions Up to Quadratics	Quadratics and Beyond	Total
Book A..............	52	69	54	57
Book B..............	45	36		42
Book C..............	53	52	51	52

Such problems, if defensible at all, are defensible as mental gymnastics, and as appeals to the interest in mystery and puzzles. As such, they are better if freed from the pretense at reality. "I am thinking of a number. Half of it plus one-third of it exceeds one-fourth of it by seven. What is the number?" is better than problems which falsely pretend to represent sane responses to real issues that life might offer. It is degrading to algebra to put it to work searching for answers which in reality would have been present as the means of framing the problem itself, save frankly as a mere exercise in sharpening one's wits and in translating a paragraph into an equation.

Of the problems which are not clearly ruled out by this criterion many concern situations or questions or both which

[1]An "Answer Known" problem is one where it is highly probable that in real life the data given would be obtained from the answer rather than the answer from the data. The totals from which the percentages are computed include problems of Type I and Type II-A. If only II-B problems were considered, the percentages of "Answer Known" problems would be much higher.

are not genuine, because in the real world the situation would probably not occur in the way described or because the answer would not be obtained in the way required.

We can make a scale for genuineness running from problems that are fantastic to problems that are entirely genuine. Zero (0) genuineness (or reality) is defined as, "Would never occur as a problem in life, in whole or in part; nor would anything at all closely like it." 10 genuineness (or reality) is defined as, "Could occur just as it is in every detail, quantities, relations, and all." 1, 2, 3, 4, etc., are defined as an arithmetical series from 0 to 10 genuineness. We may set as a very charitable criterion that a problem should be as high as 4 on this scale, 4 being the average genuineness of the following problems.

REALITY 4

A. Three men are asked to contribute to a fund. The first agrees to give twice as much as the second, and the third agrees to give twice as much as the first. How much must each contribute to make a total of $1050?

B. What angle is five times its complement?

C. A boy knows that his boat can go 6 miles per hour with the current and 3 miles per hour against the current. How far can he go and return making the whole trip in just 3 hours?

D. The diagonal of a rectangle is 102 inches and the base of the rectangle is three times its altitude. What is the length of its base?

E. The principal varies directly as the interest and inversely as the rate. If $2000 brings in $125 interest at 4%, how much principal will yield $500 at 5% for the same time?

Reality 4 is obviously not a high standard. It is doubtful whether in all the world's cases of conditional giving the problem of A has ever occurred. If B has ever occurred it probably has been in such connections that the immediate solution by $180° - \frac{1}{6}$ of $180°$ would be used. C illustrates a genuine relation but one which in reality is usually so complicated by other circumstances that only approximate estimate is made; hence the equational treatment seems rather pedantic. It is very hard to conceive cases where a person would know the proportions of a rectangle and the length of its diagonal and not already know the length of the base. The determination of the investment required to yield $500 when you don't know for how long, but do know that $2000 at 4% brings in $125 in the time in question, would not occur probably once in the lifetime of a million men.

We have counted for one of the books noted above the number of problems which passed the "Answer Known" criterion but failed to rate as high as 4 for genuineness. There were 70 out of 213.

Importance

Within the minority of problems that remain after the exclusion of bogus problems and problems whose "genuineness" is less than 4, a considerable percentage concern matters that are of importance to few people and not of much importance to them. The problems of the book in question were rated by four psychologists, three of whom were well versed in mathematics. We used as a scale for *Frequency of use* the following: 0 frequency, "Not one in a million ninth-grade graduates use it once a year," 10 frequency, "95 per cent of ninth-grade graduates use it once a month or oftener;" 1, 2, 3, 4, etc., are to represent approximately an arithmetic progression from the former to the latter. As a scale of *Importance to those who do use it* we had: 0 means "Of no use or approximately none;" 10 means "A *sine qua non* or very nearly so." 1, 2, 3, 4 are in a progression as above. The *Importance* rating was an average of that for *Frequency of use* and *Importance to those who do use it.* Importance 3 means the degree of importance possessed by the following:

IMPORTANCE 3

A. Divide $108 between A and B so that A receives eight times as much as B.

B. A has $2000 invested at 4%. How much must he invest at 6% to make the total yield equal to 5% on the total investment?

C. Mr. A paid $300 per share for some stock. At the end of five years he sold it for $550 per share. What rate of simple interest did his money produce for him during the five years?

D. Mr. A can plow a field in 6 days. Mr. B can plow it in 9 days. How long will it take them to plow the field if they work together?

E. A man does one-third of a piece of work in 10 days. He and another man finish the task together in 8 days. How many days would it take the second man to do the work alone?

F. ABC is an isosceles triangle, AD is its altitude, AD being perpendicular to BC. BD = DC. If AB is 18 inches and BC is 15 inches, find AD.

G. If a boy earns $520 during his first year of work and receives an increase of $50 a year each year thereafter, what salary does he receive the tenth year? How much has he earned in all during the ten years?

H. Each year Mr. A saves half as much again as he saved the year before. If he saved $64 the first year, how much will he save in all in seven years?

If we eliminate all the problems whose importance is less than this, we have left 96, about one-fifth of the total list. Of those so left, many are more suitably solved by mere arithmetical computation without any equation than by the organization around a symbol for the desired number and an equality sign. Omitting these also we have left 61 of the original 491.

Some of these 61 are clearly problems of Type I where no equation or formula is needed (*e. g.*, One bu. equals 32 qt. How many qt. in 4 bu.? In x bu.?) Others are clearly of Type II-A-1 where only substitution in a formula presented at the time is needed (*e. g.*, changing Fahrenheit temperatures to Centigrade). Omitting these, we have left 49, one-tenth of the original series.

These genuine problems which pass our minimum standard of importance for life seem worthy of presentation. The problems themselves show the teacher's available resources in this respect better than a description or tabulation of them would. These 49 problems appear on pages 143 to 145.[1]

Table 26 shows the facts separately for the work up to the beginning of fractions, from there up to quadraties, and from quadratics on.

TABLE 26

ANALYSIS OF THE VERBAL PROBLEMS OF A STANDARD TEXTBOOK

	Up to the Beginning of Fractions	The Beginning of Fractions Up to Quadratics	Quadratics and Beyond	Total	
1. Total numbei of verbal problems..................	268	126	97	491	
2. Number in which the answer would ordinarily not have to be known in order to obtain the data of the problem..................	129	39	45	213	(43%)
3. Number of these (2) which are rated as 4 or above for reality....................	100	16	27	143	(29%)
4. Number of these (3) which are rated as 3 or above for importance...............	73	9	14	96	(20%)
5. Number of these (4) which are not much more readily solvable by arithmetic alone and which are not clearly of Type I (where no explicit equation or formula is needed)...............	39	9	13	61	(12%)
6. Number of these (5) which are not clearly of Type II-A, where only substitution of numbers in a given formula is required........	29	9	11	49	(10%)

[1]The problems given here are not quotations, since it seems desirable to preserve the anonymity of the source, but are duplicates of the originals in general nature and form.

An analysis like that of Table 26 has been made for a second book of excellent repute, which contains work only up to quadratics. It gives the following results:

	Up to Fractions	Fractions to Quadratics	Total	
1.	283	129	412	
2.	157	82	239	(58%)
3.	125	56	181	(44%)
4.	121	23	144	(35%)
5.	39	19	59	(14%)
6.	38	15	53	(13%)

Eighteen of the problems in this second book left in class 6 are not typical problems of Class II-B. Ten are constructions of graphs, and eight are applications of simple trigonometrical facts not usually taught in algebra hitherto.

The Highest Ranking Tenth of Problems in Respect to Genuineness and Importance

1. The selling price of this book is five-fourths of its cost. Find its cost if it sells for $2.00.

2. In making a certain casting $1\frac{1}{2}\%$ of the metal is lost in the melting. How much metal is needed to make a casting weighing 86 pounds?

3. Cotton seed meal is used as a fertilizer. It contains approximately 7% of nitrogen. If a farmer wishes to put 15 pounds of nitrogen on a certain field, how much cotton seed meal must be purchased?

4. Tobacco stems contain about 8% of potash. How many pounds of tobacco stems must be bought to obtain 12 pounds of potash?

5. Divide $108 between A and B so that A receives 8 times as much as B.

6. Three men are asked to contribute to a fund. The first agrees to give twice as much as the second and the third agrees to give twice as much as the first. How much must each contribute to make a total of $1050?

7. The minimum temperature on February 2nd at Minneapolis was -15; the maximum was -4. What was the range of temperature there on that day?

8. Mr. A wishes to enclose a rectangular field, 20 rods wide. He wishes to make the field as long as he can, using 214 rods of fencing. How long can he make it?

9. If the cost of a car is p dollars and the rate of gain is 20%, what is the gain? What is the selling price? $(p+.20p=?)$

10. What was the cost of a car sold for $1320 if the gain is 10%.

11. Mr. A wishes to make 20% on some chairs. At what price must he buy them if he is to sell them at $2.00 each?

12. Mr. B wishes to sell chairs at \$7.00 each. At what price must he buy them so as to make 12% on the cost?

13. Mr. C knows that he can sell a piece of property for \$3540. How much must he pay in order to make a profit of 18%?

14. What principal must be invested at 4% to yield an income of \$600 a year?

15. How long must \$2000 be invested at 5% simple interest to produce \$375 interest?

16. The amount equals the sum of the principal and the interest. Express the amount at the end of a year when p dollars are invested at 4%.

17. What is the amount at the end of two years if p dollars are invested at 5%?

18. What sum of money invested at 6% simple interest for three years will amount to \$4000?

19. How long will it take \$1000 to amount to \$1500 if it is invested at 5%? ($1500=1000+1000\times.05y$. Solve for y.)

20. How long will it take \$1000 to double itself at 6%?

21. Let A represent the number of dollars in the amount. Let P, R, and T have their usual meanings. Show that $A=P+\frac{PRT}{100}$.

22. Use the formula above to find how many years will be required for \$7000 to amount to \$9100 at 5% simple interest.

23. Use the formula to solve this problem. Mr. A paid \$300 per share for some stock. At the end of 5 years he sold it for \$550 per share. What rate of simple interest did his money produce for him during the five years?

24. Mr. A has tea worth 65 cents and tea worth 45 cents per pound. How many pounds of each should he use to make a mixture of 100 pounds to sell at 53 cents a pound?

25. Mr. B has tea selling at 70 cents a pound and tea selling at 50 cents a pound. How many pounds of each should he use to make a mixture of 50 pounds selling at 62 cents a pound?

26. Same as the two previous, using different numbers.

27. Mr. A has \$2000 invested at 4%. How much must he invest at 6% to make the total yield equal to 5% on the total investment?

28. Mr. B has \$6000 invested at $3\frac{1}{2}$% and \$9000 at 4%. How much must he invest at 6% to make the total yield equal to 5% on the total investment?

29. A boy knows that his boat can go 6 miles per hour with the current and 3 miles per hour against the current. How far can he go and return, making the whole trip in just 3 hours?

30. A man can do a piece of work in 8 days. What part of it can he do in one day? In 7 days? In x days?

31. A man can do a piece of work in x days. What part of it can he do in 1 day? In 5 days?

32. A can do a piece of work in 6 days. B can do it in 10 days. How much can A do in one day? In x days? How much can B do in one day? In x days? How much can A and B together do in one day? In x days? How much can A do in 2 days? How much can B do in 5 days? How much can A and B together do if A works 2 days and B works 5 days?

33. A can do a piece of work in 10 days. B can do it in 5 days. How long will it take A and B together to do it?

34. A can do a piece of work in 8 hours. B can do it in 24 hours. How long will it take A and B together to do it?

35. Mr. A can plow a field in 6 days. Mr. B can plow it in 9 days. How long will it take them to plow it if they work together?

36. One machine can do a piece of work in 4 hours. Another machine can do it in 6 hours. How long will it take them both together to do it?

37. A can do a piece of work in 15 hours. B can do it in 18 hours. If A works for 7 hours, how long will it take B to complete the work?

38. A does one-third of a piece of work in 5 days. A and B complete the job by working together for 4 days. How long would it take B to do the job alone?

39. ABC is an isosceles triangle. AD is its altitude, AD being perpendicular to BC. BD=DC. If AB is 18 inches and BC is 15 inches, find AD.

40. If a boy earns $520 during his first year of work and receives an increase of $50 a year each year thereafter, what salary does he receive the tenth year? How much has he earned in all during the ten years?

41. On January 1st of each of 10 years a man invests $100 at 5% simple interest; what will principal plus interest amount to at the end of the tenth year?

42. Mr. A owes $2000 and pays 6% interest. At the end of each year he pays $200 and the interest on the debt which has accrued during the year. How much interest will he have paid off when he has paid off the debt?

43. Mr. A is paying for a $400 lot at the rate of $20 a month with interest at 6%. Each month he pays the total interest which has accrued on that month's payment. How much money, including principal and interest, will he have paid when he has freed himself from debt?

44. Mr. A plans to give his son 10 cents on his fifth birthday, 20 cents on his sixth, and each year thereafter to the eighteenth birthday, inclusive, to double the gift of the preceding year. How much will this be in all?

45. A problem in finding the height of a tower by similar triangles. A diagram is given.

46. Finding the width of a pond by similar triangles, and subtraction. A diagram is given.

47. Finding the width of a pond by similar triangles and double subtraction. A diagram is given.

48. The number of tiles needed to cover a given surface varies inversely as the length and width of the tile. If it takes 300 tiles 3 inches by 5 inches to cover a certain surface, how many tiles 4 by 6 will be needed for the same area?

49. The number of posts needed for a fence varies inversely as the distance between them. If it takes 120 posts when they are placed 10 feet apart, how many will it take when they are placed 12 feet apart?

It must be confessed that this list of what one of our standard instruments for teaching algebra offers as genuine problems to be solved by framing an equation does not support the general high estimation of problem solving of the II-B-b type. Problems 16 to 23 and 41 will not be acceptable to many because they neglect the fact that in real life the interest on the investment is almost always paid at stated intervals, not when the principal is repaid, and so can be compounded by reinvestment. Nos. 6, 27, 28, and 43 are rather fantastic. Nos. 48 and 49 require a method of finding the number of articles required which would rarely be wise, and never necessary, to use in such situations. Of the other problems, some are very probably better dealt with by the arithmetical methods which the pupils have already learned to use in such cases.

The advocate of the made-up problems will use the scantiness of this list as an argument that we must resort to the made-up, even insane, problems in order to give sufficient practice in applying principles and technique. But why should we give any practice in applying a principle or a technique to created problems when there are no sane problems to which it applies? Moreover, we must not assume that all the problem material which is genuine and of a fair degree of importance has been collected. At first thought, it would seem probable that it had, since for at least a decade progressive teachers and textbook makers have been fully aware of the need for it. A closer study of the matter, however, reveals that ingenuity and inventiveness and careful investigations do bring returns here as elsewhere. Many more such problems appear in the textbooks and teaching of 1920 than were available in 1900. Nunn has made very notable contributions. We may hope that the Nunns of the future will add more. Until we have canvassed the world's work thoroughly for problems that are genuine and important, we ought not to turn to those that are artificial and trivial.

It is a modern tendency to extend the list of genuine problems by teaching certain facts of physics, engineering, astronomy, navigation and the like so as to secure material for practice with the applications of algebra.

We very much need measurements of the time-cost of this, and of its effect upon interest in the sciences in question, in typical cases. The expectation is that often the game is not worth the candle unless it is very skilfully played, and that an undesirable attitude toward science may often result. It should be noted that the experts in teaching science rather carefully avoid algebraic and other quantitative work for pupils in high schools. High-school teachers of chemistry, geology, physical geography, the biological sciences, and economics are cautious about employing anything mathematical beyond the simplest; and this partly because they fear that it will repel students. Even in physics, descriptive work is emphasized rather than the fundamental equations; words are used instead of symbols, and sentences instead of formulas. This in spite of the fact that physics is taught in the last or next to last year of high school to a select and mature group. There is a danger that when we select problem material for algebra from the sciences we may be burdening algebra with the least attractive features of science and penalizing science by displaying its least attractive features to the pupil at the beginning of his high-school course.

There has been, so far as I am aware, no direct observational or experimental evidence published concerning the reactions of pupils to these problems taken from the sciences. Nor have we found facilities for securing such. We have, however, secured the judgments of the four psychologists mentioned previously.

They rated seventeen such problems (11 about the

principle that weight times length of lever arm equals weight times length of lever arm to make a balance, 5 about freely falling bodies, and 1 about the pressure-volume relation in gases) for reality, importance, interest, value in showing and in applying mathematical laws, excellence of statement, and value in teaching facts or laws outside of mathematics. In the combined weighted average, these problems from physics were somewhat above the average of problems in present-day textbooks.

Selection of Techniques for Application

It is hard to find psychological or pedagogical justification for the custom of concluding each topic in algebra by a series of verbal problems whose solution requires the operation of the mathematics taught under that head. The custom seems to be due partly to habits carried over from arithmetic, partly to the general fondness of intellectual persons for neat symmetrical systems, partly to a general overvaluation of the verbal problem as a means of mental training, and partly apparently to an insufficient appreciation of pure mathematics itself.

It is not likely that the arrangement of problems applying mathematical technique which is best for all children in Grades 3 to 6 will be the best for the third of them who go on to study algebra in Grade 9. Nor is it at all certain that "technique—application—technique—application" is the best arrangement in Grades 3 to 6. Good practice in the teaching of arithmetic now supplements this arrangement in Grades 3 to 6 by an arrangement by topics like "Earning and Saving," "Distances in a City," "House Plans," "A School Garden." In Grades 7 and 8, the arrangement has long been largely by topics like Insurance, Investments, Interest Given by Savings Banks, and Bank

Loans, and the like, and is developing toward an arrangement by topics like Food Values, City Expenditures, and Wage Scales.

Whatever arguments may be derived from the advantage of system would seem to be in favor of giving the main treatment of problem solving in one large unit, the general task of which would be to show that any number or numbers which can be found from certain given data, can be found by expressing the proper data in an equation or equations and solving. This chapter could well come after the pupil had learned to add, subtract, multiply and divide with literal numbers, including such simple fractions as should be mastered, and before the systematic treatment of the relation $y=ax+b$, or any treatment of quadratic equations. If a problem is suggested that leads to a quadratic (or a cubic), no harm will be done. The pupil may frame the equation, and leave it for solution until he learns the technique. This matter will be discussed further as one special problem of the order of topics in algebra. Our present purpose is simply to suggest that system does not require, or even favor, applying every technique indiscriminately in verbal problems. Of the general overvaluation of verbal problems and undervaluation of pure mathematics we shall treat in detail later.

All these are matters of minor importance for our present question if we accept as true a proposition which seems to the psychologist almost indubitable; namely, that the peculiar educative values of these verbal problems are attained by *framing* the right equations, solving them being not very greatly different from solving a similar equation framed for you by the textbook. If the problems are given primarily to train the pupil to frame the right equation or equations, we care very little about what computational techniques they happen to lead to. To take the extreme case, suppose that pupils only framed and *never* solved the equations, as in the Hotz test for problem solving. It would then be of almost no importance which techniques were required in these solutions—whether, for example, abilities with surds, quadratic equations, and certain factorizations were or were not applied. In so far as the peculiar value of problems is in framing the equations it is better *not* to give, after each technique is learned, many of the problems applying to it, because this tempts the pupil to expect that the problems will have a certain sort of equational form. He is tempted to work *toward* a certain sort of equation, instead of *from* the data given. It would then be among "miscellaneous" that a problem usually gave its best training.

We do not mean to imply that the framing of an equation and its solution are as educative if done a month apart as if done together, or that solving equations already framed for you is as educative as solving equations which you have framed yourself. We do claim that the peculiar virtue of the verbal problem is in the framing, not the solving, and that problems should be selected and arranged from this point of view rather than as exercises to show that certain algebraic computational tasks can be used in problems and to give practice in their use.

Problems as Originals and as Semi-Routines

The guiding principles in relation to this question can be briefly stated as follows:

Other things being equal, it is more educative to solve a problem as an original. Individual differences in ability need most of all to be allowed for when problems are given as originals. It is not probable that a pupil's efforts to solve problems are of great value to him when he fails with more than two out of three of them. On the other hand, pupils who are able to solve a certain type of problem as an original should certainly be excused from training in a routine method of solving it. Special training in the method of solving a certain type of problem is not desirable unless the problems in question are genuine and of some considerable importance. Clock, digit, age, and other similar problems should be given as originals, if at all. Whatever view we take of the amount of general ability developed by problem solving, one of the best ways to develop it is by trying to solve problems as originals, and, in case of failure after a reasonable effort, being given such assistance as enables one to solve them.

The Overvaluation of Verbal Problems

One reason for the great value attached to solving these verbal problems is a confusion of their value as training with their value as tests, and a misunderstanding of what they test. The ability to organize a set of facts in an equation or set of equations such that solving will produce the desired answer is very closely correlated with general intelligence of the scholarly type. The pupils who can do it well rank high in intellect and scholarship. So it is natural to infer that doing it creates and improves the ability. But this inference may be false or at least much exaggerated. Ability in supplying the missing words in sentences is also an excellent test of general intelligence. But the ability certainly has not been created or improved by supplying missing words, since that form of mental gymnastics has not been experienced by pupils save as a feature of psychological tests! The close correlation between ability in solving verbal problems and general ability is perhaps sufficiently accounted for by the fact that the task involves two abilities, each of which is closely related to general ability, namely ability in algebraic computation and ability in paragraph reading. Given a sufficient ability in algebraic computation and in paragraph reading, and pupils might conceivably solve a novel problem almost as well after two hours training in problem solving as after two hundred. Training of course improves their ability to solve the special sorts of problems they practice with, but the value of this depends largely on the genuineness and usefulness of the particular problems used.

Certain students of the teaching of algebra would agree with all this, but insist that the value of the verbal problems as training in the exact and adequate reading of paragraphs was sufficient to justify the high value attached to them.

This would conceivably be true. Solving a thousand verbal problems certainly has whatever educative value belongs to reading with great care a thousand short paragraphs and doing the thousand relevant computations. It has, indeed, the additional value that belongs to organizing the facts thus carefully read into equational forms such as will give the desired answers. The reading matter of these thousand short paragraphs is, however, so little in amount and so specialized in its nature that the training given by it seems insufficient to justify the high opinion of verbal problems or the time devoted to solving them.

The Use of Problems to Show the Need for a Certain Procedure and to Aid in Mastering It as Well as to Test and Improve the Ability to Apply the Procedure

Other things being equal, it is better for pupils to feel some need for a procedure and some purpose in learning it before they learn it. They are then more likely to understand it and much more likely to care about learning it.[1] Thus writing a real letter is now the beginning rather than the end of the lessons about "Dear Sir" and "Yours truly"; problems about the total cost of several toys or Christmas presents are the beginning rather than the end of the lessons on "carrying" in addition. "Why do we open the draughts of a stove to make the fire burn?" and, "What do we mix with the gasoline in an automobile?" are questions that introduce rather than follow the study of oxygen.

Thus in algebra problems about the average temperature of a series of days varying above and below 0, or about the total of certain credits and penalties in a rating may be excellent features in the introduction to the addition of negative numbers. Problems like the following may be useful as parts of an introduction to "− divided by − gives +."[2]

Other things are not always equal. There may be no vital, engaging problems to use as introductory material. For example, there is not, to my knowledge, any problem

[1] We do not here discuss this general educational axiom because probably it will be acceptable as stated. The whole matter of pupils' purposes in learning, and the special doctrine of "first the need, then the technique," has received a classic general treatment at the hands of Dewey. The case with algebra is much the same as with arithmetic, on which the reader may consult Chapter XIV of *The Psychology of Arithmetic* [Thorndike, '22].

[2] The illustrations here are not problems where organization in the equational form is necessary. What is said about problems in this section, indeed, concerns all problems of types I and II, not merely the II-B-b problems.

Four boys are rated for strength in comparison with the average for their age.

Arthur is −20
John is −12
Fred is − 4
Bert is − 8

Supply the missing numbers:

Arthur is....... times as far below the average as John.
Arthur is....... times as far below the average as Fred.
Arthur is....... times as far below the average as Bert.
John is....... times as far below the average as Fred.

that is vital and engaging to the average high-school pupil by which to introduce the general symbolism of fractional exponents. Such problems, though in existence, may use up more time than can be spared. Again, there is a problem almost perfectly adapted to arouse the need for knowledge of the laws of signs in multiplication, namely, the problem of measuring resemblance between a pair of measures both of which are divergences from a type or average. But it takes so long to teach the meaning of "resemblance" in such cases that probably the game is not worth the candle. The procedure may be so intrinsically valuable and interesting that mere contact with it will quickly inspire a desirable purpose and activity. For example, gifted pupils will probably learn that $\sqrt{a}\sqrt{a}=a$ as readily by straightforward consideration of $(\sqrt{4}\sqrt{4})$, $(\sqrt{9}\sqrt{9})$, $(\sqrt{16}\sqrt{16})$, $(\sqrt{2}\sqrt{2})$, and $(\sqrt{3}\sqrt{3})$, as by any introductory problem to display the need of knowing that the square root of any number times the square root of the same number equals the number.

Criteria in Selecting Problems

In this section, as in the previous one, we are concerned not alone with problems where an equation or equations are used to discover certain particular quantities relating to one particular state of affairs, but with problem material in general.

Solving problems in school is for the sake of problem solving in life. Other things being equal, problems where the situation is real are better than problems where it is described in words. Other things being equal, problems which might really occur in a sane and reasonable life are better than bogus problems and mere puzzles. Other things being equal, problems which give desirable training in framing equations from the realties or the verbal statements are better than problems which give training chiefly in solving the equations when framed. The latter training can be got easily by itself.

As was suggested in an earlier chapter, a better selection of problems will probably be secured if, instead of searching for problems which conveniently apply fractional equations, problems to apply simultaneous linear equations, and so on, we search for problems which are intrinsically worth learning to solve by algebraic methods.

If it happens that there are no genuine, important problems calling for the framing and solution of a certain technique, say simultaneous quadratics, we may simply leave that technique without application to verbal problems or we may frankly provide problems that make no false pretenses at reality as in "I am thinking of two numbers, . . . etc."

This case of simultaneous quadratics is a good one to illustrate the two points of view contrasted here. The older view, in order to have applications of simultaneous quadratics, fabricated extraordinary tasks depending on insane curiosity to know the dimensions of a field which, when altered in various ways, gives fields of certain areas, and the like. The newer view selects first the case of determining the constants in a quadratic equation from knowledge of the (x) (y) values of certain points on the curve. The ability to do this is not of great "social utility" to many of the individuals who study ninth or tenth grade mathematics, and might well be left for those who specialize further in mathematics or science. It is, however, a genuine problem. The next choice of the newer view would probably be the solution of " . . . and . . . are two curves.

Do they intersect? If so, at what points?" This again is not a question which life puts to many persons, but it is one that a sane person need not be ashamed to ask. If genuine applications of the technique of simulaneous equations are beyond the abilities and interests of high-school pupils, we may leave it without application until the abilities and interests are available.

Problems as Tests

Genuine application in the real world should be demanded in problems which are used for training; and is preferable in problems given simply to test the ability to organize a set of statements into equations to answer a question. It is preferable in the latter case because, human nature being what it is, teachers will be prone to train for the test. The nature of the examinations used has always influenced the nature of the instruction and probably always will. Except for this, we might permit as algebraic "originals" in tests, the problems about consecutive digits, hands of a watch, numerators and denominators defined by their sums, differences, products and quotients in fantastic ways, and the like, which we exclude from mathematical *training* if better material can be obtained.

Actual and Described Situations

We have noted in an earlier section that the ability to manage a problem as encountered in reality, and the ability to manage the same problem as it is described in words in an algebra book, need not be identical. Success with the former is consistent with failure with the latter, and *vice versa*.

The worst discrepancies are when, on the one hand, a state of affairs which would be very clear and comprehensible to a person experiencing it, is beclouded and confused by words, and when, on the other hand, the pupil learns to obtain correct solutions by response to verbal cues, the lack of which would cause the real situation, when encountered, to baffle him.

As an example of the first, consider this problem:

A man is paying for a $300 piano at the rate of $10.00 per month with interest at 6 per cent. Each month he pays *the total interest which has accrued on that month's payment*. How much money, including principal and interest will he have paid when he has freed himself from the debt?

In reality the man probably would be told that he had to pay $10.05 the first month; $10.10 the second month; $10.15 the third month, etc., and would easily see the progression. The difficulty with the problem in schools lies chiefly in understanding what "each month he pays the total interest which had accrued on that month's payment" means, and in the confusing use of "including principal and interest" and "debt."

It seems wiser to give more attention to providing real situations for the application of algebra. For example, it seems wise for pupils to draw a straight line and another cutting it, and find the size of all four angles by measuring one of them, as well as to solve problems in words about supplementary angles. There is not only a greater surety that the pupils are being prepared to respond effectively to situations which life will actually offer, and an insurance against the danger of unsuitable linguistic demands, but also often an increase of interest in and respect for algebra.

Like almost everything in teaching, we have to add the clause "other things being equal" to this recommendation. Too much time must not be spent in drawing, measuring, weighing, and the like. Also, the "real" situation will often be a map already drawn, a table of values already measured, a set of observations already made. Also the genuine problems to which algebra applies are, as compared with those of arithmetic, more often prophetic, foretelling what will happen if certain conditions are fulfilled or what to do in order to bring certain results to pass. The genuine task is, in many of these prophetic problems, precisely to understand a verbal description. Such for example, are problems about mixtures and alloys, about the amount of d needed to have its proportion to c the same as b's to a, and about the drawing of rectangles of specified proportions and total area.

Isolated and Grouped Problems

Problems grouped by their relation to some aspect of science, industry, business, and home life, as *Falling bodies,* or *Alloys,* or *Sliding scales for wages* or *Dietaries* have certain advantages. (1) The situations dealt with are more likely to be understood; (2) things are put together in the pupil's mind that belong together in logic or in reality or in both; (3) the data needed for all the problems can be given once for all, so that in each problem the pupil has to select the facts needed to answer it as well as to arrange them in suitable equations.

The isolated problem is indeed disappearing from arithmetic except in special exercises for particular purposes of tests, reviews, and training in alertness and adaptation. It appears that by the exercise of enough care and ingenuity, problems in arithmetic can be grouped in this way with no loss to the purely arithmetical training that is given. The same tendency is operating in algebra, and much good may be expected from it.

Problems Requiring the Selection of Data

The third advantage noted above as characteristic of grouped problems is of special importance, as was noted in an earlier chapter.

The custom is firmly fixed of giving in a problem only the facts needed to solve it, so that "there are as many distinct statements as there are unknown numbers"; and the pupil is taught to "represent one of the unknown numbers by a letter; then, using all but one of the statements, represent the other unknowns in terms of that same letter. Using the remaining statement, form an equation." Yet it seems unjustifiable. The time and thought now spent by pupils on intricate fabrications whose like they will never see again, might much better be spent in such selective tasks as are genuine and instructive.

Problems Requiring the Discovery of Data

A further step is worth consideration, namely, that of giving problems some of the data for which are lacking and must be supplied by the pupil's search. We do this to a slight extent by not including in the statement of a problem such needed facts as that 1 foot = 12 inches, or that a square has four equal sides. Is it desirable to require the pupil to find in his memory or in tables at the end of his textbook on algebra or in other reference books or from observation and measurement such facts as the inter-equivalences of inches and centimeters, the weight of a cubic foot of water, the capacity of a 4 ounce bottle, the length and width of his classroom or the area of Ohio?

There are obvious inconveniences in doing this, but there is the advantage of making problem solving in school one degree more like problem solving in science or industry or business. We might at least go so far as to assign a score of problems each with the question, "What further fact or facts must you have in order to solve this problem?", and distribute the work of discovering these facts among the pupils. The lesson that one must often supplement the facts given by the situation itself by further investigations would then be taught to all, at no great cost of time.

Such searching is, of course, not algebra; neither is the understanding of statements about rates, speeds, investments, and yields algebra. The algebra begins when statements understood are to be translated into algebraic symbols. Having already far overstepped that line in the customary work with verbal problems, we may go farther with no inconsistency.

Problems Requiring General Solutions

The most objectionable feature of problem solving in algebra today to a psychologist is the predominance of problems seeking a particular fact about some particular state of affairs—the relative neglect of problems which seek the general relation between variations in one thing and variations in something related to it.

The main service of algebra, as the psychologist sees it, is to teach pupils that we can frame general rules for operating so as to secure the answer to *any* problem of a certain sort, and express these rules with admirable brevity and clearness by literal symbolism. We take great pains to teach the pupil that pq means the product of whatever number we let p equal and whatever number we let q equal; and that if p and q equal any two numbers, the first number times the product of the two equals p^2q, and other similar facts. Then, in problems, the p's and q's or x's and y's in nine cases out of ten, mean something as unlike "any number" as could possibly be. We build up habits of computing with literal numbers and then, in problems, make almost no use thereof, reverting to an arithmetic plus negative numerals with a written x in place of the mental "What I am trying to find." Small wonder that the pupil often thinks of his algebraic computations as a mere game that one plays with a, b, c, d, $+$, $-$, $\times$, $\div$, and (). If, after a few exercises in the use of letters to mean "any number of so and so," and a few exercises in reading and framing formulas, we have him do nothing with literal symbols but play computing games, why should he think otherwise?

Why should we blow hot and cold in this way, asserting that algebra teaches us what is true of any number and then in problems, making its linear equations true of only a single number, and its quadratics of only two? Should we not alter many of our II-B-b problems into the II-B-a form, requiring the pupil to frame the general equations or formulas to solve any problem of that sort, and to obtain any particular answer by evaluating? For example, compare the two tasks I and II below:

I. A man has a lawn 40 ft. long and 30 ft. wide. How wide a strip must he mow beginning at the outside edge in order to mow half of it?

II. **1.** A man has a rectangular lawn. Make a formula to state how wide a strip he must mow beginning at the outside edge in order to mow half of it. Let l and w equal the length and width of the lawn in feet.

Let s equal the width of the strip in feet.

2. Find s if $l=40$ and $w=30$.

3. Find s if $l=100$ and $w=20$.

4. Find s if $l=80$ and $w=40$.

5. Find s if $l=80$ and $w=60$.

It seems reasonable to progress from problems of the I type to problems of the II type just as we progress from numbers to letters, and from such facts as $2\times2=2^2$, or $3\times3=3^2$ to such facts as $a\times a=a^2$, or $a(b+c)=ab+ac$.

Among problems requiring a general solution in terms of a literal formula, special importance attaches to problems of direct and inverse proportion, problems where one number varies as the square or square root of another, and other problems involving linear, hyperbolic, and parabolic relations.

Problems of Puzzle and Mystery

The earliest problems of algebra were problems of puzzle and mystery, such as Ahmes's "A hau, its seventh, it equals 18," or the finding of the age of Diophantus from his epitaph. "Diophantus passed one-sixth of his life in childhood, one-twelfth in youth and one-seventh more as a bachelor; five years after his marriage a son was born, who died four years before his father did at half the age at which his father died."

Such problems make an appeal to certain human interests. Some pupils doubtless prefer them to straightforward uses of algebra in answering questions of ordinary life. The human tendency to enjoy doing what we can do well, and especially what we can do better than others can, is often stronger than the tendency to enjoy doing what we know will profit us. Some of these problems are also arranged as strong stimuli to thought for thought's sake. By introducing an element of humor they may relieve the general tension of algebraic work, as is at times desirable. They are much more appropriate in algebra for the selected group of superior pupils who continue to high school than they are in arithmetic for all children. On the whole, however, the ordinary applications of algebra to science, industry, business, and the home will give better training to the general run of high-school freshmen and will inspire greater liking and respect for mathematics than will these appeals to the interest in puzzles and mystery.

One of the best forms of appeals to the puzzle interest is by abstract problems such as: "When will a^2 be less than a?" "When will l divided by a be greater than a?" "State a condition such that $\frac{a}{b}$ will equal $\frac{b}{a}$." "State a condition such that abc will equal a."

One of the best forms of appeal to the interest in mystery is to have pupils frame formulas[1] for such mysteries as: "Think of any number and I will tell you what it is. Think of the number. Add 3 to it. Multiply the result by 7. Subtract 20. Tell me the result. The number you thought was . . . (This result diminished by 1 and then divided by 7)." They may also make up such mysteries for the class, score being kept of the time required for pupils to find the formula for the mystery and penalties being attached to devising a "mystery" that doesn't work. $(a+b)(a-b) = a^2-b^2$ may be taught as a mystery for quickly computing products like 2998×3002, or 4980×5020. The formula for the sum of an arithmetic progression may be taught as a mystery for computing the sums of such series, either complete, or in the form "All the numbers from . . . to . . . except . . . and . . ." There are, however, better motives to use for mastering

$$S=\frac{n}{2}[2a+(n-1)d]$$

As has been so often insisted, the cardinal sin in connection with problems of puzzle and mystery is their decoration with a description of conditions and events in nature which makes the pretence that the problem is genuine when it is not; and so confuses and debauches the pupil's ideas of the uses of algebra. If they are presented in their true light and if the pupils have the option of solving them or solving problems of genuine application, they can, at the worst, do very little harm.

The Election of Problems by Students

Many of the difficulties of teaching in the case of problems are greatly lessened by arranging to have each choose a

[1] Representative problems of this sort will be found in Nunn ['13, p. 87 f.].

certain number of problems to solve from a list which contains, say, five times as many as any one pupil is to solve. We have just noted the value of election between "useful" and "puzzle" problems, if the latter are presented at all. We have noted previously that problems drawn from physics may be of very different value to pupils who are studying general science and to pupils who are not. Within the latter group, we might also differentiate between those who happen to be ignorant of science, and those who are so by their own volition. Boys and girls may well differ in their choices, though probably not so much as some theorists would expect. It may be desirable to permit and even encourage some pupils to choose the easier problems.[1] The provision of five times the number of verbal problems now given in standard textbooks would add perhaps two cents to the cost of production.

Summary

It is a worthy aim to teach pupils to organize the facts of important situations requiring numerical responses into equation form and to solve their equations. It is also worth while for pupils to learn that any quantitative question, no matter how elaborate and intricate, can be so expressed, provided adequate data are given.

Even if the educative value of this work is improved by such modifications as have been suggested in this chapter, it will still be, on the whole, less important than the framing of general equations or formulas for solving *any* problem of a certain kind. Learning to let x or q equal the unknown and to express data in terms of their relations to it is a useful lesson, but learning to express a set of relations in generalized

[1] The instructions may be "Do the ten hardest ones that you think you can do."

form is a more useful one, and, so far as psychology can prophesy, one more likely to transfer its improvement to other abilities. It is when the verbal problems of algebra advance beyond arithmetical problems in the same way that algebraic computation advances beyond arithmetical computation that they perform their chief educational service.

CHAPTER VI

THE MEASUREMENT OF ALGEBRAIC ABILITIES

We may best begin our study of the measurement of algebraic abilities by examining some actual instruments for measuring them, especially those of Rugg and Clark [1917], and of Hotz [1918].

The Rugg-Clark tests as now issued are an instrument to measure computational abilities up to quadratics, including the solution of easily factored quadratics. They are as follows:

TEST No. 1

COLLECTING TERMS

1. $5x^2+3-3x^2-2-7x^2-5=$ Answer........

2. $2a+b-6a-5b-3a-2b=$ Answer........

3. $-6y^3-8y+4y^3+3y+2y^3+7y=$ Answer........

4. $-3mn-2p+7mn+5p+4mn+3p=$ Answer........

and five more series like the above, making twenty-four tasks in all. Time, 4 min.

TEST No. 2

SUBSTITUTION

1. If $x=4$ and $y=2$ what does $2x^2-3xy$ equal? Answer........

2. If $a=3$ and $b=2$ what does $3ab+ab^2$ equal? Answer........

3. If $c=2$ and $d=5$ what does cd^2-2cd equal? Answer........

4. If $p=4$ and $q=3$ what does p^2+4pq equal? Answer........

5. If $x=3$ and $y=5$ what does x^2+2x^2y equal? Answer........

and three more similar series, making twenty tasks in all. Time, 4 min.

TEST No. 3

Subtraction

1. From $2a+3b$ take $5a+4b-c$ Answer........
2. Take $4x+2y$ from $2x-y+z$ Answer........
3. Subtract $3r-5s+10t$ from $r+s+4t$ Answer........

and six more similar series, making twenty-one in all. Time, 4 min.

TEST No. 4

Simple Equations

1. $2x+3=11$ Answer........
2. $4c=6c+12$ Answer........
3. $5x-3=-20$ Answer........
4. $13=2x-8$ Answer........
5. $12x-7-15x=10$ Answer........

and four more similar series, making twenty-five in all. Time, 4 min.

TEST No. 5

Parentheses

1. $6(3x+8)$ Answer........
2. $5(4x-2)$ Answer........
3. $-3(8x+3)$ Answer........
4. $-4(3x-4)$ Answer........
5. $9(-7x-1)$ Answer........
6. $-8(-4x-7)$ Answer........

and six more similar series, making forty-two in all. Time, 2 min.

TEST No. 6

Special Products

1. $(2x-3)^2$ Answer........
2. $(3m+n^2)(3m-n^2)$ Answer........
3. $(a-4)(a+5)$ Answer........
4. $(5x+1)(x+3)$ Answer........

and five more similar series, making twenty-four in all. Time, 3 min.

TEST No. 7

Exponents

1. $a^3 \cdot a^5 =$ Answer........
2. $5x^7 \cdot 6x^5 =$ Answer........
3. $(n^2)^3 =$ Answer........
4. $\frac{c^3}{c^2} =$ Answer........
5. $(ba^3)^4 =$ Answer........
6. $x^7 \cdot x =$ Answer........

and five more similar series, making thirty-six tasks in all. Time, 2 min.

TEST No. 8

Factoring

1. $5x^2+15x^3$ Answer........
2. a^2-64 Answer........
3. y^2-6y+9 Answer........
4. $b^2+11b+28$ Answer........
5. $5x^2+16x+3$ Answer........

and four more similar series, making twenty-five tasks in all. Time, 4 min.

TEST No. 9

Clearing of Fractions

Write without fractions but do not solve

1. $\frac{x-3}{x-4} = \frac{x+2}{x+7} + \frac{3x}{x^2+3x-28}$
2. $\frac{5}{x-4} - \frac{3}{x-9} = 0$
3. $\frac{2x-7}{5} - 2 - \frac{3x-8}{3x} = 0$
4. $4a - \frac{2a-5}{9} = -7$

and three more similar series, making sixteen tasks in all. Time, 5 min.

TEST No. 10

Fractional Equations

1. $\frac{4x+2}{3}-\frac{x-3}{4}=0$ Answer........
2. $\frac{x+1}{x-1}=\frac{5}{3}$ Answer........
3. $\frac{4}{3+x}-\frac{2}{1+x}=0$ Answer........
4. $\frac{x+8}{x+9}=\frac{x-5}{x-7}$ Answer........

and four more similar series, making twenty tasks in all. Time, 9 min.

TEST No. 11

Formulas

1. $P=ahw$ Solve for h Answer........
2. $c=\frac{E}{R}$ Solve for R Answer........
3. $E=\frac{PL}{K}$ Solve for P Answer........
4. $L=\frac{Mt-g}{t}$ Solve for M Answer........
5. $I=\frac{bd^3}{3}$ Solve for b Answer........
6. $E^2=\frac{JWhr}{t}$ Solve for h Answer........

and three more similar series, making twenty-four tasks in all. Time, 5 min.

TEST No. 12

Quadratic Equations

1. $x^2-81=0$ Answer........
2. $y^2+y=6$ Answer........
3. $n^2-7n=-12$ Answer........
4. $x^2+5x=6$ Answer........

and six more similar series, making twenty-eight tasks in all. Time, 7 min.

TEST No. 13

Simultaneous Equations

1. $2x+y=10$
   $3x-2y=1$ Answer........
2. $2x=3y+3$
   $5x+3y=39$ Answer........
3. $4n-2r=0$
   $3n+5r=13$ Answer........

and four more similar series, making fifteen tasks in all. Time, 12 min.

TEST No. 14

Radicals

Leave answer in *simplest* radical form.

1. $\sqrt{8}$ Answer........
2. $\sqrt{a^3b^4}$ Answer........
3. $\sqrt{\frac{2}{3}}$ Answer........

and six more similar series, making twenty-one tasks in all. Time, 3 min.

A test on graphs and a test on quadratic equations with irrational roots, included originally, are not now issued as a part of the standard examination.

It should be noted that each test gives a possibility of from four to seven trials for each unit task, so that casual errors due to carelessness may be distinguished from real inabilities.

For comparison with certain achievements the tests are given with these time limits[1]:

Tests 1 to 4: 4′, 4′, 4′, 4′, respectively.
Tests 5 to 8: 2′, 3′, 2′, 4′, respectively.
Tests 9 to 12: 5′, 9′, 5′, 7′, respectively.
Tests 13 and 14: 12′, 3′, respectively.

They are then clearly tests of speed and accuracy in combination. They may, however, be given with a long time limit

[1] It may be noted here that for convenience in administering tests it is almost always best to set time limits.

and instructions to take sufficient time for perfect work, or with the announcement of very heavy penalties for wrong answers so as to reduce the relative weight of speed upon the score to any desired minimum.

These tests are noteworthy in what they omit. For instance, parentheses within parentheses, division by a polynomial, multiplication by a polynomial other than the case of $(ax+b)$ $(cx+d)$, factorization other than of monomial factors, x^2-y^2, and products of $(ax \pm b)$ $(cx \pm d)$, complex fractions. In general the aim is obviously to measure the mastery of a few of the more widely used tools of algebra, in their simpler uses, rather than the extent of the pupil's computational repertory or his ability to use it to handle elaborately complex simplifications, or to see the possibilities of ingenious short-cuts.

It may be questioned whether the ability to perform these computations one at a time in the form in which they are given in the Rugg-Clark test implies the ability to perform them when the circumstances are changed or complicated. For example, will a pupil who does $(3m+n^2)$ $(3m-n^2)$ correctly in Test 6 do $(3m-n^2)$ (n^2+3m) correctly? Will a pupil who does $6(3x+8)$ and $-3(8x+3)$ correctly in Test 5 do $6(3x+8)-3(8x+3)$ correctly?

We know that in general any change in the situation no matter how slight has some disturbing effect upon the connection from that situation to its proper response. For example, the ability to add

$$\begin{array}{r} 6 \\ 8 \\ \underline{3} \end{array}$$

in column does not ensure ability with $3+8+6$; ability with $7 \times 9 = 63$, and with $7 \times 6 = 42$, and with $42+6=48$, and knowledge of the process to be used, do not ensure

$$\begin{array}{r} 69 \\ \underline{7} \\ 483. \end{array}$$

The pupil

may be unable to get his 42 when he has the added burden of holding in mind the 6 to be added. For a test to be a perfect inventory of algebraic abilities it is perhaps necessary that each ability be tested not only in isolation but also in any commonly used context, if the presence of that context has a possibility of disturbing the ability. This is not intended at all as a criticism of the Rugg-Clark tests, but as a warning against improper inferences from them and as a general principle to be considered in measuring algebraic abilities.

Another general principle is suggested by the directions, and by the titles which serve as directions. The Rugg-Clark tests measure a compound made up (a) of certain abilities to operate, and (b) of certain trade secrets of algebra. Such are in Test 1, the understanding of "Collecting terms"; in Test 4, knowledge that "Answer to $2x+3=11$" means "Find what x equals"; knowledge that "." means multiply; in Test 11 knowledge that "Solve for h" means "Find what h equals"; in Test 13 knowledge that when *Simultaneous Equations* is at the top of the page, the "answer" to $\begin{Bmatrix} 2x+\ \ y=10 \\ 3x-2y=\ \ 1 \end{Bmatrix}$ is the value of x and the value of y which will make both equations true statements.

In general, it seems desirable to keep measures of operating ability as free as possible from measures of knowledge of these trade secrets. We should all agree, for example, that Test 11, if put as:

$P=ahw$. Change the subject of the formula to h,

$C=\frac{E}{R}$. Change the subject of the formula to R, etc.,

would baffle many pupils who nevertheless had mastery of the operating. Yet "Solve for" is as truly a trade secret as "Change the subject of the formula to," though it is much more widely known. In practice for mastery, of course, the claims of brevity justify terms like "simplify" or "solve," and the assumption that the pupils will know what to do with, say, simultaneous equations. In tests of mastery, however, where brevity is far less important than clearness, and where pupils are to be compared who may have been taught different customs as to terms and procedures, it seems better, as said above, to test the power of operating separately from the power of understanding what operation is desired.

The Hotz tests are an instrument to measure elementary algebraic abilities in general, including abilities with graphs and problem solving as well as computation. They are shown in part below.

SPECIMEN HOTZ TESTS

ADDITION AND SUBTRACTION

1. $4r+3r+2r=$

5. $7x-x+5-4=$

10. $8c-(-6+3c)=$

15. $\frac{1}{a-x}-\frac{3x}{a^2-x^2}=$

20. $\frac{1}{a+1}-\frac{a}{a^2-a+1}-\frac{a-4}{a^3+1}=$

24. $\frac{a}{a-2}-\frac{a-2}{a+2}+\frac{3}{4-a^2}=$

MULTIPLICATION AND DIVISION

1. $3 \cdot 7y=$

5. $\frac{2}{3}$ of $9m=$

10. $\frac{4x^4}{5}\div 2x^2=$

15. $\frac{m+n}{a}\cdot\frac{b}{m^2-n^2}=$

20. $\frac{p^2+4p-45}{p^2+2p+4}\cdot\frac{p^3-8}{p^2-81}\cdot\frac{1}{3pr-15r}=$

23. $\frac{3\sqrt{6a}}{2a\sqrt{18}} \cdot \sqrt{12a} =$

EQUATION AND FORMULA

1. $2x = 4$

5. $7n - 12 - 3n + 4 = 0$

10. $\frac{2x}{3} = \frac{5}{8}$

15. $\frac{4}{3-x} = \frac{2}{1+x}$

20. $\frac{2}{x^2+4x+3} = \frac{3}{x^2+3x+2}$

25. $\sqrt{x^2-1} - x = -1$

PROBLEMS

1. If one coat cost x dollars, how much will 3 coats cost?

5. The distance from Chicago to New York by rail is 980 miles. If a train runs v miles an hour, what is the time required for the run?

10. A tower casts a shadow of 20 feet. A man, 5 feet 9 inches high, who is near at the same time, casts a shadow of 2 feet 6 inches. Find the height of the tower.

14. An open box is made from a square piece of tin by cutting out a 5-inch square from each corner and turning up the sides. How large is the original square, if the box contains 180 cubic inches?

GRAPHS

1. The following diagram represents the length of certain rivers.

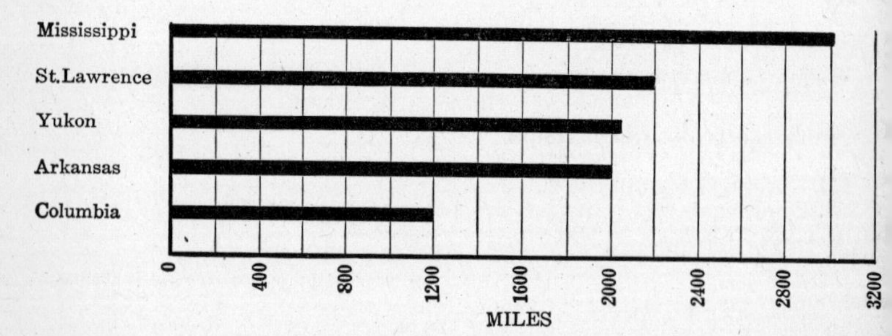

How many miles long is the Arkansas river as represented in this diagram?

. . . .

5. The following graph is used to convert degrees of temperature from the Fahrenheit scale (F) to the Centigrade scale (C), and from the Centigrade scale to the Fahrenheit scale.

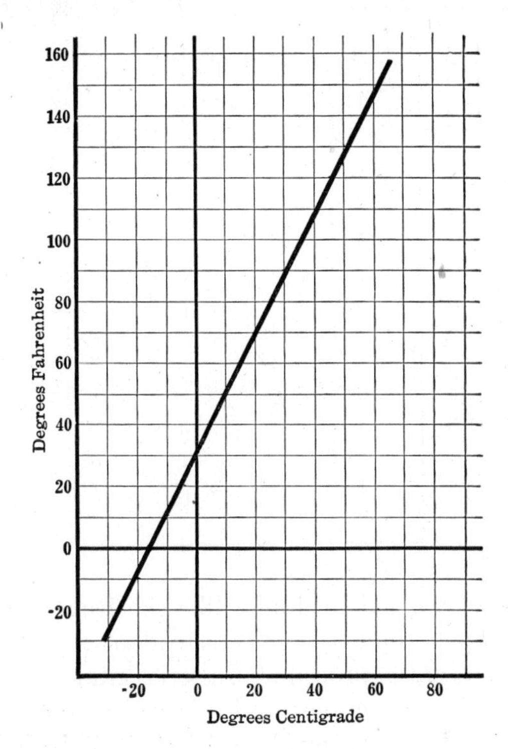

When it is +20° on the F scale, what is the temperature in degrees on the C scale? . . . .

11. A boy begins work with a weekly wage of \$9 and receives an increase of 25 cents every week. Another boy starts with a weekly wage of only \$6 but receives an increase of 50 cents every week.

Draw a graph which shows the wage of each at the beginning of every week for 15 weeks.

According to this graph when will their wages be the same? . . . .

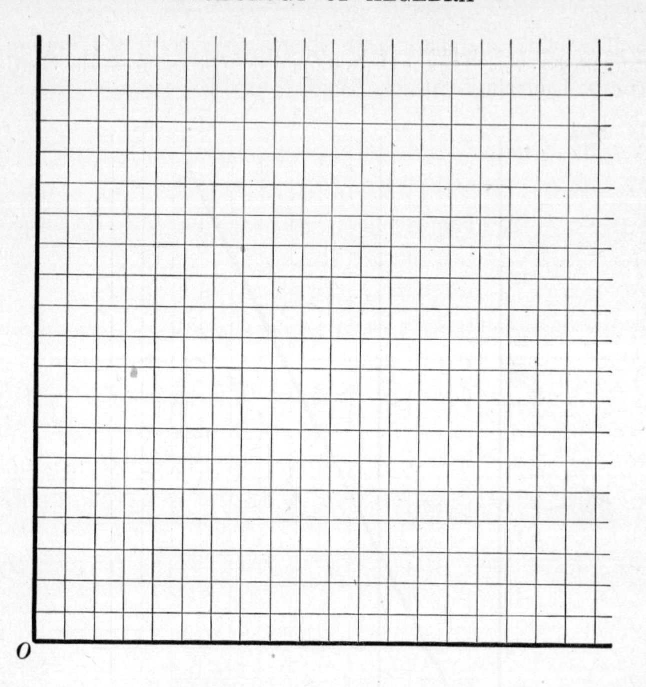

They are arranged by topics like the Rugg-Clark tests, but within each topic there is instead of a series graded to cover the essential procedures, a series graded in difficulty and including both simpler and easier and more elaborate and harder tasks than the Rugg-Clark tests include. The topics are also different.

This graduation is empirical, difficulty being measured by the percentage of a group of pupils who fail task by task. The difficulty is thus a function of the nature of the training had by the pupils of certain communities at a certain date [1917, 1918], as well as of the intrinsic essential difficulty of the task. The order of difficulty may be, and probably will be, greatly altered by such differences in the teaching of algebra as each new decade shows. For example, at the time and in the place where Hotz standardized these tests, No. 17 below was a much harder task than No. 14

14. $3m+7n=34$
$7m+8n=46$

17. In the formula RM = EI, find the value of M

but it would almost certainly be the easier of the two for a group who had been taught the content of Nunn's *Exercises*, Book I, by the methods which Nunn advocates.

Any empirical gradation as to difficulty must of course be thus a function of the nature of the teaching in vogue; and be itself subject to change. The graded or "ladder" form of test has, none the less, great merits. First, it helps us to state how hard a task a class or an individual can perform with substantial mastery. Suppose, for example, that in the test for addition and subtraction the percents correct for the class were:

Task	Percent
1	95
3	90
5	100
9	95
10	95
12	100
13	95
15	80
18	75
22	65
23	45
24	20

It becomes at once clear that this class has a substantial mastery of addition and subtraction through difficulty No. 13. There is a great difference between knowing $\frac{n}{2}$ things perfectly and half knowing n things. The graded or "ladder" test informs us concerning how many things a class knows and which they are.

Second, it measures with fairness and economy of time, over a wide range of ability. Individual differences in the capacities of pupils to learn algebra are such that if the pupils spend approximately the same time, say 400 hours, in study and class-work, some should have acquired far more ability than others. If ten tasks of about equal difficulty are set as the means of measuring them, many of them should do all ten perfectly and many of them should fail on all ten. The "ladder" test avoids such "undistributed perfects" and "undistributed zeros."

Third, the "ladder" test helps us to distinguish lack of ability from lack of care. Suppose, for example, that John and William have taken the Hotz test for addition and subtraction, extended to include five tasks at each level of difficulty, with the following results:

Level of Difficulty	Number Right	
	John	William
1	5	5
3	3	5
5	4	5
9	4	5
10	5	5
12	4	5
13	5	0
15	0	0
18	0	0
22	0	0
23	0	0
24	0	0
Total Right	30	30

John is obviously much more careless than Will, who is indeed an Admirable Crichton in this subject. Carelessness may be defined objectively as a mixture of failing on tasks easier than those one succeeds with and failing on tasks which one usually succeeds with. Both features are important. Graded tests specially reveal the former as the Rugg-Clark tests do the latter.

Some features of the Hotz tests may be criticized on the ground that the situations involved are unreal and unlikely ever to be met with by a sane person in a real world. For example:

$$\frac{3-2x}{(x-1)^3}+\frac{x+1}{(x-1)^2}-\frac{1}{x-1}=$$

$$\frac{c^4-d^4}{(c-d)^2}\cdot\frac{c-d}{c^2+d^2}=$$

$$\frac{p^2+4p-45}{p^2+2p+4}\cdot\frac{p^3-8}{p^2-81}\cdot\frac{1}{3pr-15r}=$$

$$\frac{6x-2}{x+3}-3=\frac{3x^2+13}{x^2-9}$$

$$\sqrt{x^2-1}-x=-1$$

A gold watch is worth ten times as much as a silver watch, and both together are worth $132. How much is each worth?

The total number of circus tickets sold was 836. The number of tickets sold to adults was 136 less than twice the number of children's tickets. How many were sold of each?

The area of a square is equal to that of a rectangle. The base of the rectangle is 12 feet longer and its altitude 4 feet shorter than the side of the square. Find the dimensions of both figures.

In general we should certainly prefer to test ability in computation and in solving problems with kinds of computing and problems such as real life may offer.

A test of algebraic abilities is much increased in usefulness if it exists in a number of alternative forms of approximately equal difficulty, so that it can be used to measure improvement as well as status, and so that special coaching upon its particular tasks will be of little or no profit save as general practice in algebra. Where the measurement of algebraic abilities is used as a measure of merit, as for promotion, the assignment of marks, and the like, a large number of alternative forms is necessary. The Rugg-Clark tests 1 to 14 could be easily extended to such alternative forms. The Hotz tests could be so extended, but more ingenuity and labor would be required.

As a third type of measuring instrument we may examine that printed below, which is one of a set of many alternative forms[1] any one of which (1) makes an approximate inventory of algebraic abilities, (2) is closely equal to any other in difficulty, (3) is scored with absolute objectivity, and (4) can be given as an examination either without access or with full acccss to textbooks. The time allowed is 180 minutes, the score is $\frac{1}{2}$ of the sum of the credits earned by the following schedule:

PART I

Task	Credit	Task	Credit	Task	Credit	Task	Credit
1	1	11	2	21	2	31	4
2	1	12	2	22	2	32	4
3	1	13	2	23	2	33	3
4	1	14	2	24	2	34	4
5	1	15	1	25	2	35	5
6	1	16	2	26	2	36	5
7	1	17	2	27	2	37	5
8	1	18	2	28	3	38	3
9	2	19	2	29	4	39	5
10	1	20	2	30	6	40	5

PART II

Task	Credit	Task	Credit	Task	Credit	Task	Credit
1	2	6	6	11	6	16	8
2	3	7	8	12	6	17	8
3	3	8	7	13	6	18	8
4	3	9	7	14	8	19	8
5	5	10	6 if 3 are right 3 if 2 are right 0 if 1 (or none) is right	15	8	20	8

[1] These examinations were constructed by the Institute of Educational Research of Teachers College for experiment, and possible adoption by the College Entrance Examination Board. No two are exactly alike in the abilities tested, but any one gives a rough test of almost all important abilities. Being intended for purposes of promotion, credit, exemption, and the like, these examinations are designed to be especially effective toward the lower end of the scale of ability. They are not intended to give accurate measures of great, very great, and extraordinary mathematical gifts, though they seem to do this better than one might expect.

I. E. R. ALGEBRA TEST

PART I. TO QUADRATICS. NINETY MINUTES. FORM A

Write your answers on the dotted lines. Use the blank pages and empty spaces for your work

Write answers here

1. $(5a^2+4-3a^2)+(a^2-2-7a^2-5)$ **1.**............

2. $(-2a^3-10a-4a^3)+(5a+3a^3-4a)$ **2.**............

3. From $3a+4b$ subtract $5a-9b-3c$ **3.**............

4. From $5a-b-2c$ subtract $3c-3a$ **4.**............

5. $8a+8b-(3a+6b)$ **5.**............

6. $(5d-e)-(7e+2f)$ **6.**............

7. $7d \times 2de^2$ **7.**............

8. $de^2 \times d^2e$ **8.**............

9. $8-5\ (d+2)$ **9.**............

10. $4e^2+e\ (-4e-3)$ **10.**............

11. $5np-3p\ (4n+3p)$ **11.**............

12. $m+\frac{8\ m^2n^2}{mn}$ **12.**............

13. $4m+\frac{2m^2np^3}{mp}$ **13.**............

14. $\frac{m^2np}{mn}-\frac{m^2n-mp}{m}$ **14.**............

15. $(m^5n)\ (m^2n^3)$ **15.**............

16. $(2a-7)^2$ **16.**............

17. If $a=2$, and $b=3$, what does $5a^2-2ab$ equal? **17.**............

18. If $a=.7$, and $b=1.2$, what does $2a^2-5ab$ equal? **18.**............

19. If $a=1$, $b=2$, $c=.4$, and $d=100$, what does a^2-b+cd equal? **19.**............

20. If $a=12$, $b=6$, $c=5$, $d=3$, and $e=1$, what does $\frac{22}{7}\ [ab+c\ (d-e)]$ equal? **20.**............

21. If $d=2$, $e=3$, $f=4$, what does $\dfrac{\frac{df}{d+e}}{ef}$ equal? **21.**............

Write your answers on the dotted lines. Use the blank pages and empty spaces for your work

22. $d=\frac{ef}{g}$. What does f equal? **22.**............

23. $\frac{e}{W}=\frac{r}{R}$. What does W equal? **23.**............

24. $\frac{PV}{T}=\frac{P_1V_1}{T_1}$. What does V equal? **24.**............

25. $4q=7q+5$. What does q equal? **25.**............

26. $15=7w-4$. What does w equal? **26.**............

27. $\frac{V}{4}=a-2$. What does V equal? **27.**............

28. $\frac{6}{2-u}-\frac{11}{3-u}=0$. What does u equal? **28.**............

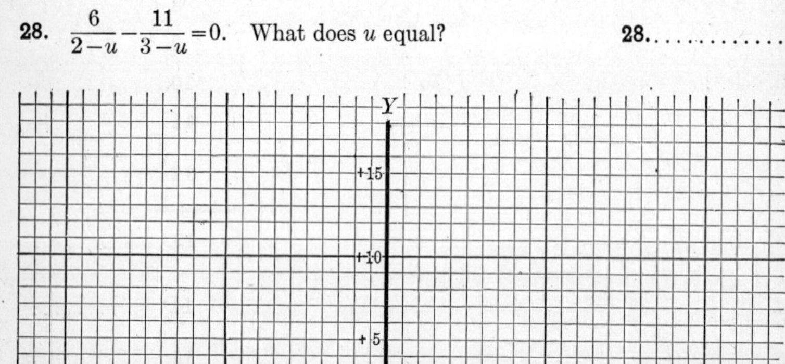

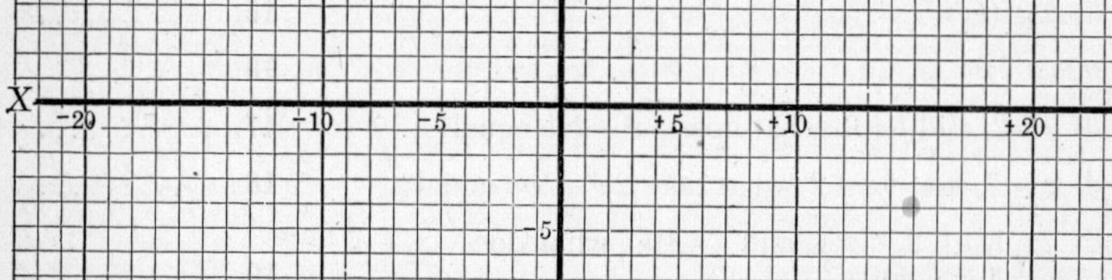

29. $P=\frac{b_1b_2}{c_1+c_2}$. b_1, b_2, c_1, and c_2 are all positive.

Write "*larger*" or "*smaller*" or "*you cannot know*" on the dotted lines.

If b_1 becomes larger, P will become ....

If b_2 becomes larger, P will become ....

If c_1 becomes larger, P will become ....

If c_2 becomes larger, P will become ....

30. $N=\frac{ab}{c}-d\ (e-f)+\frac{1.414}{g+\frac{1}{h}}$. a, b, c, d, e, f, g, and h are all positive.

Write "*larger*" or "*smaller*" or "*you cannot know*" on the dotted lines.

If a becomes larger, N will become ....

If b becomes larger, N will become ....

If c becomes larger, N will become ....

If d becomes larger, N will become ....

If e becomes larger, N will become ....

If f becomes larger, N will become ....

If g becomes larger, N will become ....

If h becomes larger, N will become ....

31. If $a=1$, $b=10$, and $c=100$, express 216 in numbers and letters. ....

Express 2.16 in numbers and letters. ....

32. To two times a certain number 2 is added. From 3 times the number 7 is subtracted. The two results are equal. What is the number? ....

Use cross section paper on opposite page for Nos. 33, 34, 35.

33. Make a little cross (x) at the point for which $x=2$ and $y=3$.

Make a little circle (○) at the point for which $x=-4$ and $y=1$.

Make a little triangle (△) at the point for which $x=1\frac{1}{2}$ and $y=-4$.

34. Draw a graph of $y=1\frac{1}{3}x-5$. Mark it No. 34.

35. Draw a graph of $y=\frac{1}{x}+8$. Mark it No. 35. Draw enough of it to show clearly that you understand it.

Write your answers on the dotted lines. Use the blank pages and empty spaces for your work

36. Find two numbers such that:—
Twice the first, if added to three times the second, equals 2.
Six times the second, if added to ten times the first, equals 7. **36.** ....

37. What is the length of the diagonal of a rectangle, if the sides of the rectangle are 8 ft. and 6 ft.? **37.** ....

38. Multiply $\sqrt{\frac{a}{b}}$ by $\frac{b}{\sqrt{a}}$ **38.** ....

39. Write a plus sign (+) on the dotted line if the statement is true.
Write a minus sign (−) on the dotted line if the statement is false.

$\sqrt{\frac{ab}{cd}}=\frac{\sqrt{abcd}}{cd}$ ....

$\sqrt{a}\,\sqrt{b}\,\sqrt{c}=\sqrt{a+b+c}$ ....

$a^{\frac{1}{2}}=\frac{1}{a}$ ....

$a^{\frac{1}{3}}\times a^{\frac{1}{3}}=a\sqrt[3]{2}$ ....

$\sqrt{\frac{1}{5}}=\frac{\sqrt{5}}{5}$ ....

Write words or numbers on the dotted line to make a true statement.

40. Under what conditions will $a+b+c=abc$?
If.......................................$a+b+c$ will equal abc.

PART II

Quadratics and Beyond. Ninety Minutes. Form A

Write your answers on the dotted lines. Use the blank pages and empty spaces for your work

1. Multiply g^3hj^{-1} by g^5hj^{-6}. Ans............

2. Divide $pq^{\frac{2}{3}}$ by $p^{\frac{2}{3}}q^{\frac{1}{2}}r$. Ans............

3. Write 7.03×10^8 as an ordinary number. Ans............

4. Write the square of $p^{-3}+6q^{\frac{1}{3}}$. Ans............

5. What are the factors of $6x^2-x-12$? Ans............

Write your answers on the dotted lines. Use the blank pages and empty spaces for your work.

6. Solve the equation $x^2-3\frac{1}{2}x+1\frac{1}{2}=0$. Ans............
("Solve the equation" means find its roots, the values of x which satisfy the equation.)

7. Solve the equation $\frac{x^2}{c}+\frac{x}{2}+d=0$. Ans............

8. Write the middle term of $(m+0.2n)^6$. Ans............

9. Write the middle term of $\left(\frac{m}{2}+n^{\frac{1}{2}}p^{-3}\right)^6$ Ans............

10. Examine each of these. If it is an arithmetical progression, mark it A. If it is a geometrical progression, mark it G. If it is neither, mark it No.

8.4	7.6	6.8	....
$\frac{3}{4}$	$1\frac{1}{8}$	$1\frac{11}{16}$	....
$5f-g+3h$	$f+3h$	$g+3h-3f$	....

11. What is the sum of the first six terms of the series beginning 1, $1\frac{1}{2}$, $2\frac{1}{4}$? Ans............

12. Multiply $3\sqrt{a}-2\sqrt{b}$ by $\sqrt{a}+5\sqrt{b}$. Ans............

13. If $p-3\sqrt{p}+2=0$ what does p equal? Ans............

Do any Four of Numbers 14 to 21

14. A body falls 16 feet in the first second, three times as far in the second second, five times as far in the third second, and so on. How far will it fall in a minute? Ans............

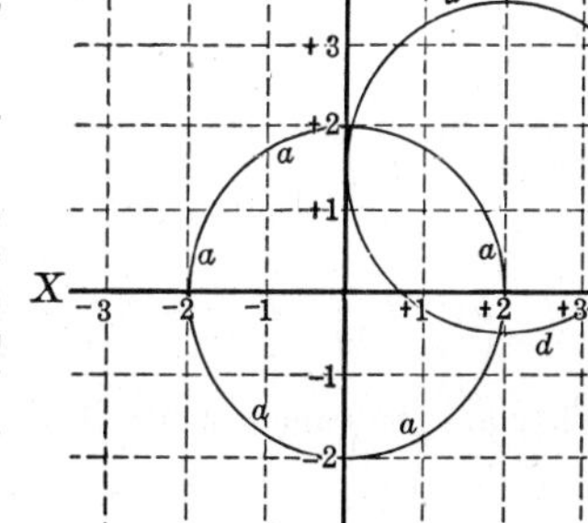

15. A man gives to his daughter each year as many dollars as she earns and the square of the number of dollars she saves. How much must she earn in a year to get $400 in all, if she saves half of what she earns? Your answer need be correct only to the nearest dollar.

Ans............

Write your answers on the dotted lines. Use the blank pages and empty spaces for your work

16. The equation of circle *a a a a* is $x^2+y^2=4$. What is the equation of circle *d d d d* whose center is at $+2, +1\frac{1}{2}$? (See previous page for figure). Ans.............

17. Examine these equations: $y+\frac{4}{x}=0, y=\frac{x}{4}+2$. If there are real values of x and y which satisfy both equations (that is, if the two lines intersect) write *Yes*, and state approximately what the values are. An answer will be called right if it is within 10 per cent of the exact answer. If there are no such values of x and y (that is, if the two lines do not intersect) write *No* and make a rough sketch showing the two lines. Ans.............

Use logarithmic tables for 18 and 19

18. $\frac{.88078\times 21.277}{.065925}$ Ans.............

19. $\frac{\sqrt[3]{3.21}\times\sqrt[3]{.0043725}}{.12111}$ Ans.............

20. In the right triangle ABC, sin A=0.17, $c=25.6$. Find a. Ans.............

21. A man walks 1,000 yds. up a slope of 20°. How high is he then above his starting point? Ans.............

The writer may be permitted to illustrate the change in our notions of what should be measured in a test in algebra and how it should be measured, by a test planned and given by himself in a number of schools about 1900. It was as follows:

Do these examples as quickly as you can. Do not copy them but put the work right under each example. Take the quickest way you can to get the correct answers:

1. Simplify

$$\left(\frac{a^2-b^2}{x-y}\right)\left(\frac{x^2-y^2}{a-b}\right)\left(\frac{c^2}{x+y}\right)$$

2. What are the values of x and y if $5x+3y=8$ and $7x-3y=4$?

3. A shepherd being asked how many sheep he had in his flock, said: "If I had as many more, half as many more, and seven sheep and a half, I should then have 500." How many sheep had he?

4. What are the values of x and y if $1-\frac{x+y}{x-y}=\frac{3x}{x-y}$ and $\frac{7x-3y}{23}=3$.

5. Simplify

$$\frac{\frac{m+n}{m-n}+\frac{m-n}{m+n}}{\frac{m-n}{m+n}-\frac{m+n}{m-n}}$$

6. If to the double of a certain number 14 be added, the sum is 154. What is the number?

This represented the orthodox content and method of examinations of that day. It illustrates almost every fault that a test can have! Only two of the six tasks would be tolerated by experts in educational measurements today.

Other instruments for measuring achievement in algebra are as follows:

Algebra tests *A*, *B*, *C*, *D*, *E*, and *F* by Walter S. Monroe [1915 and 1917] are tests of certain computational abilities.[1]

Test A, 2 minutes, to multiply $\pm a(\pm bx \pm c)$ where a, b, and c in no case were greater than 9, and in no case were all positive.

Test B, 3 minutes, to reduce fractions to a common denominator.

Test C, 1 minute, to solve equations of the type $\pm ax = \pm b$.

Test D, 2 minutes, to transpose terms in equations such as $4x-6+5=7x-4-2$.

Test E, 3 minutes, to collect terms in expressions such as $-5x+6x-11x+8-3-9$.

Test F, 12 minutes, 13 equations to be solved, Nos. 1, 4, 7, 10 and 13 being:

$$\frac{-3x-2}{4}=\frac{x+2}{6}$$

$$1-\frac{x+5}{6}=\frac{-4x+9}{8}$$

$$\frac{-3\,(2x-5)}{5}+\frac{-2\,(3x+4)}{7}=0$$

$$5+\frac{3x+1}{8}-\frac{-1+2x}{12}=0$$

$$\frac{2x+3}{10}-\frac{1+2}{12}+\frac{-4x-5}{15}=1$$

[1] Monroe was the first to provide tests for general use, with a statement of results obtained in a number of schools. The need of tests graded in difficulty had been pointed out by Thorndike [1914].

Dalman [1920] prepared tests of the inventory type, with each topic represented by four levels of difficulty or range of ability or both, for use in gradation and promotion in algebra. Mastery meant a grade of C, B, A or A+ according to the level mastered. There were two sets of tests, one for the first and one for the second semester. This recognition of the distinction between knowing one half of a subject and half knowing the whole subject had, as might be expected, a very beneficial effect upon learning.[1] Many of the difficulties and defects of grading were also avoided or lessened by it.

[1] The actual content of the tests would, however, not meet modern standards. The A+ test for the second semester, for example, has as its even-numbered tasks the following:

Find the prime factors of:

2. $2x^3-8$

4. $6b^2c^2-25bc+14$

6. y^4+3y^2+4

8. $27y^6+64m^3$

Simplify:

2. $\frac{a+b}{a^2+ab+b^2}+\frac{a-b}{a^2-ab+b^2}+\frac{2b^2}{a^4+a^2b^2+b^4}$

FRACTIONAL EQUATIONS

2. A's age is one-third of B's age, and 6 years ago it was one-fifth of B's age. Find their present ages.

SIMULTANEOUS EQUATIONS

2. The sum of two digits of a number is 14; and if 18 be subtracted from the number, the remainder equals the number obtained by reversing the digits. Find the number.

QUADRATIC EQUATIONS

2. Find the dimensions of a rectangle whose area is 720 square feet if the sum of the base and altitude is 54 feet.

The University of Wyoming Algebra Test, by C. E. Stromquist, is a 36-minute test, comprising seven parts as stated and illustrated below:

I. APPLICATION OF FORMULAS. FOUR MINUTES

1. Solve for t: $v=gt$

16. Solve for b: $T=\frac{1}{2}a\ (b+b')$

II. DIVISION. FOUR MINUTES

1. $c-3d\ \overline{|2ac-6ad+5bc-15bd}$

6. $2a^2+7b\ \overline{|6a^5+17a^3b+8a^2b-14ab^2+28b^2}$

III. SIMPLE EQUATIONS IN ONE UNKNOWN. FOUR MINUTES

1. $5x-12=x+23-3x$

12. $1.5-.3x=.3-.6x$

IV. SIMPLE SIMULTANEOUS EQUATIONS. FOUR MINUTES

1. $3x+y=5$
$2x+y=2$

6. $7x+2y=8$
$8x+3y=9$

V. FACTORING. FIVE MINUTES

Factor to simplest factors.

1. $36mx+48my=$

12. $15x^2+x-6=$

VI. NUMERICAL SUBSTITUTION. FIVE MINUTES

1. $5a+3b-4c$ for $a=3$, $b=2$, $c=-\frac{1}{2}$

10. $\dfrac{-b+\sqrt{b^2-4ac}}{2a}$ for $a=3$, $b=-11$, $c=+6$

VII. FORMING EQUATIONS. TEN MINUTES

Form the equations for the following problems but do *not solve them*. Indicate first what the letter you use stands for.

1. Twice a number increased by $\frac{1}{3}$ of the number is 28. Find the number.

10. The combined area of the floors of two rooms is 405 sq. ft. The first room is in the form of a square and the second a rectangle. The length of the second room is the same as that of the first and its width is 12 ft. Find the length of the first room.

The Douglass tests [1921] are of the same general nature as the Hotz test, but contain no work with evaluation, formulas, graphs, or verbal problems.

"They comprise a series of four sets of ten exercises each devised to test the four formal operations of elementary algebra as it is usually taught in the first year of secondary schools, which operations were selected as the essential fundamentals of elementary algebra by the combined judgment of competent judges. The exercises are selected so as to provide tests for all the principal teaching units and phases of each operation and at the same time to provide problems of a wide range of difficulty. The first part of this statement may be verified by an examination of a number of algebra elementary texts; the latter part may be seen from the tabulated results showing the relative difficulty of the problems. The tests are to be given with a uniform procedure and under natural conditions. From giving the tests, no measure of speed may be obtained further than the fact that the extremely slow are penalized. The resulting scores are a measure of power, not of speed. Each problem has an assigned weight, to be regarded in scoring papers, which weight is determined by, and is directly in proportion to, the relative difficulty of the problem." [1921, p. 33.]

The first two and last two elements of each of the four tests are:

TEST

1. Add

$$\begin{array}{r} 15m \\ \underline{12m} \end{array}$$

2. Add

$$\begin{array}{r} 8ab \\ -7ab \\ -5ab \\ \underline{2ab} \end{array}$$

Collect Terms

9. $a^2+8a+7a^2+x+7+9x+4+3a=$

10. $2x-xy+3y-3x+2xy+7y+8=$

TEST II

1. Multiply

$$\begin{array}{r} 9m \\ \underline{2} \end{array}$$

2. Multiply

$$\begin{array}{r} 4x^2 \\ \underline{3x^2} \end{array}$$

9. Multiply

$$\begin{array}{r} 3a^2bc+5b^2cd \\ \underline{-4ax+2} \end{array}$$

10. Multiply

$$\begin{array}{l} 7by^2z-4a^2cx^2+9ab^2y \\ \underline{9byz^2+8a^2bc^3} \end{array}$$

TEST III

1. Divide $12a^4$ by $2a^3$
2. Divide $16x^3$ by $-4x^2$
9. Divide $6a^3-18a-11a^2+20$ by $2a-5$
10. Divide $r^3-19r+84-6r^2$ by $r-7$

TEST IV

1. Solve for x $4x=12$
2. Solve for a $4a=3$
9. Solve for x $\frac{2x}{3}+8=\frac{3x}{2}+\frac{x}{6}+14$
10. Solve for x $\frac{3x}{5}-\frac{x}{2}+\frac{7x}{10}=\frac{1}{2}+2x$

In all these tests some time limit will be set, and we have always the problem of equating quality of work with speed. Even if sufficient time is given to allow each pupil to do all that he can do, this issue is not avoided; for some pupils will then finish ahead of time and the question arises whether they shall receive any extra credit therefor. The general principle should be to give credit for speed in so far as it is a symptom of mastery. Intrinsically, it is of little value. Algebraic computation is used so seldom in life that the mere utility from time saved by speed is of almost no consequence in comparison with the abilities themselves. Algebra is not like reading or writing or simple arithmetical computation, where speed means a substantial daily saving of time.

As a symptom of mastery speed does deserve consideration. If a pupil has sure command of the treatment of signs when a parenthesis is removed, for example, he will not only do work like $4a-(2-a)$ with precision but will also do it rapidly. Moreover if one pupil can do harder things than another pupil, he will in general do easier things, which both can do, more rapidly than the other pupil. This correlation between how hard things a pupil can do and how rapidly he can do easy things is close enough so that even rather high credits for speed are not notably unfair—do not notably disturb the order in which a group of pupils is put.

The Rugg-Clark and the Hotz tests and the I. E. R. test differ notably from the customary school, state, or college entrance examination in being composed of many more tasks and including easier tasks. This will be found generally true of the examinations devised by psychologists and by scientific students of education.

What is gained and what, if anything, is lost by having pupils spend their time on sixty or more units of work including "easier" tasks, rather than on eight or ten tasks which are more elaborate or more difficult or both?

There is a gain first of all in the objectivity of the scoring. If only a dozen or so tasks are given, it becomes in practice necessary to give partial credits for work which, though wrong, is not as wrong as it might be. Different scorers will judge differently about these, and, unless they work under an elaborate uniform scheme of what credit to give to this, that, and the other, the measures will be tainted by individual caprice.

There is a further gain in the reliability of the scores as measures of the pupils in question. The effects of having had the same problem in his textbook, of misunderstanding a word, of being misled by an error in copying, and the like will tend to equalize themselves amongst individuals in sixty tasks, but may be causes of serious variations when there are only a dozen.

Nobody would advocate measuring a pupil in comparison with other pupils by one problem only. Sixty have the same sort of advantage over ten that ten have over one. This is more serious than examiners have realized. Wood [1921], in a study made for the College Entrance Examination Board, found that the official marks for the June 1921 Math. A paper (100 candidates taken at random) had a reliability coefficient of only .76. The correlation of the score for the odd-numbered questions with the score for the even-numbered was only .61. A coefficient of .76 means that the error of estimate by the examination is over six-tenths as large as if the grades had been assigned by chance.

Using the same method, the reliability coefficient of the examination shown on pages 181-186 is, for a group of about four hundred pupils in ten schools who had studied algebra at least one year, .95½. The two halves (one being 1 to 5, 11 to 15, 21 to 25, 26, 27, 32, 33, 35, 37, and 39, the other being the balance) correlate .91½. These correlations should not be compared with those obtained by Wood without making allowance for the probably smaller variability of his group. If in our group only such pupils as would attempt the College Entrance Board examinations were used, the reliability coefficient would be lower.[1] On the other hand, the examination used by Wood was of three hours' length, whereas the one reported here was of only one and a half hours' length. It is reasonable to estimate that in a three-hour examination the reliability coefficient of the new type of examination would be .10 higher than that of the old type for the same group of pupils.

We lack measures of the unreliability obtained directly by having the same pupils do a number of different forms of each sort of examination. The examination composed of many elements, graded in difficulty, and all scored objectively would probably show an increased superiority, since its several forms are very nearly equal in difficulty, while

[1] Using as a group the boys who were in four excellent private schools in New England, we find a coefficient of .80, the correlation between the two halves being .67. This group probably has even less variability than the group of candidates studied by Mr. Wood.

examinations of the ordinary type, even when prepared with great care, as by the College Entrance Examination Board, are known to vary greatly in difficulty, and this adds greatly to the error of estimate unless the gross measures are transmuted on some assumption about the nature of the group tested.

There is further gain, in that the inclusion of a wider variety of tasks gives us to some extent an inventory test, and tells us which abilities the pupil has and which he lacks. Finally, the inclusion of easy tasks gives us to some extent a "ladder" test which measures algebraic abilities over a wide range, and tells us to some extent the degree of mastery he has within each sort of ability.

The loss, if any, is likely to be in the lack of measures of a pupil's ability to organize and use in new situations, various combinations of these abilities, selecting the appropriate ability or combination of abilities according to his needs. Thus to solve $\frac{5}{x-7}-\frac{7}{x+3}=\frac{5x+1}{x^2-4x-21}$, the pupil must organize and use many elementary abilities in adding, subtracting, multiplying, and dividing, and seeing possibilities of factoring. To factor and simplify

$$\frac{a^2-11a+30}{a^3-6a^2+9a}\times\frac{a^2-3a}{a^2-25}\div\frac{a^2-9}{a^2+2a-15}$$

he must play a sort of selective game that utilizes many relations together. In both cases he must make no mistakes in the course of a rather long series of operations. It may even be said that the essential ability sought in algebra is not to operate algebraic abilities singly or in their common combinations, but to master a complex and novel task wherein these abilities must coöperate, the manner of their coöperation being selected by algebraic insight. Such tasks as "Find by a short process 2^{19}, given $2^{10}=1024$," are then esteemed features of an examination in algebra.

The three issues thus brought up are each worthy of consideration. In what respects is one task which requires the successful operation in sequence of ten abilities, better for measurement than ten tasks each testing a single ability, or than five tasks each testing two abilities? In what respects is a task which requires rather elaborate choice and organization of abilities, better than one which uses them more simply? In what respects are tasks requiring novel applications of algebraic theory and technique, what we may call algebraic "originals," better than routine applications?

Tasks Where a Correct Result Requires the Correct Action of Many Abilities

Consider as a sample this case:

Find the H. C. F. and L. C. M. of x^3-125, $5x^3-125x$ and $x^2-10x+25$. For an ordinary pupil this is substantially a test of whether he can do *all* of these:

(1) Know what H. C. F. means.
(2) Know what L. C. M. means.
(3) Know that 125 is the cube of 5.
(4) Know that $x^3-(5)^3=(x-5)\ (x^2+5x+25)$.
(5) Know that the promising thing to do first with $5x^3-125x$ is to change it to $5x(x^2-25)$.
(6) Know that $x^2-25=(x+5)\ (x-5)$.
(7) Know that $x^2-10x+25$ to be factored means $(x-?)$ $(x-?)$.
(8) Find that $(x-5)\ (x-5)$ will do the trick, with or without experimentation with other combinations.
(9) Remember or see that he has $(x-5)\ (x^2+5x+25)$, $5x(x+5)\ (x-5)$ and $(x-5)\ (x-5)$.

(10) Apply his knowledge of what H.C.F. means to select $(x-5)$ from them.

(11) Apply his knowledge of what L.C.M. means to select $(x-5)(x^2+5x+25)5x(x+5)(x-5)$ from them.

Suppose that it were replaced by:

Factor 125.
Factor x^3-5^3.
Factor $5x^3-125x^2$.
Factor x^2-25.
Factor $x^2-10x+25$.
What is the H.C.F. of x^3-125, $5x^3-125x^2$, and $x^2-10x+25$?
What is the L.C.M. of x^3-125, $5x^3-125x^2$, and $x^2-10x+25$?

A pupil then might score from 0 to 7, whereas before he could score 0 or 1.

Suppose that the mastery of each of these seven component abilities is such that it operates correctly nineteen times out of twenty. The pupil would get a score of 7 about two times out of three, of 6 about once out of three, and, but very rarely, scores of 5 or lower. In the other plan he would score 1 about two times out of three, and 0 about once out of three.

Suppose that the mastery of each of these seven component abilities is such that it operates correctly nine times out of ten. The pupil would get a score of 7 about once out of three, a score of 5 or 6 about twice out of three, and scores of 4 or below very rarely. By the other plan, he would score 1 about once out of three and 0 about twice out of three.

Suppose that the mastery of each of the seven is such that it operates correctly four times out of five. The pupil would then almost never obtain other than 0 as his score by the second plan.

The "long" task, that is, attaches no weight to the component abilities until they have reached a rather high degree of strength. This is reasonable in so far as the abilities begin to be useful in life only at a high degree of strength. It leaves always as undistributed zeros individuals who vary greatly in ability. This makes it desirable to have similar "short" tasks to secure differentiation amongst these, and to discover which of the component abilities are specially weak and in need of training.

Tasks Where a Correct Result Requires Elaborate Organization of Abilities

The "simplify" tasks are the commonest case. In so far as they are worth training these selective and organizing abilities are worth measuring. It would be inadequate to measure algebraic abilities singly and to neglect the ability to organize them. The tasks should, however, be genuine and not much more elaborate than life itself offers.

Algebraic "Originals"

Tasks requiring novel applications of algebraic theory and technique measure one of the most important abilities that the study of algebra can give to those who have the capacity. A set of a thousand such originals which possess significance for pure or applied mathematics would be useful in testing and also in teaching. Some such are given below:

1. Let $a=$ any integral number. Write an expression that will always represent an even number. Write an expression that will always represent an odd number.

2. State a condition such that $a+b+c=abc$. State a different condition that will produce the same result.

3. Under what conditions will increasing the value of a increase N if $N=ab-\frac{b}{a}$?

4. State a condition such that $abc=\frac{ac}{b}$.

5. Draw the graph of $x=5$.

6. What would be the graph of $x^2+y^2=0$?

It must be borne in mind, of course, that when such originals are made the subject of explanation and drill, and when certain types of them are prepared for as the theory of signs or of exponents is prepared for, their value as originals is lost.

THE MEASUREMENT OF MORE GENERAL ABILITIES

It is probable that, in spite of the emphasis in recent years upon direct specific algebraic abilities, of use in science, technology, and business, four out of five teachers of algebra think that its service as an improver of more general abilities is the greater. These more general abilities have never been rigidly defined, but may be thought of under these five heads:

The ability to deal with symbols.

The ability to deal with relations, especially the more "intellectual" relations like resemblance, cause and effect, proportionality, etc.

The ability to generalize and deal with generalizations.

The ability to select elements and features as needed.

The ability to organize ideas and habits, using a number of them together to good effect.

In the following pages an instrument is presented with which to measure the status reached and the improvement made in a composite of these more general abilities. The instrument is arranged in parts, each of which emphasizes one of the abilities listed. One part does not, however, measure one ability exclusively. A test which did so would be likely to be so artificial as to be of value only to psychologists. The critical reader, perhaps, should neglect the division into parts and simply consider the entire examination as an instrument to measure a composite of abilities.

The description given here is of only one-half of the instrument, a second examination being available so that pupils may be measured before and after their study of algebra.

The reader who is acquainted with instruments for the measurement of intelligence will recognize the similarities between this examination and the examinations by which a psychologist measures intelligence, or, more exactly, intelligence as operating with ideas, abstractions, and symbols. The five abilities of our list would indeed be tolerable as a rough definition of this abstract intelligence — of the thing which the Stanford Binet at upper ages, the Army Alpha, the National Intelligence Test, and the Thorndike Intelligence Examination for High School Graduates, do measure. Conversely, it would not be grossly unfair to test the gain in general abilities hoped for from algebra by one of these standard intelligence tests. It would be somewhat unfair, since certain tests which do in fact correlate with intelligence, such as knowledge of the vernacular, range of general information, and memory and comprehension of directions, and which are included in most of the instruments for measuring it, could not well be assumed to test the abilities which algebra is expected to improve. The examination shown here is thus preferable for our purpose.

Certain cautions are needed with respect to its use. By merely giving Form A at the start of the course in algebra and Form B at its close, certain information accrues. We have then measured the gross gain, and can compare the gains of individuals, making such interpretations as seem best. If we take the trouble to arrange for classes equal in ability at the start, and treated alike in all respects save some point to be investigated, say a difference in the subject matter of the algebra course, or a difference in methods of teaching, we can measure the effect of this difference. We can compare the gains in the case of the more algebraic with the gain in the case of the less algebraic data.

By giving Form A and Form B to various groups under various conditions we can obtain estimates of the amount of improvement to be expected in a year in individuals of various sorts who spend the year in various ways. For example, we can compare students who study algebra with those who do not. The control of the conditions in such an experiment is, however, a matter requiring special facilities and great care.

In all cases it must be remembered that a certain gain in tests of this sort is to be expected from the general mental growth and training of a year, and that the second trial of the test may show a rather large gain over the first trial as the result of familiarity with the form of the test.

We have not been able as yet to carry on the experimentation necessary to determine the comparative difficulty of the A and B series, element by element. We hope to do this later, but may not. As totals they are presumably very closely equal in difficulty; and a collection of all the elements of A having mathematical content will probably be very closely equal to a corresponding collection from B. The same holds good for a collection of the elements having specially verbal content. Knowledge of the difficulty of corresponding elements in the A and B halves is not essential, since the comparison is between the gains of the group having the training in question and those of the control group lacking the training.

THE I. E. R. TESTS FOR ABILITY WITH SYMBOLS. FORM A

TEST 1

1. If $a \times a = a^2$
$a \times a \times a = a^3$
$a \times a \times a \times a = a^4$
How will you express, $n_1 \times n_1 \times n_1 \times n_1 \times n_1 \times n_1$? Ans.............

2. How will you express $a \times a \times a$ etc., if there are n a's in the row? Ans.............

3. Let pp^n mean any flat surface enclosed by straight lines, the n denoting the number of sides it has. Let e mean equiangular. What is the common name for $e\ pp^4$? Ans.............

4. Express this surface using letters and numbers as described in 3. Ans.............

5. Express in briefest form "any number times a constant divided by that number equals the constant." Ans.............

6. Let a = the number of months a man has lived. How many years has he lived? Ans.............

7. Express in brief form, using I, B and D: "The illumination varies directly as the brightness of the light and inversely as the square of the distance." Use = K times for "varies as". Ans.............

8. A man's annual income is $2,000 plus the profits from his business. He plans to put $\frac{1}{5}$ of his income in the savings bank, to spend $1,500 plus $\frac{1}{4}$ of the profits from his business for necessary expenses, to spend $\frac{1}{4}$ of the profits of his business for refinements and luxuries, and to use the rest of his income to enlarge his business. Using P, S, Ne, Rl and B, show how he plans to use his income. Ans.............

TEST 2

Write the answers on the dotted lines. Use the bottom of the page to figure on.

1. Let n = any number
Let n_r = 1 divided by n
Let n_R = 10 divided by n
Let n^s = the number raised to the same power as itself.

What does $\left(\frac{a_R}{a_r}\right)^s$ equal? Ans.............

2. Let m, m_2, m_3, etc., be any numbers.
Let n be their number; that is, n tells how many m's there are.
Let S ( ) mean "the sum of".
What name will you give to $\frac{S\ (m\text{'s})}{n}$? Ans.............

3. Let *A. D.* = the average of the deviations of a set of numbers from their average, disregarding the signs of the deviations.
Find the *A. D.* of **6**, **9**, **10**, **11**, **14**. Ans............

4. Let *S. D.* = the square root of the average of the squares of the deviations of a set of numbers from their average. Find the *S. D.* of **11, 13, 14, 15, 17.** Ans............

TEST 3

Suppose that | stands for *he*
y stands for *yes* or *is* or *are*
< stands for *increasing*
o. stands for *outside* or *out*
□ stands for *house*
g stands for *good*
gg stands for *better*
ggg stands for *best*
b stands for *bad*
l stands for *long*

1. Write what should stand for *worse* ....
2. *longest* ....
3. *inside* ....
4. *no* ....
5. *decreasing* ....
6. *she* ....

Write what you think each of these means.

7. | *y* < *gg* ....
8. □ < *y* ....
9. | *y* < *ll* ....
10. | – *y* □. ....
11. *b* □ *y gg n* □ ....
12. *ll y bb y* ....

TEST 4

Read the first sentence carefully to see what it means. Then check the two of the four sentences which have the same meaning. Check only two.

1. Better be a big frog in a little puddle than a tadpole in a lake.
......Better the head of an ass than the tail of a horse.
......I had rather be a door-keeper in the house of my God than to dwell in the tents of wickedness.
......Better to reign in hell than serve in heaven.
......Better to be a beggar in Rome than a prince in a village.

2. Don't cross the bridge before you come to it.

......Look before you leap.
......Don't borrow trouble.
......Don't lock the barn after the horse is gone.
......Take care of today and tomorrow will take care of itself.

3. Great men may jest with saints; 'tis wit in them, but, in the less, foul profanation.—(Shakespeare.)

......The rank is but the guinea's stamp.
......Great men tremble when the lion roars.
......What's sin for the servant is saintly in the master.
......That in the captain's but a choleric word which in the soldier is flat blasphemy.

4. Solon compared the people unto the sea, and orators and counselors to the winds: for that the sea would be calm and quiet, if the winds did not trouble it.—(Bacon.)

......Orators and counselors are puppets of public opinion.
......Orators and counselors are responsible for the unrest of the people.
......Solon feared spontaneous uprisings among the masses.
......He believed the people to be essentially passive and inert when left alone.

[Twenty more similar tasks follow.]

Test 4 is from the Psychological Examination of the Carnegie Institute of Technology.

THE I. E. R. TESTS FOR SELECTIVE AND RELATIONAL THINKING

TEST 1

Twenty-one Arithmetical Problems

"Find the answers to these problems. Write the answers on the dotted lines. Use the blank sheets to figure on." Nos. 1, 4, 7, 10, 13, 16, and 19 are:

1. What is the cost of four tickets at 50 cents each? Ans.............

4. How much will 24 lemons cost at 30 cents a dozen? Ans.............

7. At 6 for 25 cents, what is the cost of 3 dozen? . Ans.............

10. What number minus 7 equals 23? Ans.............

13. 4 per cent of $600 equals 6 per cent of what amount? Ans.............

16. A family spends $600 on rent, $3,000 on other expenses and saves $200. If they increase their total expenses to $4,200 and their savings in the the same ratio, how much will they save? Ans.............

19. Jofas are 4 for 25 cents. Kelas are $2\frac{1}{2}$ cents each.
A Jofa costs.............as much as a Keia.

TEST 2

Sixteen Absurdities (Thorndike after Woodworth)

"Write '*imp*' before each statement that could not possibly be true. Write '*poss*' before each statement that might possibly be true (even if it is not probable)." Nos. 1, 4, 7, 10, 13, and 16 are:

........**1.** Each singer shouted at the top of his voice, but the big fat man with the red necktie could be heard above all others.

........**4.** The poor wanderer, finding himself without means of lighting his camp fire, made a fruitless search through his equipment by the light of a single candle.

........**7.** Using a field glass, the captain now clearly perceived what he had previously surmised — a group of mounted men moving cautiously along the river bank on their hands and knees.

........**10.** By the light of a dim lantern, the farmer found the source of the nauseating odor.

........**13.** He stood on the dry grass watching the rain, which had been falling steadily for two days and nights.

........**16.** Starting half way between two posts, he walked slowly all around the field and each post that he met was shorter than any he previously had passed.

TEST 3

Five-line Arrangements

1. Draw 6 triangles using only 5 lines. All the triangles in your drawing may be counted to make up the six. For example, in the upper drawing there are three triangles: ABC, ACD, and ABD. In the lower drawing there are six rectangles, ABGH, BCFG, CDEF, ACFH, BDEG, and ADEH.

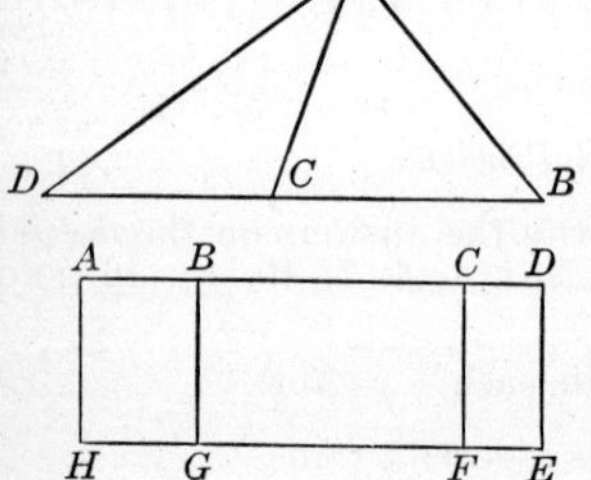

2. Draw 5 squares using only 6 lines.

3. Draw a hexagon, 2 squares, and 4 triangles, using only 9 lines. (Hexagon means a 6-sided figure.)

4. Draw a square surrounded by 4 triangles, using only 6 lines.

5. Draw 10 triangles, 2 squares and a hexagon, using only 11 lines.

TEST 4

Thirty-five Wylie Opposites

"Look at each of the words in the list below. Then write a word after each one which means just the opposite and which also begins with the letter *b*. If you come to any word which you cannot do, then go on to the next one. These three samples are given as they should be:"

girl — boy
covered — bare
upset — balance

Numbers 1, 6, 11, 16, 21, 26, and 31 are:

1. Good **6.** White **11.** Straight **16.** Loose
21. Cultured **26.** Exhume **31.** Spiritual

TEST 5

Twenty Number-series Completions (*Thorndike After Rogers*)

"In the lines below, each number is gotten in a certain way from the numbers coming before it. Study out what this way is in each line, and then write in the space left for it the number that should come next. The first two lines are already filled in as they should be."

Samples: 2, 4, 6, 8, 10, ...**12**...
11, 12, 14, 15, 17, ...**18**...

Lines 1, 5, 7, 10, 13, 16, and 19 are:

5, 10, 15, 20, ...
103, 95, 87, 79, ...
$5\frac{1}{2}$, 7, $8\frac{1}{2}$, 10, ...
$21\frac{1}{5}$, $21\frac{2}{5}$, $21\frac{3}{5}$, $21\frac{4}{5}$, ...
10, 12, $12\frac{1}{2}$, $14\frac{1}{2}$, 15, 17, ...
$1\frac{3}{16}$, $1\frac{3}{8}$, $1\frac{9}{16}$, $1\frac{3}{4}$, ...
7, 9, 10, 11, 13, 14, 15, 17, ...

TEST 6

Geometrical Relations

Two sets of ten each (*Thorndike*)

"In lines 1 to 10 draw a fourth figure in each line such that the fourth figure is to the third as the second is to the first, as shown in lines A and B."

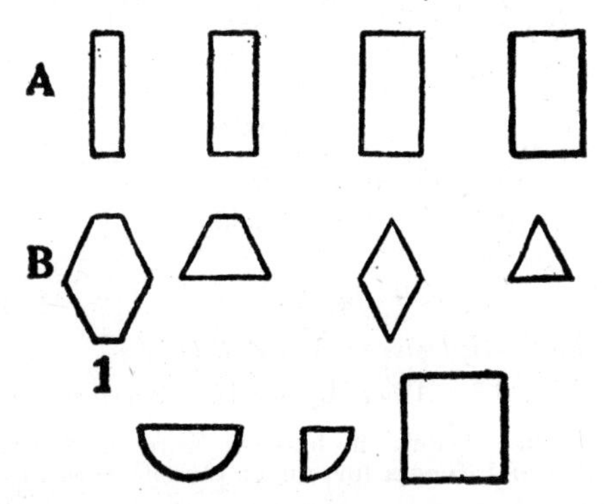

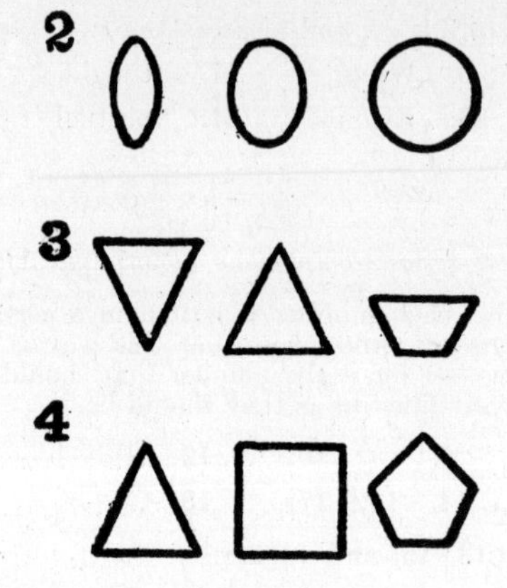

TEST 7

Thirty-two Verbal Analogies or Relations

(Thorndike, Selection and Extension from Briggs)

"Write on the dotted line a word which fits the third word in the same way that the second word fits the first word. The samples show what you are to do."

Samples	long	longer	bad	*worse*
	push	pushed	run	*ran*
	boy	boys	ox	*oxen*

Nos. 1, 6, 11, 16, 21, 26, and 31 are:

1.	child	child's	Poll and Hunt	....
6.	cook	cook's	Burns	....
11.	wings	wing	they	....
16.	driver's	drivers'	my	....
21.	stones	stone	strata	....
26.	prepare	preparation	flee	....
31.	fit	fitness	young	....

TEST 8

Forty Analogies or Mixed Relations

(Thorndike, After Army Alpha, After Woodworth-Wells)

"In each of the lines below, the first two words have a certain relation. Notice that relation and draw a line under the *one* word in the parenthesis which has that particular relation to the third word. Begin with No. 1 and mark as many sets as you can."

Samples:
- sky—blue: grass (grow, *green*, cut, dead)
- fish—swims: man (boy, woman, *walks*, girl)
- day—night: white (red, *black*, clear, pure)

Nos. 1, 10, 20, 30, and 40 are:

1. eat — bread: drink (water, drunk, chew, swallow)

10. tiger — wild: cat (dog, mouse, tame, pig)

20. poison — death: food (eat, bird, life, bad)

30. birth — death: planting (harvest, corn, spring, wheat)

40. advice — command: persuasion (help, aid, urging, compulsion)

THE I. E. R. TESTS FOR GENERALIZATION AND ORGANIZATION

TEST

One-half of the Pressey Moral Judgment Test Modified

(12 lines). "In each line cross out the word that does not belong."
Lines 1, 4, 7, and 10 are:

1. borrowing, gambling, overcharging, stealing, begging

4. stinginess, carefulness, generosity, charity, economy.

7. stupidity, dullness, foolishness, dishonesty, ignorance.

10. meekness, vanity, self-confidence, self-esteem, self-respect.

TEST 2

Twenty Selections, Each of a Member of a Class Defined by Three Samples (Thorndike, After Otis)

"Look at the words in line 1. Find the way in which the things named by the first three are alike,— the quality or feature which they have in common. Then look at the other four words on line 1, and draw a line under the name of the thing that is most like all the first three,— which has some quality or feature which the first three all have. Do the same for lines 2, 3, 4, 5, etc."

Lines 1, 5, 10, 15, and 20 are:

1.	fat grease butter	melt lard burn fry
5.	clam scallop limpet	shell beach salt oyster
10.	sapphire amethyst ruby	ring sparkle topaz costly
15.	football baseball golf	chess tennis whist bat
20.	north south east	compass wind turn up

TEST 3

Ten Selections of a Member of a Class as in Test 2, But with Pictures (Thorndike, After Otis)

Lines 1, 4, and 7 are:

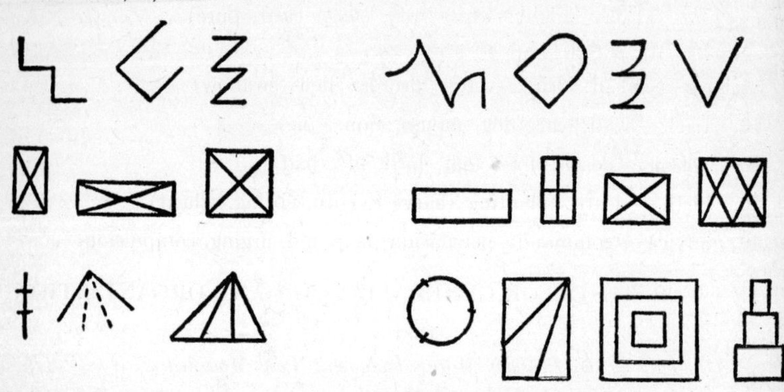

TEST 4

Twelve Selections of a Member of a Class as in Tests 2 and 3, But With Numbers (Thorndike, After Otis)

Lines 1, 4, 7, 10, and 12 are:

21	49	700	800	63	12	94
$\frac{6}{4}$	$\frac{15}{10}$	1.5	$\frac{2}{3}$	$2\frac{1}{2}$	3.0	$\frac{4}{2\frac{2}{3}}$
25	64	10000	1000	9	640	64025
7	26	63	121	122	123	124
27	125	1000	270	1250	64	10000

TEST 5

Trabue Completions J and L (or K and M)

"Supply the missing words to make each sentence true and sensible. Write only one word in each blank space."

Lines 1, 5, 10, and 15 are:

1. The boy will......his hand if......plays with fire.

5. The poor little......has......nothing to......; he is hungry.

10. To......many things......ever finishing any of them......a...... habit.

15. It is......that a full-grown man should......a ghost......he is.......

TEST 6

Cutting a Surface so as to Produce Given Surfaces
(Thorndike, After Army Beta Test 7)

"Think how you would cut the triangle so as to make the pieces shown in 1. Draw a line or lines to show how you would cut it. Do the same for the triangles in 2, 3, 4, 5, etc."

The test Beta 7 is so well known that we do not reproduce samples of the surfaces and parts here.

TEST 7

Rearranging Data to Form a True Equation
Twenty Equations (Thorndike)[1]

"Write the numbers and signs in each line in the proper order, so that they make a true equation as shown in the three sample lines. Use the loose sheets of paper to figure on if you need to."

Sample lines	3 3 6 = +	$3+3=6$
	4 7 8 20 = + ×	$7\times4=20+8$
	2 3 3 7 18 = + − × ( )	$7+2=18-(3\times3)$

Lines 1, 5, 10, 15, and 20 are:

3 3 4 10 = + −
3 3 8 48 = + ÷
1 3 4 4 12 = + − × ( )
1 3 3 3 3 21 = + − × ÷ ( )
2 $2\frac{3}{4}$ 3 4 5 = + × × ( )

To the tests shown above we may well add a test in reading difficult paragraphs and answering questions about them such as require discrimination and a selective organization of the ideas presented in each paragraph. The harder half of the Thorndike McCall test in paragraph reading and the paragraphs in Part III of the Thorndike Intelligence Examination for High School Graduates will serve this purpose adequately.

Such tests in symbolism, selective and relational thinking, generalization and organization, and comprehension of paragraphs give a definite meaning to these terms. If the abilities measured by the tests described above are accepted

[1] Some further information about these tests may be found in Thorndike [1922a].

as the abilities which we are trying to improve by algebra, we can proceed to measure these subtler and more general results just as we measure the obvious and specific abilities to add literal numbers or read a graph. In general, teachers of mathematics do accept them as reasonable definitions by sample. In general, they expect that the general improvement of thinking due to the learning of algebra will show itself in such acts of thinking as these tests require.

A complete measurement of the improvement produced by algebra would obviously be a very elaborate matter. It would measure not only the separate abilities as in the Rugg-Clark tests but also their organized working in tasks such as are given in the Hotz test, the I. E. R. test and typical school examinations so far as these are tasks which life may be expected to set. It would measure not only ability at routine tasks, but also the ability to apply algebraic theory and technique to original problems. It would measure not only strictly algebraic abilities, but also the more general powers of dealing with symbols and relations and abstractions and generalizations, selecting essential elements and organizing ideas and habits. It would report how hard a task the pupil can do with substantial mastery along each line of ability. It would probably give three or more examples of each sort of task so as to distinguish with surety lack of ability from lack of care. It would include a record of the time required in some cases as additional evidence of the degree of mastery attained.

Such a measurement could be used as an inventory of a pupil's achievements; as a record of his progress; as a diagnosis of where his difficulties were, and what training he needed; and as a means of measuring the effects of methods of teaching. It would need fifteen to twenty hours of a pupil's time. A selection of two or three hours' worth of the tasks would be adequate for graduation, promotion, college certification, and the like.

Selections from its easier levels could be used to measure progress in the early parts of the algebra course. Selections by topics could be used to find weak spots in preparation for reviews. The instrument could be used as a bill of specifications of the learner's job.

NEW TYPES OF EXAMINATION QUESTIONS

Certain new types of examination questions deserve attention from the teacher of algebra.

The first is the *true-false* type, a sample of which follows:

39. Write + on the dotted line if the statement is true.
Write − on the dotted line if the statement is false.

$a^{\frac{1}{3}} = \sqrt[3]{\frac{1}{a}}$ ....

$a^{\frac{1}{3}} \times a^{\frac{1}{3}} = \sqrt[3]{a^2}$ ....

$\sqrt{\frac{a}{b}} = \frac{ab}{\sqrt{b}}$ ....

$\sqrt{a^5b} = 2a\sqrt{b}$ ....

$\sqrt{\frac{1}{2}} = \frac{1}{\sqrt{2}}$ ....

This sort of test is useful in algebra as a stimulus to teach pupils to examine and, if necessary, check their results before leaving them as valid.

The second is the *selection* test, illustrated here in the case of knowledge of the meaning of Arithmetical Progressions and Geometrical Progressions.

Examine each of these. If it is an arithmetical progression, mark it A. If it is a geometrical progression, mark it G. If it is neither, mark it N.

$\frac{5}{12}$	$1\frac{2}{3}$	$6\frac{2}{3}$
5.8	6.6	7.4
$\frac{5}{8}$	1	$\frac{11}{8}$
1.7	2.0	2.3

NEW TYPES OF EXAMINATION QUESTIONS—Continued

$4\frac{1}{2}$	$5\frac{1}{3}$	$6\frac{1}{4}$
5	5	5.5
abc	bc	c
rs^2t	st^2	$\frac{t}{r}$
$\frac{e}{cd}$	e	cde

This sort of test is especially useful in the case of meanings of terms, and for training and testing the power to detect gross errors.

A third variety is the *matching* test, illustrated here in the case of the understanding of the relations represented by certain equations.

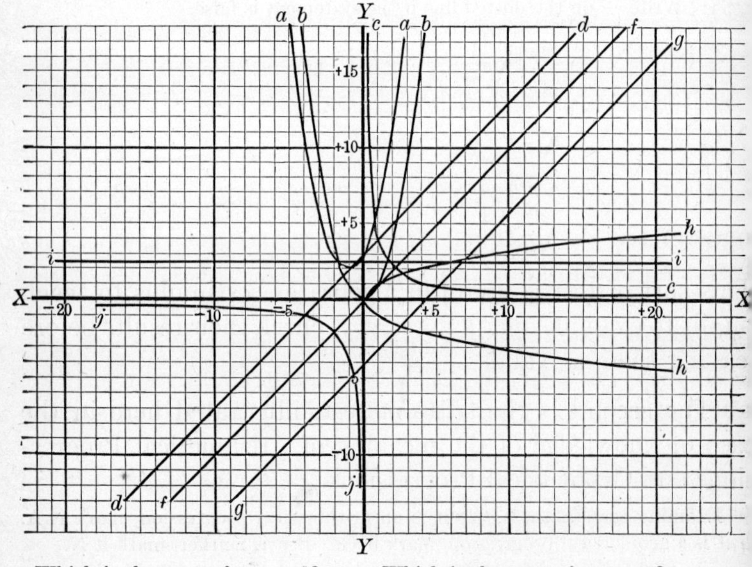

Which is the curve for $y=\frac{5}{2}$?
Which is the curve for $y=x$?
Which is the curve for $y=x+3$?
Which is the curve for $y=x-4$?
Which is the curve for $xy=4$?
Which is the curve for $y=x^2$?
Which is the curve for $y=x^{\frac{1}{2}}$?
Which is the curve for $y=x^2+2x+3$?

This form of test is often capable of gradation from very easy to very difficult distinctions, and has, in a very high degree, the merit of testing whether a pupil can adapt his abilities to a form of task differing from the tasks by which he acquired them. The matching form of test also often organizes or reorganizes pupils' abilities in valuable ways.

A special variety of the matching test is the *ranking* test, as where a series of events are ranked in order for importance, or in chronological order, or where a series of definitions are ranked for merit, or a series of cities for size. For example:

Write 1 before the largest of these.
Write 2 before the next largest of these.
Write 3 before the third largest, and so on.

a. 1.17×10^3
b. The number whose logarithm is 3
c. $1000000^{\frac{1}{2}}$
d. $(10+\frac{1}{10})^3$
e. The number whose logarithm is 4
f. 2^{20}
g. 4^{11}

All these tests arouse interest in the pupil by their novelty and employment of familiar facts and as a change from computation. They are quickly scorable by the pupils themselves. However, they are not relatively so important in algebra as in subjects like English history, civics, and economics, where the ordinary examination attaches too much weight to literary ability and is very hard to grade accurately.

A still more important type is the completion test, in which parts of a statement, or picture, or diagram, or map are left to be supplied by the pupil. This type is already in universal use in algebra; for the solving of an equation is a completion test, in which x or some other symbol for the unknown quantity is used instead of an empty space. It may be used to good effect also in the form of interpolation and extrapolation with curves, and perhaps occasionally in sentences about definitions, rules, and proofs.

MEASUREMENTS OF THE CAPACITY TO LEARN ALGEBRA

Rogers [1918] has constructed a series of tests to prophesy in advance how well a pupil may be expected to succeed in comparison with other pupils in the study of algebra. They include:

(1) A test in geometrical conclusions, reasons, and proofs, the necessary assumptions and data being all given.

(2) Two tests in algebraic computation, tasks 1, 5, 10, of the 11 of Test I, and tasks 1, 4 and 7 of the 7 of Test II being as follows:

TEST 1

1. If $a=2$, $b=3$, $c=5$, and $d=1$, find the value of each of the following:

(a) $5a$ *Answer:* . . . .

(b) $2a-d$ *Answer:* . . . .

(c) $\frac{a+b+d}{d}$ *Answer:* . . . .

(d) $\frac{2a+c}{3d}$ *Answer:* . . . .

(e) $\frac{2c}{a}-\frac{4b}{3d}$ *Answer:* . . . .

5. If $2x+3=15$, what is the value of x? *Answer:* x. . . . .

10. If l stands for the number of feet in the length of a room, what is the number of feet in the length of a room 4 feet longer? *Answer:* . . . .

TEST 2

1. Multiply the following as indicated:

(a) $4\,(3x-4)$ (b) $-5\,(-4x-6y)$ (c) $-4\,(x+2)$

Answers: (a) . . . . (b) . . . . (c) . . . .

4. Find the value of x:

$$\frac{-7x-2}{6}=\frac{x+1}{8}-x$$

Answer: x . . . .

7. Find the values of x and y:

$$5x+2y=34$$
$$7x-3y=7$$

Answer: x .... y...........

(3) Two tests in supplying numbers in arithmetical series. Tasks A, F, K, P and T of Part I and of Part II being as follows:

PART I

A.	1	3	5	7	...	11	13	15	17	...	21
F.	5	13	...	29	37	45	53	61	...	77	85
K.	3	...	...	18	...	...	33	...	...	...	53
P.	...	...	13	...	...	...	29	...	...	...	45
T.	7	...	...	...	...	...	31	...	...	...	47

PART II

A.	1	8	15	22	...	36	43	50	...	64	71
F.	5	17	29	...	53	65	77	...	101	113	125
K.	1	...	...	7	...	...	13	...	...	...	21
P.	...	...	16	...	...	...	44	...	...	...	72
T.	5	...	...	...	...	...	83	...	...	...	135

(4) A superposition test of the following type with forty-eight tasks:

DIRECTIONS FOR SUPERPOSITION TESTS 1 AND 2

Suppose that the figure with a circle in it is a small card with one of its edges painted black and with a hole in one corner.

If this card is moved around so that its black edge lies upon the long heavy black line, it will fit one of those two figures shown.

Decide which it fits and then with your pencil *draw a circle where the hole would be.*

Try the three following examples.

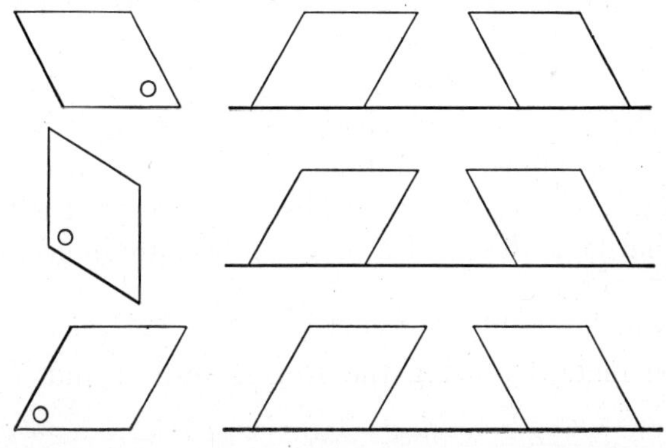

(5) A sentence completion test with 14 tasks ranging from

Boys and......soon become......and women

to

One can......do his......at one......while......of another.

(6) A mixed relations test of 43 tasks, Nos. 1, 11, 21, 31, and 41 being:

eye — see	ear — ....
little — less	much — ....
past — present	present — ....
growls — dog	roars — ....
mice — cat	worms — ....

Dr. Rogers' tests are the best so far published to select according to promise of ability in algebra and geometry. We have evidence, however, that their value consists chiefly in their being a good measure of abstract intellect, and that any reliable measure of abstract intellect would prophesy success in algebra and geometry nearly as well. Also it seems probable that algebraic ability and geometrical ability differ nearly if not quite as much as do ability in algebra and ability in any other abstract subject such as physics or Latin. Consequently tests specialized for numerical and spatial data may be found to do this prognostic work even better than the Rogers tests.

If these tests could be arranged so as to be given before any algebra or geometry had been studied, it would, of course, be an added advantage. The Rogers tests presuppose a few months' study of algebra, and are much influenced by familiarity with the idea of and procedure in geometrical proofs.

It seems probable that such a prognostic test for algebra could be made by using the Rogers tests 3 and 5 with a test in arithmetical problems and a test in disarranged numerical equations.

MEASUREMENTS OF THE CHANGES IN INTEREST AND ATTITUDE DUE TO ALGEBRA

The important work in this direction is that of Kelley [1920]. His determination of values imputed to the study of mathematics and of tests to measure these includes all the important general values and so measures, amongst others, the abilities tested by our tests of Selective and Relational Thinking, Generalization and Organization, and Symbolism. His tests of what we have called interest and attitude are of especial interest, because they are unique of their kind.

The values imputed to high-school mathematics were found to be:

Mathematical Values

1. Provides entertainment for recreational hours.

2. Prepares for advanced sciences.

3. Prepares for advanced mathematics.

4. Leads to an understanding of ancient and modern accomplishment of mankind.

5. Establishes algebraic method of solving problems.

6. Establishes the habit of reasoning in terms of symbols.

7. Leads to an understanding of formulas.

8. Leads to use of formulas in solving scientific and social problems.

9. Leads to an understanding of graphic methods.

10. Leads to use of graphic methods.

11. Establishes the habit of proving results.

12. Establishes the habit of considering situations from their quantitative (instead of merely qualitative) aspects.

13. Establishes the habit of differentiating between the known and unknown elements in a situation.

14. Develops religious sense and appreciation.

15. Develops high ideals of life.

16. Develops mathematical ideals.

17. Leads to self-discovery and guidance.

18. Develops respect for truth (honesty).

19. Develops self-reliance.

20. Develops originality.

21. Develops powers of clearness in statement (definiteness).

22. Develops powers of concentration (sustained attention).

23. Develops powers of generalization.

24. Develops powers of inference (constructive imagination).

25. Develops powers of analysis.

26. In addition to the assigning of each problem to one or more of the preceding 25 values, the judges were asked to indicate such of the problems as test transfer of training.

27. Measures accuracy in fundamental operations.

28. Measures knowledge of arithmetic.

29. Measures accuracy in thinking.

The instrument by which they are to be measured is the following:

MATHEMATICAL VALUES TEST ALPHA

1. How would you find the value one year hence of a W. S. S. for which you now pay $4.16?

2. It has been claimed that there is an algebraic type of thinking. What does this mean to you?

3. A certain professor gave a lecture in a town some distance away, for which he was to receive $100 and his expenses. The note-book in which he kept track of his expenses read as follows:

Feb. 2.	Ticket	$19.10
	Dinner	1.00
	Cab	1.50
	Hotel Bill	18.20
	Liberty Bond	50.00 paid first installment
Feb. 5.	Couple dollars of tips and Pullman 2.00	
	Return fare 17.10	

One morning the professor tossed this note-book to his wife and asked her to make out a bill covering all moneys due him. She did her best.

(*a*) Express your opinion with reference to each item in the note-book as to whether or not it is complete and satisfactory.

(*b*) Make out the best statement or bill that you can from the note-book.

In your mind how is algebra related to

4. Religion?

5. Life in the home?

6. Life out of doors?

7. Other school subjects (*a*) home economics? (*b*) sciences? (*c*) history? (*d*) any other school subjects?

8. Write down just what you have been told higher mathematics deals with. Who told you this?

Now think of mathematics that is still more advanced than this that you have just described. Make a good guess and describe what you think it deals with.

9. Where did our present numerals originate and about when were they first used in Europe?

What system of numerals did people in Europe generally use before they used our present one, and how did they perform such operations as addition, subtraction, multiplication, and division?

10. The area of a circle is πa^2. What does π stand for? a^2?

11. Suggest three or four problems other than those discussed in your algebra class that could be made clear by means of a graph.

12. The algebraic method is one of supposing the unknown quantity known, making a statement (an equation) which relates this to the known quantities, and then determining the unknown in terms of the known (solving the equation). Can you think of some problems suggested by your other school work, or your life outside of school, in which this method applies? Explain.

13. In the solution of what kind of problems is it desirable to use logarithms?

14. The expression of a physical law by means of a mathematical formula is probably the most powerful tool in modern science for the interpreting of scientific facts. For example, $S = \frac{1}{2}gt^2$; in which S = the space passed over by a falling body, g = the force of gravity, and t = the length of time that the body is falling. The simple formula $S = \frac{1}{2}gt^2$ tells more about the law of gravity than a whole volume could tell without it.

Write down any other formulas that you know.

Write down for each of the following fields two relationships which you think have been, or are capable of being, expressed by means of formulas:

15. Electricity (Sample answer: There is probably a formula which gives the relation between the amount of current which can flow through a wire and the size of the wire).

16. Air (Sample answer: There is probably a formula which gives the relation between the temperature of the air and the rapidity with which sound travels through it).

17. Mention some recreational activities in a person's life that algebra can make more enjoyable.

18. Do you know any game in which algebra is used? If so, describe it.

19. Two men meet and the one says to the other "My father is your father's son." What kin were they?

20. What do any of the following names suggest to you in connection with mathematics?

Pythagoras; Newton; Euclid; Pascal; Archimedes; Leibnitz; Leonardo of Pisa; Descartes.

Add the names of any others whom you know who have made important mathematical contributions, and tell what they did.

21. Certain great concepts or ideas have been discovered as humanity has evolved. Some of these are listed below. Number them 1, 2, 3, etc., in the order of their momentousness or importance. *Consider that one which has revolutionized procedure most to be the most important;* for example, (*a*) below is of tremendous importance, for the present number system displaced the old cumbersome Roman system composed of X's, L's, C's, V's, etc., and it is extremely difficult to do so simple a thing as to multiply in the Roman system.

a. The present number system.

b. The idea or concept of the letters x and y as unknowns in an equation and of a and b as knowns.

c. The concept of 0.

d. The concept that $a^2a^5=a^7$.

e. The concept that $x^2-y^2=(x+y)\ (x-y)$.

f. The concept that an equation can express the path of a comet or planet around the sun.

g. The concept that an equation of this type, $4y-x=7$, is a straight line.

h. The concept that \$1.00 at 5 per cent interest compounded annually $=\$1.10\frac{1}{4}$ at the end of 2 years.

i. The concept that if $2x^2-10x=28$, then $x=7$.

j. The concept that in many problems the unknown quantity can be dealt with as though known, in building up equations, the solution of which gives the value of the unknown.

k. The concept that $ax^2+bx+c=0$ is the same as

$$x=\frac{-b\pm\sqrt{b^2-4ac}}{2a}$$

22. If a man 5 ft. tall weighs 110 lbs., a man 6 ft. tall should weigh how much?

If a dwarf 3 ft. tall weighs 30 lbs., a man 6 ft. tall should weigh how much?

If a new born baby 20 inches tall weighs 8 lbs., a man 6 ft. tall should weigh how much?

23. The area of a triangle equals one-half the product of the base and altitude. If the area of a triangle $=ab$, what does a stand for, and what does b stand for?

24. If F represents *force* of attraction, G is a constant, M represents the *mass* of one body, m the *mass* of a second body, and D the *distance* between the two bodies, what does $F=\frac{G\,M\,m}{D^2}$ mean?

Is there any mathematical symbol or operation that makes you think of, or seems in some way similar to:

25. God's all pervading power — that is, his presence in nature around us?

26. The intimacy of God's power — that is, his presence in our own natures?

27. Living creatures can be divided into two fundamental groups, male and female. Mention all the mathematical symbols or operations that show an equally fundamental division into two aspects. There are many illustrations, so give more than one.

28. $R=\frac{nr}{1+(n-1)r}$ $\quad R=.90$ $\quad r=.45$

What is the unknown in this equation? Solve for it.

29. The accompanying graph gives the amount of U. S. tonnage for a number of years. How do you account for the low spot in the curve?

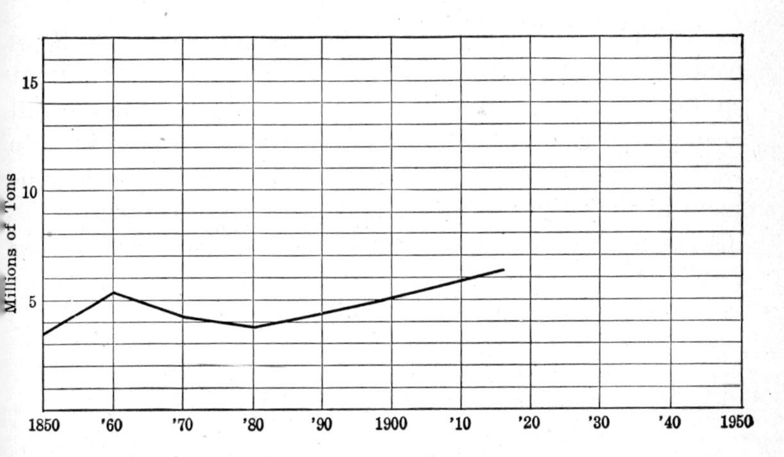

30. In January, 1918, the U. S. expected to build about 3 million tons during 1918 and 6 million in 1919. It was feared that submarines, storms, etc., would sink $\frac{1}{6}$ as much as was built. Assuming these estimates to be correct, draw the curve from 1918 on to 1920. Make a guess and draw it from 1920 on to 1950. What effects have the important wars of the U. S. had upon tonnage?

31. The accompanying graph gives the percentages of married graduates of a certain girls' college who married at different ages. At what age did the most marriages occur? At what ages were there just 10 per cent of marriages?

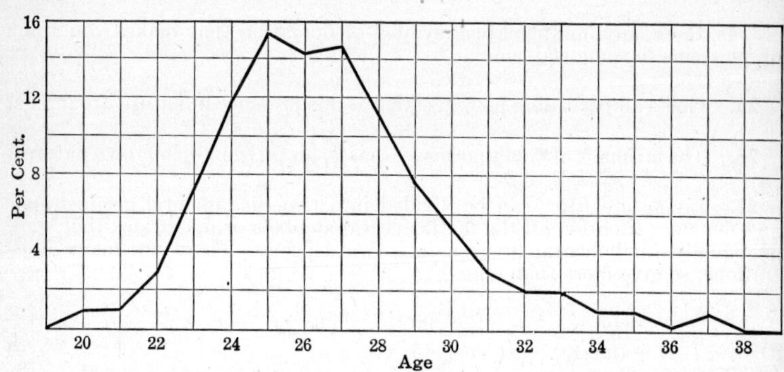

32. Mrs. A has a skirt pattern cut for a 40 inch hip measure. Her own hip measure is 44. The pattern is in two parts, one measuring 25 inches at the bottom and the other 30 inches at the bottom. What should be the bottom measures for the two parts in order to conform to Mrs. A's hip measure?

The use of the responses to these questions as evidence concerning the values in question is naturally somewhat subtle and elaborate. The student should consult the full account by Kelley. It is sufficient for our purpose to note that differences between two classes in respect to the attainment of the values in question can be well measured by the instrument (provided, of course, that neither teachers nor pupils prepare especially for the examination).

CHAPTER VII

The Constitution of Algebraic Abilities

The abilities most worth acquiring in a year's course in algebra seem to us to be the following:

(1) Ability to understand formulas, to the extent of being able to answer questions as hard as, or a little harder than those below.

$P=\frac{b_1 b_2}{c_1+c_2}$. b_1, b_2, c_1, and c_2 are all positive.

Write "*larger*" or "*smaller*" or "*you cannot know*" on the dotted lines.

If b_1 becomes larger, P will become............

If b_2 becomes larger, P will become............

If c_1 becomes larger, P will become............

If c_2 becomes larger, P will become............

$N=\frac{ab}{c}-d\,(e-f)+\frac{1.414}{g}$. $a, b, c, d, e, f,$ and g are all positive.

Write "*larger*" or "*smaller*" or "*you cannot know*" on the dotted lines.

If a becomes larger, N will become............

If b becomes larger, N will become............

If c becomes larger, N will become............

If d becomes larger, N will become............

If e becomes larger, N will become............

If f becomes larger, N will become............

If g becomes larger, N will become............

(2) Ability to translate into a formula any clear statement of a thoroughly understood quantitative relation which a reasonable person might need to put in a formula.

(3) Ability to "evaluate" for any letter or other significant unit in formulas, such as the following, or even more intricate.

$$d=\frac{60\ Aln}{231}$$

$$v=\frac{4}{3}\pi r^3$$

$$v=\frac{h}{6}(A_1+4A_m+A_2)$$

$$A=\pi\frac{Dd}{4}$$

$$C=\frac{nE}{R+nr}$$

$$\frac{1}{R}=\frac{1}{r_1}+\frac{1}{r_2}+\frac{1}{r_3}$$

$$\frac{PV}{T}=\frac{P_1V_1}{T_1}$$

$$\frac{W_0}{W_1}=\frac{R_1{}^2}{R_0{}^2}$$

$$F=\frac{Wv^2}{gR}$$

$$t=\pi\sqrt{\frac{l}{380}}$$

$$t_0=\frac{m_1t_1+m_2t_2}{m_0}$$

$$L=\frac{20w\sqrt{n}}{PK}$$

$$S=v_1t+\frac{1}{2}at^2$$

$$l=n\sqrt{c^2+h^2}$$

$$S=1.35\ D^2+111D-111$$

$$P=\frac{2\pi sI}{Dy}$$

$$D=60^2+8H\sqrt{\frac{T-t}{491}}$$

$$A=\pi r\sqrt{h^2+r^2}$$

$$v=\frac{Ah}{2}+.5236\ h^3$$

$$Fs=Rs+\frac{Wv^2}{2g}$$

(4) Ability to "solve for" or "change the subject to" any letter or other significant unit in such formulas as those just quoted.

(5) Ability to frame an equation or set of equations expressing any quantitative problems which high-school graduates meet on the average at least once in five years, provided the problem situation or description is itself clearly understood, and the data are adequate.

(6) Ability to solve such equations or sets of equations if they are linear or quadratic.[1]

[1] There may be a few exceptions in the form of genuine problems of the importance specified which will lead to simultaneous quadratics not solvable by the method of quadratics.

(7) Ability to understand any graph representing by Cartesian coördinates the relation of one variable to another. Ability to make a graph for a table of values relating one variable to another, or for any clearly understood description of the relation of one variable to another.

(8) Understanding of the elementary facts concerning the relations expressed by

$$y=ax$$
$$y=ax+b$$
$$y=\frac{a}{x}$$
$$y=\frac{a}{x}+b$$
$$y=x^2$$
$$y=ax^2+b$$
$$y=ax^2+bx+c$$

and perhaps of those expressed by

$$y=\sqrt{x}$$
$$x^2+y^2=a$$
$$y=a^x$$

(9) Ability to solve for the constants in such equations, given the x, y values of two points on the curve.

(10) The abilities in algebraic computation required to deal with formulas and equations as stated above.

(11) Ability to understand and use negative and fractional exponents.

(12) Ability to use logarithms with multiplications, divisions, powers and roots.

(13) Such insight into the use of algebra for formulas

expressing numerical relations themselves as comes from the study of

(a)

$$a \times 0 = 0$$
$$0 \div a = 0$$
$$a^m a^n = a^{m+n}$$
$$\frac{a^m}{a^n} = a^{m-n}$$
$$ax + bx = (a+b)x$$
$$\frac{-b \pm \sqrt{b^2 - 4ac}}{2a}$$

for the roots of a quadratic and the other essential formulations of rules for the computations mentioned above.

(b) Certain formulas useful in computation or in understanding approximations, especially such as

$$(a+b)\ (a-b) = a^2 - b^2$$
$$a^2 - b^2 = (a+b)\ (a-b)$$
$$(a+b)^2 = a^2 + 2ab + b^2$$
$$(a-b)^2 = a^2 - 2ab + b^2$$
$$(a+b)^3 = a^3 + 3a^2b + 3ab^2 + b^3$$
$$(a-b)^3 = a^3 - 3a^2b + 3ab^2 - b^3$$

(c) The formulas for arithmetical progressions, geometrical progressions, and the binomial theorem.

(14) Certain informational abilities, especially knowledge of the meaning of *ratio*, *varies directly as*, *varies inversely as*, *reciprocal*, *hypot.*$^2 = S_1^2 + S_2^2$, *constant*, *variable*, *tangent*, *sine*, and *cosine*, the laws for corresponding dimensions in similar figures (in so far as these have not been learned in arithmetic), and the use of tables of roots, powers, reciprocals, logarithms, tangents, sines, and cosines.

These abilities may be constituted in very many different ways. A mathematician might think of the pupil as acquiring these abilities by learning a few principles of notation, the laws of signs, the theory of exponents, the axioms, and the general rule that you can operate with literal numbers as with ordinary numbers. A mathematician who knew nothing of the conventional treatment of algebra in schools might begin to teach algebra as follows:

Let a, b, and c represent any three numbers. Let $+$, $-$, $\times$, $\div$, the fraction line, $\sqrt{\ }$, $\sqrt[3]{\ }$, ^{2} and ^{3} be used as in arithmetic. Let ab or $a \cdot b$ mean $a \times b$. Let ( ) mean that the expression within is to be treated as one number.

$$+a=0+a \qquad +b=0+b \qquad +c=0+c$$
$$-a=0-a \qquad -b=0-b \qquad -c=0-c$$

Adding $-a$ is the same as subtracting $+a$

Subtracting $-a$ is the same as adding $+a$

$$(+a)\ (+b)=+ab$$
$$(+a)\ (-b)=-ab$$
$$(-a)\ (+b)=-ab$$
$$(-a)\ (-b)=+ab$$
$$\frac{+a}{+a}=+1$$
$$\frac{+a}{-a}=-1$$
$$\frac{-a}{+a}=-1$$
$$\frac{-a}{-a}=+1$$

If equals are added to equals the results are equal.

If $a+b=c \qquad a+b+d=c+d$

If equals are subtracted from equals the results are equal.

If $a+b=c \qquad a+b-d=c-d$

and so on.

Such a straightforward general treatment has a good deal in its favor, in the way of brevity and dignity. The experience of teaching, however, shows that algebraic abilities are not constituted, in the minds of the pupils, out of a few general sweeping laws. No textbook or teacher of today, for example, would dare to assume that pupils who had been taught as above would be sure to understand that $(c-d)-(e+f)$ means "Subtract $e+f$ from $c-d$," or that $\frac{cd}{a}-\sqrt{a+b+c}$ means "Subtract $\sqrt{a+b+c}$ from $cd \div a$," or even that $7cd^2-4cd^2$ means "Subtract $4cd^2$ from $7cd^2$."

The customary approved teaching of today builds up these abilities out of many detailed abilities. The pupil learns the meanings of about two hundred and fifty terms, such as:

Abscissa
Absolute term
Absolute value
Aggregation
Algebraic addition
Algebraic expression
Algebraic number
Algebraic product
Algebraic solution
Algebraic subtraction
Antilogarithm
Applying a formula
Ascending powers
Axes
Axiom
Base (distinct from power)
Binomial
Binomial theorem
Brace

He learns about one hundred and fifty rules, such as:

Like roots of equals are equal.

To add a positive number to a negative number take the difference of their absolute values and prefix the sign of the numerically greater number.

$$a-0=a$$
$$0-a=-a$$

To add similar monomials find the algebraic sum of the coefficients of the common factor and prefix this sum to the common factor.

He forms many habits either as applications of these rules or as accessories acquired in the course of computation and problem solving. For example, the principle *We can represent numbers by letters*, develops into at least ten distinct habits of thought, namely,

1. A letter may mean a particular number of things, like men, boys, or eggs.

2. A letter may mean a particular number of units, like cents, quarts, feet.

3. A letter may mean any one of a number of numbers, like the number of dollars in the cost of any number of suits of clothes of a certain sort, or the number of square feet in any rectangle.

4. A letter may mean any number, as in $(p+q)\ (p-q)=p^2-q^2$.

5. If you call a certain number p, you may call 3 times that number q or r or s or any letter except p that you please, but it is commonly useful to call 3 times that number $3p$.

6. If you call a certain number p you may call 3 more than that number any letter except p that you please, but it is commonly useful to call it $p+3$.

7, 8, and 9. The same principle of consistency and utility with $p-3$, $\frac{p}{3}$, and $\frac{3}{p}$.

10. If we call a certain number (say, the profit Mr. A. made in January, 1922) p we don't call it anything else and don't call p something different so long as we are thinking about the problem to answer which we called that number p.

In spite of all our experience in teaching algebra we do not seem to have found the optimum constitution of these algebraic abilities. Reformers like Rugg and Clark and Nunn not only eliminate certain abilities, change the emphasis on those which they retain, and add new ones; they also constitute the retained abilities in different ways. They would be the first to expect that other desirable changes will be found by further experimentation and improved insights.

To the psychologist who tries to follow through the mental operations of pupils from their first solutions of simple formulas to their comprehension of the parabolic relation, of the binomial theorem, and the like, there seem to be many promising invitations to experiment, and even many cases where improvements can be made at once by a straightforward application of the laws of learning. In the remainder of this chapter we shall note some of the general features of the constitution of these abilities, namely the provision for habits now neglected, the elimination of unnecessary habits, and the use of adaptable, even of loose habits, in place of inflexible rules which in actual learning have to be broken. In the next chapter we shall treat the psychology of learning the meanings, operations, and principles of algebra, and shall present detailed suggestions concerning particular items of learning to be added or omitted.

PROVISION FOR HABITS NOW NEGLECTED

The hundred and fifty rules do not cover all that the pupil must know and do. For example, throughout algebra the pupil has to decide when to indicate an operation only and when to carry it through according to some known routine.

Suppose that he is faced with the need of dividing a by b, a^2 by a, a by a^2, and $\sqrt{625}$ by $\sqrt{10}$. In the first case he must only indicate $\frac{a}{b}$; in the second he must subtract with with the exponents and write a; in the third case he must (according to usual methods of teaching), not subtract with the exponents, but indicate the division as $\frac{a}{a^2}$ and then divide each term by a; in the last case he is again customarily taught to indicate the division as $\frac{\sqrt{625}}{\sqrt{10}}$ and then proceed to $\sqrt{\frac{625}{10}}$ or $\sqrt{62.5}$, or to $\frac{25}{\sqrt{10}}$, or still worse to $\frac{\sqrt{625} \times \sqrt{10}}{10}$ or $\frac{\sqrt{6250}}{10}$. It would seem worth while to provide definitely for each useful habit, where there is some one procedure that is best, and to provide also a clear understanding that in certain cases you can do one of a number of things, any one of which is right so far as it goes, but some one of which will be best according to circumstances.

To take another illustration, in the teaching of division of a monomial by a monomial it is customary to use almost or quite exclusively tasks like $\frac{a^2}{a}$, $\frac{ab^2c^3}{ab^2c}$ and $\frac{27a^3b^3c}{9a^2b}$ where the quotient is integral — where no factor is left below the fraction line. $\frac{a}{a^2}$, $\frac{ab^2c}{ab^2c^3}$ and $\frac{9a^2b}{27a^3b^3c}$ commonly do not occur at all in the systematic treatment of division, and occur very rarely in the reduction of fractions to simplest form. In a census of all the work, oral and written, computation and problem solving, provided for the first year's study of algebra in three standard instruments of instruction, we find the following enormous disparity between samples of the two types of task:

Let a, b, c, etc., represent numerals.

Let x, y, z, etc., represent literal numbers. Then the averages for the three textbooks are:

$\frac{ax}{b}$ and $\frac{axy}{b}$ occur 848 times; $\frac{a}{bx}$ and $\frac{a}{bxy}$ occur 2 times;

$\frac{ax}{x}$, $\frac{x^2}{x}$, and $\frac{x^3}{x}$ occur 245 times; $\frac{x}{ax}$, $\frac{x}{x^2}$, and $\frac{x}{x^3}$ occur 8 times.

This is an inadequate preparation for the actual use of algebra in later mathematical study, science, or the general work of life. The larger numeral, the persisting literal factors, and the higher power probably do go in the numerator oftener than in the denominator, but not in the ratio of 110 to 1! If fifteen year old boys and girls have a certain result happen 110 times as often as its opposite, they tend to think of that opposite as impossible or wrong. They will feel no surety when any division comes out with a balance in the denominator.

The customary procedure is due to the tradition that considered fractions as a difficult matter, not to be touched in any way until all operations with integers have been mastered, and to the effort to fit the pupils' operations to the two rules of dividing by subtracting exponents and of reducing fractions by dividing both numerator and denominator by the same factor. The tradition is simply nonsense, a fallacious notion due to analogy with arithmetic and the idea that because fractions as a whole are hard, everything about them is hard. The fraction form is universally used in the division of a monomial by a monomial. The plan of fitting the pupils' operations to the two rules has much in its favor, and can reasonably be retained, the defect in the pupils' habits being remedied by extending the notion of division when reduction of fractions is learned, and by giving suitable practice then and thereafter.

It would, however, perhaps be still more effective to make the procedure general from the beginning by introducing -1, -2, and 0 as exponents, teaching their meanings, and permitting either a^{-1} or $\frac{1}{a}$ as an answer to $\frac{a}{a^2}$ or $\frac{a^2}{a^3}$, and similarly for all factors remaining below the fraction line. Teachers in general will be shocked at the proposal to teach anything about negative exponents at this stage; but the most acute teachers may not be. For some of them have perceived by intuition and experience, what psychologists infer from general principles, that it is often very advantageous to get used to a few elements of a topic or doctrine long before the doctrine as a whole is systematically taught. Thus in arithmetic we now teach certain facts about the addition of fractions two or three years before the general treatment of the addition of fractions, and teach addition, subtraction, multiplication, and division with United States money long before the general treatment of decimals, and do so with great profit, both to the early work and to the general treatment when it comes.

THE ELIMINATION OF UNNECESSARY HABITS

We teachers and learned people are obsessed by two tendencies: to learn everything about anything and to teach everything about anything that attracts our intellects. The former tendency is, in spite of certain pedantries, perhaps the greatest blessing the world has. The latter is also a blessing on the whole, but it can be made a greater blessing if we control it. It has needed control in algebra. The educational reformers have had to work hard to convince teachers of mathematics that it is not wise or humane or scientific to burden the learning of algebra by children with all the factorizations ingenuity can devise, or to hide x^2-y^2 under all the disguises through which mathematical acuity can penetrate. This tendency to indiscriminate teaching still needs control.

Consider the case of the teaching of radicals as a sample. Suppose that we eliminated all the customary work with radicals and the customary general systematic treatment of powers during the first year, and instead taught the facts now taught much later concerning fractional exponents and negative exponents, plus the following:

1. $\frac{a^m}{b^m}=\left(\frac{a}{b}\right)^m$. If $a=b$, $a^m=b^m$. The rule for principal roots.

2. $\sqrt{a}$ means $a^{\frac{1}{2}}$, $\sqrt[3]{a}$ means $a^{\frac{1}{3}}$, $\sqrt[4]{a}$ means $a^{\frac{1}{4}}$, etc.
   $\sqrt{a^3}$ means $a^{\frac{3}{2}}$, $\sqrt{a^4}$ means $a^{\frac{4}{2}}$, $\sqrt{a^5}$ means $a^{\frac{5}{2}}$, etc.
   $\sqrt[3]{a^2}$ means $a^{\frac{2}{3}}$, $\sqrt[3]{a^3}$ means $a^{\frac{3}{3}}$, $\sqrt[3]{a^4}$ means $a^{\frac{4}{3}}$, etc.

3. If in any task you meet an expression with $\sqrt{}$ or $\sqrt[3]{}$ or $\sqrt[4]{}$, change it to an equivalent expression with the proper exponents (and with parentheses if necessary).

4. If after multiplying, dividing, raising to powers and finding roots by proper use of the exponents, any arithmetical number still has an exponent other than 1, evaluate by using tables of powers and roots, or by logarithms, or by trial and correction, if you cannot get the necessary tables.

5. If, however, you see some way of shortening the work, as by $(5^{\frac{1}{2}}+7^{\frac{1}{2}})\,(5^{\frac{1}{2}}-7^{\frac{1}{2}})=5-7$, take it.

This suggestion will, like the one on page 232, shock many teachers. They will think, or rather, feel, that the general procedure with exponents is too hard to teach thus early, that $\sqrt{\ }$ should precede, not follow $^{\frac{1}{2}}$, and that it will seriously mutilate algebra to omit such stimuli to ingenuity as $\sqrt[3]{9}\,(27)\sqrt{3}$, or $(\sqrt{8}+\sqrt{7})\,(\sqrt{8}+2\sqrt{7})$, or $\sqrt[4]{\frac{2}{3^3}}$, or $\frac{2-\sqrt{5}}{3+\sqrt{2}}$, or to replace them by equivalents with fractional indices whose simplifications are aided by tables.

These objections are instructive. The difficulty pupils have in learning the general procedure with fractional and negative exponents is largely of our own manufacture. In the ordinary course of events what they learn about roots, radical signs, surds, and rationalizing hinders them in learning the general treatment. "If it is 2 in $\sqrt{\ }$ how can it be $\frac{1}{2}$? If it requires two distinct operations in $\sqrt[3]{8^2}$ how can you express it as a single fraction $8^{\frac{2}{3}}$? And why does it turn bottom side up? When I had to be sure to turn $\sqrt[3]{a^4}$ into two a's $(a\sqrt[3]{a})$ as soon as I saw one, is it reasonable for me to call it $a^{\frac{4}{3}}$ now? In any case why not leave me to do it in peace in the old way that I learned at such cost?" Such is the unconscious argument of the pupil's nervous system. In the ordinary type of "explanation" of the procedure with exponents, we are prone to add to its difficulties for all save the gifted pupils. We explain how it can be true, justify it, and present it in the highly abstract rules:

$a^m \times a^n = a^{m+n}$ whatever m and n may be
$a^m \div a^n = a^{m-n}$ whatever m and n may be
$(a^m)^n = a^{mn}$ whatever m and n may be

and quiz the pupils lest they be not convinced.

Now the difficulty is not that the pupils feel logical objections, or any intellectual shock at the innovation. Unfortunately many of them would not rebel intellectually if they were told that hereafter a^2 would always mean $10a$, a^3 would mean $100a$, and so on. Nor are they helped greatly to understand the reasonableness of the new system by $a^m \times a^n = a^{m+n}$, etc. Their chief difficulty is that they are not used to it and do not realize what they are doing. They feel lost and dizzy with these strange exponents. They need first just to become at home with $(a^{\frac{1}{2}} + b^{\frac{1}{2}})(a^{\frac{1}{2}} - b^{\frac{1}{2}})$ and the like. They need, second, to see how the results do turn out by this new procedure, that $a^{\frac{5}{2}}$ means $a^{2\frac{1}{2}}$ and is about halfway between a^2 and a^3, that 10^{-4} is $\frac{1}{10000}$.

Such series as the following are useful for this:

$10^4 = 10000$	$4^3 = 64$	$9^3 = 729$
$10^3 = 1000$	$4^{\frac{5}{2}} = 32$	$9^{\frac{8}{3}} = 350+$
$10^{2\frac{1}{2}} = 316+$	$4^{\frac{4}{2}} = 16$	$9^{\frac{7}{3}} = 168+$
$10^2 = 100$	$4^{\frac{3}{2}} = 8$	$9^{\frac{6}{3}} = 81$
$10^{1\frac{3}{4}} = 56.2$	$4^{\frac{2}{2}} = 4$	$9^{\frac{5}{3}} = 38.9$
$10^{1\frac{1}{2}} = 31.6$	$4^{\frac{1}{2}} = 2$	$9^{\frac{4}{3}} = 18.7$
$10^{1\frac{1}{4}} = 17.8$		$9^{\frac{3}{3}} = 9$
$10^1 = 10$		$9^{\frac{2}{3}} = 4.3267$
$10^{\frac{3}{4}} = 5.6234$		$9^{\frac{1}{3}} = 2.0800$
$10^{\frac{2}{3}} = 4.6416$		
$10^{\frac{1}{2}} = 3.1623$		
$10^{\frac{1}{3}} = 2.1544$		
$10^{\frac{1}{4}} = 1.7782$		

They need in the third place to have enough practice in the operations so that they can do them correctly and have their minds free to consider what they are doing, and what comes of it, and why it is a useful thing to do, and how the formulas, $a^m \times a^n = a^{m+n}$ and the rest, sum up in a beautiful set of rules a great many operations which they have learned to perform and to trust as sure means of obtaining correct results. The very pupils to whom these formulas are baffling verbal edicts at the beginning of the learning may, after enough instructive practice and checking, find them admirable summaries of what they have learned.

The second objection to the effect that $\sqrt{\ }$ and $\sqrt[3]{\ }$ and $\sqrt[4]{\ }$ should precede $^{\frac{1}{2}}$ and $^{\frac{1}{3}}$ and $^{\frac{1}{4}}$ can hardly have any other origin than mistaking what is familiar to us as what is easy and natural.[1] $^{\frac{1}{2}}$, $^{\frac{1}{3}}$, $^{\frac{1}{4}}$ are better symbols than $\sqrt{\ }$, $\sqrt[3]{\ }$, $\sqrt[4]{\ }$ in every way whatsoever save two, for all pupils whatsoever. Their two demerits are, of course, the likelihood of confusing them with common fractions, and the more common use of $\sqrt{\ }$, $\sqrt[3]{\ }$, $\sqrt[4]{\ }$.

Pupils in high school are protected against the confusion with fractions by long habituation to a^2, a^3, a^4, etc. They should learn to understand both types of symbol, but $^{\frac{1}{2}}$, $^{\frac{1}{3}}$, etc., should precede. These should indeed be used in the first treatment of roots in algebra, whenever that may be, the pupil being taught that the square root of a in algebra is written $a^{\frac{1}{2}}$, or, sometimes, $\sqrt{a}$.

The third objection was that the rule permitting students to get rid of certain surds by direct use of tables, logarithms, or computation instead of searching for clever ways to rationalize would rob algebra of a certain portion of its intellectuality. It would, but the general addition of intellectuality, by making the treatment of radicals as a whole

[1] It seems probable that the antipathy to fractional exponents, founded on their unfamiliarity, is made much worse by the strain that commonly attends reading them because the type used is so small. The traditions of the printer in this respect are very bad, and the authors of textbooks have weakly given way to him. Fractional exponents should, of course, be printed in type that is easily legible.

a logical and straightforward application of the theory of exponents, far outweighs the loss. Radicals as a whole as now taught are a confused set of maxims plus a bag of clever tricks.

On the whole it seems conservative to estimate that at least three-fourths of the time now given to radicals could be saved with no loss, but a net gain to the pupil's development.[1]

ADAPTABLE HABITS *versus* INFLEXIBLE RULES

The rigor of its definitions and the universality of its rules are among algebra's chief merits, and we should cherish them. We do not, however, dignify algebra by claiming a universality when it is really lacking, nor by asserting something rigorously and evading it soon thereafter. When the definition works only part of the time and the rule is helpful for only some of the cases, it may be better to organize the learning as a set of rather loose and opportunistic habits than to insist on the rule and then qualify and amend and even break it.

Consider these two definitions and these two rules:

(1) If an expression is separated into two factors, either factor is called the coefficient of the other.

(2) Terms which differ in no respect or only in their coefficients are called like terms.

[1] It may be noted that an appreciation of the desirability of using the general procedure with exponents in the computations with radicals is shown by certain teachers and authors of textbooks who insert more or less of the general treatment along with the older teaching as an *alternative* mode of handling the computations. This, however, seems to be of little or no avail in lightening the pupil's load. We do not wish to have him learn the inferior way and that there is a better way, but to be blissfully ignorant of the inferior way. We wish him only to turn $\sqrt{\ }$ into exponential form and to know the common equivalents so that he can read books using $\sqrt{\ }$. Apart from that, the less he does with $\sqrt[3]{a^2}$, etc., and the more exclusively he works with $a^{\frac{1}{2}}$, $a^{\frac{2}{3}}$, $a^{\frac{3}{2}}$, etc., the better pleased we should be.

(3) To add like terms add the coefficients of the common factor and prefix this sum to the common factor.

(4) To add polynomials we write like terms in the same column and add these terms, writing their sums as a polynomial.

Consider $4ab^2c$. By the definition (1) b^2 is called the coefficient of $4ac$, and it would be reasonable to call it so, but in actual fact it isn't so called. $4ab^2$ by the definition is the coefficient of c.

Consider $5a^2b^3c$, regarding one factor as $5a^2b^3$ and the other as c. By the definition (2) $4ab^2c$ and $5a^2b^3c$ are called like terms, and it would be reasonable to call them so, but teachers would not call them so in one case out of fifty, and might rebuke pupils if they so called them. Consider: Add $3abc+4ab^2c+5a^2b^2c$ to $6adc+16\frac{abc^2}{c}+7cef+8abc$. Applying Rules (3) and (4), which shall we put in the same column — all of them or only the first and last, or the first, last and second from the last?

In fact, it is not the definitions and rules, but the habits developed by experience with particular numbers and expressions that teach a pupil what to regard as the coefficients, and what terms to collect together in addition. A practical joker could arrange a set of exercises in early addition that would force a conscientious pupil who tried to follow the definitions and rules into utter bewilderment. Would it not be better to build up useful habits of addition and subtraction without the pretense that they are logical deductions from the definitions and rules?

"An algebraic expression in which the parts are not separated by + or − is called a monomial" is given as a rigorous definition; but later the pupil has to consider

3$(ab-cd)$ as a monomial; and later he must treat $\frac{12a^2b^2c}{3abc}$ as two monomials though there is no separation by + or −.

A special case of importance is the erection into fixed rules on a level with the most general axioms and principles, of technically convenient procedures like arranging in ascending or descending powers before multiplying or dividing, writing like terms under one another, or beginning a monomial with its numerical factor. Every such rule which a pupil finds that he can break (as he can all such) without getting wrong answers means a risk that he will lose respect for and confidence in the really imperative rules. If we leave to habit everything that can be done as well by habit, we gain an added dignity for the matters that really are matters of principle. For example, it seems sound psychology to teach "If equals are added to equals the results are equal," as a rule, but to teach "transposing," universally valid though it is, as a convenient habit. It is not because he does not value rules and principles in algebra that the psychologist often prefers to use habits instead; it is rather because he does value principles and does not wish them to be misused and cheapened.

CHAPTER VIII

THE CONSTITUTION OF ALGEBRAIC ABILITIES (Continued): LEARNING ALGEBRAIC MEANINGS AND PROCEDURES

LEARNING MEANINGS

It used to be customary to teach literal number, coefficient, exponent, term, factor, radical, surd, root of an equation, and the like, by giving a definition and a few illustrations, and then proceeding to use the new fact or symbol in computation.

For example, a typical first lesson in algebra would teach the meaning of literal numbers as follows:

In arithmetic, numbers are represented by the digits, 1, 2, 3, 4, 5, 6, 7, 8, 9, and 0. In algebra numbers are also represented by letters. Numbers represented by letters are called *literal numbers*.

Example 1: A girl earns 10 cents per day. How much does she earn—
(a) in 2 days?
(b) in 8 days?
(c) in any number of days?

The result in (c) may be expressed thus: "as many cents as you obtain by finding the product of the number of days and 10."

In algebra it may be expressed thus:

Let n = the number of days
Then $10 \times n$ = the number of cents earned
So, if n is 3, $10 \times n = 10 \times 3$ or 30
if n is 7, $10 \times n = 10 \times 7$ or 70.

Example 2: How many quarts are there—
(a) in 4 bushels?
(b) in any number of bushels?

Let x = the number of bushels
Then $32x$ = the number of quarts in x bushels.

The assumption seems to have been that the learner got the meaning substantially from the definition, and made sure of it from the illustrations.

Whatever the merits of the method of teaching, this assumption was almost certainly not true save perhaps for a very few of the most gifted pupils.

Most of the pupils do not master the meanings once for all from the definitions and illustrations, and then fix them in memory by using them. They obtain vague, partial, and even inaccurate notions from the definitions and illustrations. These they use as best they can, and, in the course of using them, improve them. The lesson quoted above, for example, will give most pupils only a weak predisposition to think of n days, f dollars, m apples, p cakes, and d weeks, as numbers of dollars, apples, etc., rather than as misprints or nonsense. It will not guarantee that they are sure that this is the case, still less that you can multiply a figure-number by a letter-number, still less that since literal numbers are numbers, you can add, subtract, multiply, and divide with them.

If pupils who have studied diligently the lesson quoted above are then given the test below, there will be very few perfect scores.

1. How many cents are there in f dollars?
2. How many dollars are there in f cents?
3. How many quarts are n quarts less 4 quarts?
4. How many cents are x cents plus 7 cents?
5. How many cents are there in g dollars and h dollars together?
6. John has m apples. Fred has n apples more than John. How many apples has Fred?
7. Mary bought p cakes and ate q of them. How many cakes had she left?
8. How many weeks will it take Alice to earn b dollars, if she earns c dollars per week?
9. How much will Nell earn in d weeks if she earns e dollars a week?

By taking such tests, and by being told when he is right and when he is wrong, and what the right answers are, the pupil will come in time to an understanding of what literal numbers are. The process is not usually to get the meaning all at once from the definitions and illustrations, but rather to get it gradually from operating with the things in question — literal numbers, coefficients, exponents, equations, or whatever they are.

It is true that algebra is a most favored case for learning meanings from definitions. The concepts to be acquired are beautifully clear, can be rigidly defined, are free from variations and exceptions, do not depend upon time or place or environment, and have been prepared for by the pupil's study of arithmetic, of whose concepts they are in the main only extensions. Definitions are relatively more efficacious in algebra probably than in any other high-school study except geometry. Yet observation of the learning of algebra shows that the pupils learn by their concrete experiences with letters, coefficients, exponents, etc., more than by analytic scrutiny of their definitions. They learn by what they *do with* algebraic facts, and what the results are, more than by what they *are told about* them.

So Rugg and Clark, after their extensive and careful study of algebraic learning under class-room conditions, largely replaced the study of definitions by graded exercises whereby the pupil's own activities lead gradually to a comprehension of terms. Compare, for example, their treatment of radicals [1918, pp. 337-346] with the older type of treatment.

The need of actually working with symbols and abstract ideas in order to understand them, beginning with partial and crude notions which are extended and refined as one uses them, is demonstrable even in learning by highly trained adults of notably superior ability. Let the reader, for example, if he has not studied the theory of variable measurements, work through such a book as Yule's *Theory of Statistics*, and observe his learning. He will find that he learns what variability, mean square deviations, correlation coefficients, regression equations, partial correlations, and the like are by working the problems as truly and as much as he learns how to work the problems by the definitions and illustrations and discussions of these terms. Yet it is probably much easier for a person of the ability of the average reader of this book to understand these statistical concepts by direct analysis and the inspection of a few illustrations than for the average first-year high-school pupil to understand literal numbers, polynomials, signs of aggregation, the relations of signs to the four quadrants of the Cartesian coördinate system, fractional exponents, and the like.

Pupils learn meanings by direct experience of the facts in question just as truly in the case of coefficient, exponent, and surd, as in the case of ampere, volt, and ohm, or in the case of oxygen, hydrogen and sulphur, or even in the case of amœba, paramœcium and amphioxus, diphtheria and smallpox. The difference is one of degree; the verbal description helps much more with the mathematical ideas, which deal with a narrow content of number and quantity, in a few relations, than it does with the biological or medical ideas which involve size, color, shape, and chemical composition, in many intricate relations and with elaborate serial changes.

Pupils also learn meanings gradually in algebra as they do in the less formal sciences. Consider what would happen if, after the first few lessons which assume to teach pupils that you can represent any number by a letter, and that a formula states a fact about literal numbers, and another set of lessons which state and illustrate the axioms that you can add equals to equals, subtract equals from equals, etc., the pupil was confronted by the following task:

$ax+b=0$

State a formula for finding what x equals in any such equation.

Find what x equals in $\frac{3x}{8}+p=0$ by substituting in your formula.

Find what x equals in $\frac{ax}{b}+c=0$ by substituting in your formula.

Find what x equals in $\frac{bx}{a}+(c+b)=0$ in the same way.

The pupils (save a very few, very gifted individuals) do not from these early lessons really understand what a literal number is, or what a formula is. They grow into that understanding by adding, subtracting, multiplying, dividing, by transposing, evaluating, and solving. Computation with literal numbers is not simply a technique learned so as to enable pupils to find certain answers; it is one chief means by which they get the notions of algebra itself, of a "generalized arithmetic," of the possibility of stating a quantitative relation in a formula or equation, of directed numbers, or of the system of exponents.

LEARNING ALGEBRAIC COMPUTATION

The usual method of teaching the operations in algebra is to state the rule or procedure with a certain amount of explanation that it is a right rule and why it is a right rule, then have the pupil apply the rule, and finally give him practice so that he can do the operation in question almost without thinking of how he does it or why he does it in that way. The principle or rule with its justification comes first, then the operating in accordance with it. There is a rather wide variation in the amount of explanation that the procedure is right and why it is right, but there is rather general agreement in assuming that the pupil first learns how he is to operate by a general rule, and eventually comes to operate with little or no thought.

Modern psychology, however, is suspicious of all cases where habits are supposed to be easily derived from principles. It so often happens that the really effective principle is the product of the habits, not their producer. A man's conduct seems to determine his conscience more than his conscience his conduct. Principles are not, as a rule, general but rather are limited to the fields where they have had habitual operation. Young notes that pupils asked to find the value of, say, e in $a=1.4b+\frac{c}{e}$, will proceed to "Let $x=e$; then $a=1.4b+\frac{c}{x}$, etc." In the learning of arithmetic in the common schools the understanding of how and why to carry in adding, to manipulate the partial products in multiplication, to place the decimal point, and the like, seems to come to many pupils only after they have done these things many times. Many of them learn to operate by imitation, in the first instance, the general statements of how and why being little more than nonsense words to them, and acquire their understanding of the general statements by first acquiring habits which the general statement describes or justifies.

There is, of course, a better chance for learning by grasping a principle or rule and applying it so as to form certain habits, in the case of high-school pupils than in the case of pupils in general; and in the case of algebra than in the case of, say, a foreign language. The high-school pupils are, with few exceptions, from the top two-thirds of the population for abstract intellect. The operations in algebra are clearly and fully describable, and the reasons why one should operate in such and such ways are rather easily seen and appreciated. $a^m \times a^n = a^{m+n}$ and $(a^m)^n = a^{mn}$, for example, are perhaps as teachable as any rules found in all secondary teaching.

Nevertheless we maintain that the pupil learns the rules of algebra by operating as truly as he learns how to operate by being taught the rules. We maintain, for example, that the formation (even by mere imitation and unthinking acceptance) of special habits like $a^2 \times a = a^3$, $a^2 \times a^2 = a^4$, $a^2a^3 = a^5$, $b^2 \times b = b^3$, etc., is a natural and useful step toward learning the general modes of multiplying. The pupil builds up or integrates his habits into rules, as well as derives new habits from rules. Learning to compute algebraically is not only, or chiefly, learning rules and how to apply them; it is also building up a hierarchy of habits or connections or bonds which clarify, reinforce and, in part, create the understanding of what the rules mean and when to apply them.

Algebra appears to a competent abstract thinker after he has learned it as a series of deductions from certain definitions, axioms, and very general laws (such as that $a+b+c=b+a+c$ or $c+b+a$, or that $abc=bac$, or bca or acb). Algebra to most learners, however, is in large measure forming more or less particular bonds or connections, such as $a \times ab = a^2b$, $a(a+b) = a^2 + ab$, a means $1a$, $-a \times -b = +ab$, learning to operate several of these together as needed, organizing them further into more inclusive habits and insights, summing up what one has learned to do in rules, and thus gradually attaining a sense of what it is right to do with literal numbers and why.

We may become convinced of the importance and primacy of this process of association, connection, or bond forming by considering (1) the changes in the teaching of algebra made decade by decade, and (2) the stock errors made by learners.

In the course of selecting methods of teaching that give good results, less and less emphasis is put by teachers, textbooks, and courses of study on the learning of rules.

Our fathers were supposed to learn how to multiply algebraically by learning to answer such questions as those shown below [Davies, 1866, p. 56 f].[1]

MULTIPLICATION

41. **1.** If a man earns a dollars in 1 day, how much will he earn in 6 days?

Analysis.—In 6 days he will earn six times as much as in 1 day. If he earns a dollars in 1 day, in 6 days he will earn $6a$ dollars.

2. If one hat costs d dollars, what will 9 hats cost? Ans. $9d$ dollars.

3. If 1 yard of cloth costs c dollars, what will 10 yards cost? Ans. $10c$ dollars.

4. If 1 cravat costs b cents, what will 40 cost? Ans. $40b$ cents.

5. If 1 pair of gloves costs b cents, what will a pairs cost?

Analysis.—If 1 pair of gloves cost b cents, a pairs will cost as many times b cents as there are units in a: that is, b taken a times, or ab; which denotes the product of b by a, or of a by b.

Multiplication is the operation of finding the product of two quantities.

The quantity to be multiplied is called the *multiplicand;* that by which it is multiplied is called the *multiplier;* and the result is called the *product.* The multiplier and multiplicand are called *Factors* of the product.

6. If a man's income is $3a$ dollars a week, how much will he receive in $4b$ weeks?

$$3a \times 4b = 12ab.$$

If we suppose $a = 4$ dollars, and $b = 3$ weeks, the product will be 144 dollars.

NOTE.—It is proved in Arithmetic (Davies' School, Art. 48. University, Art. 50), that the product is not altered by changing the arrangement of the factors; that is,

$$12ab = a \times b \times 12 - b \times a \times 12 = a \times 12 \times b.$$

MULTIPLICATION OF POSITIVE MONOMIALS

42. Multiply $3a^2b^2$ by $2a^2b$. We write,

$$\begin{aligned} 3a^2b^2 \times 2a^2b &= 3 \times 2\ a^2\ a^2\ b^2\ b \\ &= 3 \times 2\ a\ a\ a\ a\ b\ b\ b; \end{aligned}$$

in which a is a factor 4 times and b a factor 3 times; hence (Art. 14),

$$3a^2b^2 \times 2a^2b = 3 \times 2a^4b^3 = 6a^4b^3$$

[1] The entire treatment is quoted so as to show the questions in their proper relations and also for its general interest. The last exercise in application of the rule is "Multiply 70 $a^8b^7c^4d^2fx$ by 12 $a^7b^5c^3d\ x^2y^3$!"

in which *we multiply the coefficients together, and add the exponents of the like letters.*

The product of any two positive monomials may be found in like manner; hence the

RULE

I. *Multiply the coefficients together for a new coefficient:*

II. *Write after this coefficient all the letters in both monomials, giving to each letter an exponent equal to the sum of its exponents in the two factors.*

41. What is multiplication? What is the quantity to be multiplied called? What is that called by which it is multiplied? What is the result called?

42. What is the rule for multiplying one monomial by another?

In good schools today one seldom hears the order "State the rule for" in the beginning stages of the learning of an operation. The illustrations are given more attention; instructive exercises which aid in the formation of the mental connections are given still more. Thus it is now customary to give analogous exercises from arithmetic which recall the mental connections that the pupil already has as a result of his past learning and to modify these connections or bonds to fit the uses of algebra. The exercises applying the rule are more and more such as a pupil might do without first understanding the rule.

For example, "To add a positive number to a negative number, take the difference of their absolute values and prefix the sign of the numerically greater number" is followed by problems about an elevator going up and down, an engine going back and forth, a man's property and debts, and a game with credits and penalties. Common sense does not need psychology to tell us that these problems help a pupil to learn the rule much more than the rule helps him to solve the problems.

The rules more and more take the form of convenient statements of what you do rather than laws as to how you should think. Such a rule as,

"A minus sign before a number means that the number is to be subtracted, *e. g.*,

-4 means that 4 is to be subtracted
$-a$ means that a is to be subtracted
$-mp^2$ means that mp^2 is to be subtracted,"

is obviously designed to prevent the pupil from efforts to think out the distinction between − as a sign of the direction of the number and − as a sign for the operation of subtraction. It hides that issue deliberately so as to give the pupil help in the actual operating connections to be formed with −.[1]

The exercises applying the rule more and more take the form of graded series which clear up what the rule means. The custom, once fairly common, of making the exercises chiefly a series of probes to find weak spots in a pupil's knowledge of the rule, is vanishing. Such exercises are fewer in number and are given late in the development of the topic.

Mr. Symonds found [1922] that certain pupils' errors were in large measure explainable by inadequate mental connections, and inability to operate several connections (each adequate in certain simple situations) together in proper coöperation. It is important to remember that these pupils were chiefly pupils of good achievement in other school subjects. They were then not, in the main, pupils

[1] We do not imply that this clever attempt at unification of the habits of response to − (the minus sign) is justifiable. The point here is simply that it is a recognition (possibly unconscious) of the importance of connection-forming,— a sacrifice of the "logic" of the rule to its serviceability in establishing useful mental bonds. The very same point is illustrated by those who would *accentuate* the difference between the two uses of − by writing the sign above the letter to mean the direction of that number from zero, and by writing the sign before the letter to mean only the nature of the operation. That is, they would use $+\overset{-}{a}$, $-\overset{-}{a}$, $+\overset{+}{a}$ and even $-\overset{+}{a}$. A chief value of this would be that good mental connections or habits would not interfere with each other, as they now do.

who were unable to learn the general rules of algebra, who had to work by mere habituation, and consequently had to make both their successes and their failures on the habit level. On the contrary, if these pupils, intelligent in general, fail often in algebraic work by lack of proper direction, strength, and organization of algebraic connections or bonds, it is probable that the other pupils fail still more.

Learning algebraic computation is, and should be, in large measure the formation and organization of a hierarchy of mental connections or bonds. The science of teaching algebra must consider what these bonds are, the amount of practice each should have, how this practice should be distributed, how the order and methods of formation of these bonds may provide the maximum of facilitation and the minimum of interference among them.[1]

LEARNING GENERAL PRINCIPLES

The general psychology and pedagogy of associative thinking in relation to analysis, abstraction, and reasoning is the same for algebra as for arithmetic. The matter has been fully treated by one of us elsewhere (Thorndike, 1922,

[1] This section may have given the impression that psychology regards the announcement of a rule at the beginning of a topic as essentially bad pedagogy. We have, however, been careful to say nothing of the sort. Indeed we believe that in certain cases the early announcement of a rule is desirable. The rule may, for example, be a stimulating way of putting a question. For example, the contemplation of "Like powers of equals are equal" and "Like roots of equals are equal, if you treat the signs properly," may be a good way of arousing the question, "What can we do in the way of squaring and cubing and getting square roots and cube roots and still be sure that our results are true?" A pupil who gets from these two rules at the outset only a confused sense that people square and cube and extract roots with the members of equations and that he is expected soon to learn something appertaining thereto, may still have profited from them. A rule may be useful by suggesting a job, even if it does not specify it, or by specifying a job even if it does not help do the job, or by helping do the job, though only as one minor factor. Rules have many uses. In general, however, it seems desirable to use rules less as edicts at the beginning, and more as convenient verbal summaries of what a pupil *has learned to do.*

Psychology of Arithmetic, chapters 3, 4, 9, and 10); and we shall make here only a bare statement of the relevant conclusions. Readers who find difficulty in understanding or in accepting these conclusions should consult the discussion in the case of arithmetic.

In recognizing the importance of association or connection-forming or mental habit in the learning of algebra there need be no sacrifice of the so-called "higher powers" of abstraction, generalization, and reasoning. These higher powers are in reality the coöperation of many connections or bonds selected and given proper weight for some purpose. An equal or greater mastery of the general principles of algebra is secured by reaching them gradually and in the progress of the learning of algebra as a result of learning it, rather than by demanding that they be mastered all at once at the beginning as a means to learning it. For very gifted pupils there is perhaps a loss of time and of training in reasoning, but for most pupils there is a gain. Instead of demanding that a pupil master certain meanings, rules, and principles from verbal descriptions, and making it difficult for him to learn algebra unless he does, we form bonds which provide him with useful algebraic abilities in any case, and stimulate him to understand principles so far as he has the ability.

CHAPTER IX

THE CONSTITUTION OF ALGEBRAIC ABILITIES (Continued): THE SELECTION OF THE PARTICULAR MENTAL CONNECTIONS OR BONDS TO BE FORMED

DESIRABLE MENTAL CONNECTIONS NOW OFTEN NEGLECTED

Bonds should be formed with tasks extending algebraic symbolism to include Roman letters, capital letters, primes, and subscripts, and using other letters than a, b, c, x, y, and z. This is of real, though not of great importance. There are three reasons for it. First, the generality of algebraic symbolism is more surely taught. Second, the fact that any literal number does in actual use stand for some number of something is more surely kept in mind if we let n_1 and n_2 rather than x and y equal the smaller number and the larger number, if we let l', l'', and l''' or p_1, p_2, and p_3 equal the length of the three pendulums in inches, etc. Third, the pupil is better fitted for reading algebraic discussions. Customary usages should also be followed in the symbols for angles, masses, distances, times, etc., other things being equal.

Bonds should be formed between the situation "A formula or equation to be read" and a realization of the relations which the formula states. Really reading $P=\frac{AV}{M}-K$ means understanding that as A increases P increases, as V increases P increases, as M increases P decreases, in each case proportionately; and that a discount of K is always made from $\frac{AV}{M}$ to obtain P. Professor Hedrick has emphasized the importance of these bonds of insight into relations. They have not been formed by the standard teaching of the past, and will not be formed, save in the very gifted pupils, unless teaching takes pains to form them. Of pupils who had been taught algebra for one to over two years in excellent schools, but with the older restriction of work with equations to solving them, half failed with such exercises as:

$P = \frac{b_1 b_2}{c_1 + c_2}$. b_1, b_2, c_1, and c_2 are all positive.

Write "*larger*" or "*smaller*" or "*you cannot know*" on the dotted lines.

If b_1 becomes larger, P will become ....

If b_2 becomes larger, P will become ....

If c_1 becomes larger, P will become ....

If c_2 becomes larger, P will become ....

From 182 pupils the numbers of correct completions of these four sentences were 110, 91, 93, and 87. Some of these, of course, were due to chance.

These same pupils did very well with tasks to which they were accustomed, three out of four obtaining right answers to such tasks as:

To two times a certain number 2 is added. From three times the number 7 is subtracted. The two results are equal. What is the number?

Find two numbers such that: Twice the first, if added to three times the second, equals 2. Six times the second, if added to ten times the first, equals 7.

What is the length of the diagonal of a rectangle, if the sides of the rectangle are 8 ft. and 6 ft.?

The bond between "a table of values of a variable to be read" and "the consideration of each in relation to the others, and of the whole as the story of a relation" should be formed, or at least started on the way toward formation. Exercises in interpolating, in drawing the graph of the table, in finding an equation which expresses the table (where such is a reasonable task for high-school pupils), and in such completions as "The number in column 2 is about . . . times the corresponding number in column 1"; "Add . . . to each entry in column 2 and each will be about . . . times the corresponding number in column 1," are much to be desired in algebra. Much of the work in ratio and proportion may well be correlated with table-reading.

Bonds should be formed between the sight of any literal equation and the expectation that it tells the story of the way certain things occur in the world. Two great aids towards this are: first, the extension of work with formulas and their evaluation, always evaluating each formula studied for several cases of the relation; second, obtaining the general answer to a verbal problem as well as the special answer that fits one case of the relation in question.

After the formal general treatment of $y=ax+b$ has been given and its graphic representation is understood, bonds should be formed between these general facts and the solution for x of equations in the form of $y=ax+b$, given values of y, and their solution for y, given various values of x. The latter is the more important. The solution of $ax+b=0$ should be connected as a subordinate special case of solving for x when y is given. To emphasize it without its proper setting is to interfere with the pupil's learning of what a linear equation is.

Bonds should be formed between the graph of the relation of one variable to another, and the graph formed by the ends of a series of rectangular columns. These columns should become thinner and thinner. There need be and should be no talk about limits, but the pupil should see the ends of the columns each expressing a table-entry grow into an approximately smooth line as the table entries become of narrower and narrower range.

Bonds should also be formed later between the graph of the relation of one variable to another, and the graph formed by a moving point. Let one boy move the x crayon along the x axis, while another boy moves the y crayon so as to make $y=x$, or $2x$, or $\frac{1}{2}x$, or $4+x$, etc.

The first of these two supplements to the standard procedure of locating points and joining them is needed to connect the algebra of graphs with the relevant facts of common life, which are mostly to the effect that when x is from 0 up to 1.0, y averages so and so; when x is from 1.0 up to 2.0, y averages so and so, etc. The second is needed to enforce the lesson of dependence and relation, that the y varies with the x. The standard method has an important defect. The x variable does not seem to do anything or suffer any change, or in fact be in the game at all. A red chalk moving for x and a yellow chalk moving synchronously for y, help to supply the deficiency.

Bonds should be made between the facts about ratio and the verbal forms in response to which we use our knowledge of ratio. The most important is: p is.... as many (much, heavy, long, large, etc.) as q. When p is known to be larger than q, we usually think the word "times" before "as many." The next in importance is "p and q are in the ratio of.... to....," the "fillers" to be made simple for memory's sake, and for convenience in calculation.

The third is, "The coefficient of so and so (or the specific gravity, or the nutritive ratio, or the visibility, or what not) is the ratio of the such and such to the such and such." One must know that the first such and such is the numerator, and the second such and such the denominator. Science should learn to change this usage to "is the such and such divided by the such and such," which would save probably one hundred thousand hours a year to students of algebra and first year general science alone. In the meantime, we can save part of it by engraving on every mind "The ratio of *this* to *that* means $\frac{this}{that}$."

Unless these bonds are made, it is hard to learn what ratio means, and still harder to use your knowledge of ratio when it is needed.

Bonds between computing and checking the computation are of great importance. A truly algebraic computation in life is, except for evaluations, often a momentous matter, since some general relation or law is often involved, not merely the money of some individual, or the size of some one farm. The pupils are also, in many cases, soon to leave school and be without a teacher to inspect their work. They are old enough to assume responsibility for accurate work. They do not, as things now are, do accurate work.[1]

The simplest check of all for ordinary computations and evaluations is not often urged by teachers. That is, to do the work all over again independently after, say, a half hour spent on something else. Yet for many pupils it would be an admirable practice to do, say, three-fourths of the exercises, repeating after a half hour, instead of a full assignment once. The customary checks for the solution of equations should be not only advised or ordered, but made profitable to pupils by the attachment of heavy penalties for wrong answers. A good way to provide practice with numerical computation, and drill on the laws of signs and the removal of parentheses, is through checks on the literal computation.

UNDESIRABLE MENTAL CONNECTIONS NOW OFTEN FORMED

It is important to find and eliminate any unnecessary or wasteful bonds now being formed in the learning of algebra.

[1] The evidence for this will be presented in Chapter XII.

For we need to make room for the work with formulas, graphs, and the study of the relations between two variables, which every line of evidence shows are of notable educative value, and for the brief treatment of logarithms, certain other aids to computation, and the tangent, sine, and cosine, which are also strongly recommended. Moreover, the first-year course in high-school mathematics should be made a model first-year course in all respects, so that pupils who ought to study mathematics further will be led to do so, and so that communities which ought to provide further mathematical instruction will be led to do so. Mathematics ought not to be content to live on the reputation made when it was almost the only respectable science available for young learners.

Elaborate Computations

In a previous chapter we noted the general demand from experts in general education and in the teaching of mathematics that computations so elaborate as to be seldom or never used should be omitted. No competent person disputes the wisdom of this; but, though converted, we are far from being reformed. For example, in an algebraic contest between schools, arranged by progressive teachers, in one of our most enlightened cities, in 1916, the competing teams were tested with tasks a majority of which required mastery of the special products and factorizations which are the least used features of algebra. The tasks were required to be of the following type forms:

TYPE FORMS OF PROBLEMS

I. FACTORING

$m^2x^2 \pm 2mnxy + n^2y^2$	$ax^2 \pm bx \pm a$
$x^2 \pm ax \pm b$	$a^2x^2 - b^2y^2$
$m^2 \pm 2mn + n^2 - a^2k^2$	$x^3 \pm a^3$

TYPE FORMS OF PROBLEMS—Continued

II. Special Products

$(ax \pm by)^2$	$(m \pm n \pm a)\ (m \pm n \mp a)$
$(m+n)\ (m-n)$	$(x \pm a)\ (x \pm b)$
$(a \pm b \pm c \pm d)^2$	$(ax \pm by)\ (cx \pm dy)$
$(a \pm x)^3$	

III. Fractions

(a) Problems in addition and subtraction.

(b) Problems in multiplication and division.

These problems shall not involve any type of factoring or special product not shown in I and II, and no problem shall contain more than two fractions. The object is to illustrate the simple types of fractions.

IV. Equations

(a) Solution by factoring:

$$x^2 - a^2 = 0$$
$$x^2 \pm 2ax + a^2 = 0$$
$$ax^2 \pm bx \pm c = 0$$

(b) Simultaneous equations:

$$ax + by = c$$
$$a^2x^2 + b^2y^2 = c^2$$

[Boughn, E. T., 1917, p. 330]

Unreal and Useless Problems

In a previous chapter it was shown that about half of the verbal problems given in standard courses were not genuine, since in real life the answer would not be needed. Obviously we should not, except for reasons of weight, thus connect algebraic work with futility. Similarly we should not teach the pupil to solve by algebra problems which in reality are better solved otherwise, for example, by actual counting or measuring. Similarly we should not set him to solve problems which are silly or trivial, connecting algebra in his mind with pettiness and folly, unless there is some clear, counterbalancing gain.

This may seem beside the point to some teachers, "A problem is just a problem to the children," they will say,

"The children don't know or care whether it is about men or fairies, ball games or consecutive numbers." This may be largely true in some classes, but it strengthens our criticism. For, if pupils do not know what the problem is about, they are forming the extremely bad habit of solving problems by considering only the numbers, conjunctions, etc., regardless of the situation described. If they do not care what it is about, it is probably because the problems encountered have not on the average been worth caring about save as *corpora vilia* for practice in thinking.

Another objection to our criticism may be that great mathematicians have been interested in problems which are admittedly silly or trivial. So Bhaskara addresses a young woman as follows: "The square root of half the number of a swarm of bees is gone to a shrub of jasmine; and so are eight-ninths of the swarm: a female is buzzing to one remaining male that is humming within a lotus, in which he is confined, having been allured to it by its fragrance at night. Say, lovely woman, the number of bees." Euclid is the reputed author of: "A mule and a donkey were going to market laden with wheat. The mule said, 'If you gave me one measure I should carry twice as much as you, but if I gave you one we should bear equal burdens.' Tell me, learned geometrician, what were their burdens." Diophantus is said to have included in his preparations for death the composition of this for his epitaph: "Diophantus passed one-sixth of his life in childhood one-twelfth in youth, and one-seventh more as a bachelor. Five years after his marriage was born a son, who died four years before his father at half his father's age."

My answer to this is that pupils of great mathematical interest and ability to whom the mathematical aspects of these problems outweigh all else about them will also be interested in such problems, but the rank and file of pupils will react primarily to the silliness and triviality. If all they experience of algebra is that it solves such problems they will think it a folly; if all they know of Euclid or Diophantus is that he put such problems, they will think him a fool. Such enjoyment of these problems as they do have is indeed compounded in part of a feeling of superiority.[1]

Unnecessary Vocabulary

Other things being equal, the learning of algebra should not be made dependent on knowledge of rare and unimportant words. Pupils should not be asked to solve problems which they can read only with recourse to the dictionary, or to study explanations whose language is in part unknown.

New technical terms whose learning is desirable, must, of course, be learned. These should be defined by the teacher or textbook as they occur. Certain semi-technical and non-technical words (like approximate, asset, asterisk, assumption, axes) which a considerable percentage of pupils will not know, must be learned because their learning will be worth while for the sake of algebra and general education. As a rule, however, the difficulties of algebra itself should not be complicated by an elaborate or recondite vocabulary.

These precepts should be obvious, but teachers probably violate them again and again. They are violated occasionally by authors of textbooks who have greater ability and much more time to plan what they write than the teacher has to plan what he says.

[1] Teachers do not sufficiently realize that certain pupils are only too ready to think of school studies as pedantic, freakish pursuits, beneath the serious attention of a person who can earn money, play football, or be taken to dances!

For example, within the first fifty pages of three standard textbooks for first-year algebra, we find these fifty words:

aggregation	Ahmes	algebraist	buoyancy	complementary
consecutive	debit	deflection	Demosthenes	denominate
Descartes	Diophantes	displacement	dissimilar	distinctive
elementary	equilateral	et	evaluation	facilitate
frustum	Harriot	haw	Herigone	Hindu
identical	initiation	Leibnitz	neutralize	nitrogen
notation	Oughtred	papyrus	Pell	polygon
potentia	prism	projectile	Pythagoras	reintroduce
resultant	scalepan	specific	Stifel	substitution
subtrahend	supplementary	trinomial	Vieta	Widmann

The use of some, perhaps a majority, of these in their particular contexts may be advisable, but they will need explanation by the teacher or special study by the pupils. Some of the fifty are almost certainly better replaced by commoner and better known words.

Two first-year books, both written with great care and intelligence, make use of 580 words that are not among the first ten thousand of the language in importance. Many of these are desirable technical terms. Others are proper names used in problems and involve no appreciable burden on the learner. Some, like *acceptable*, *accumulation*, *adaptation*, *alternative*, and *applicable*, are probably worth learning even at the cost of some interference with algebraic learning itself. Some, however, are hard to justify. *Calculus*, *commutative*, *concept*, *conjugate*, *criteria*, *cryptologist* and *curriculum*, for example, almost certainly do more harm than good in an algebra for Grade 9.

Undesirable Terms and Definitions

In connection with learning to compute algebraically it is customary to learn the meanings of many technical terms such as *coefficient*, *exponent*, *monomial*, *trinomial*, *solve*, *evaluate*, and the like. There are also non-technical or semi-technical words and phrases which a considerable per cent of ninth-grade pupils probably will not know. Such are, for example, *assign*, *axiom*, *brace*, *bracket*, *constant*, *convergent*, *elimination*, *identity*, *inverse*.

Four general principles may be noted. First, other things being equal, the fewer technical terms used, the better. But, second, it is desirable to have a name for any fact or principle or operation, if it needs to be treated as a unit in thought. When well chosen, such hooks on which to hold units of knowledge are not burdens but aids. For example, with or soon after experience of negative numbers they should receive a name, such as *negative* or *directed*. Third, these names should be, so far as possible, descriptive and unambiguous. *Negative numbers* is thus descriptive and unambiguous. *Exponent* is not descriptive but is unambiguous. *Term* is neither descriptive nor unambiguous. Fourth, the abilities required are not to define these terms, but to respond correctly to appropriate uses of them.

A useful though not infallible procedure for distinguishing which terms should be taught and which terms should be dropped or replaced by more useful ones is to test pupils at the end of the course and thereafter, noting those cases where the fact or principle or operation is retained but the name for it is forgotten, and also those cases where the name has been retained but the fact or principle or operation has been forgotten. In both cases the name has been of dubious value. Thus, few persons remember what *minuend* and *subtrahend* mean, though they subtract competently. Thus many persons remember the words *parabola*, *hyperbola*, *sine*, and *cosine*, but not what the facts are.

The commonest error of teachers is to use too many technical terms. Bonds should not be formed between technical mathematical terms and their meanings unless the knowledge in question saves more time than is required to attain it. Samples of such wasteful learning are: *abacus, accumulating errors, aeq., aequalis, affected quadratic equation, alternation, antecedent, associative law, commutative law, comparison, composition.*

It may be objected that this verbal erudition may be worthy as a part of general education, even if it does not pay its way in the learning of algebra. This may be so, and in a few cases it probably is so, but in general if a technical mathematical term is not worth while for the learning of mathematics, it will not be worth while otherwise.

The evil is real and important. Courses of study and textbooks by gifted authors use technical terms whose values for the learning of algebra are slight or even negative. Individual teachers probably indulge in them even more. For it is natural for anybody to think that something which is useful to him will be useful to his pupils; it is also natural to think that a word or phrase which one has devised is useful; and, except for a very keen critical sense, it is natural to think that whatever we know is good for others to learn.

An inspection of five standard elementary textbooks reveals about five hundred technical terms (including, of course, ordinary words or phrases used with a technical meaning).[1] Some of these are obviously needed for the teaching of algebra (or are needed until better terms are devised and made customary). Some will be of little or no help, will detract from interest, and will encourage the pupil to think of algebra as a burden and a folly.

For the convenience of any student who wishes to experiment with algebraic terminology we present here a list of the technical and semi-technical terms or words used with

[1] Many are, however, due to the geometry which is taught along with the algebra in some of the books. Some of the terms, of course, are put in these books with no requirement that they be learned.

a special meaning in algebra, together with the ratings given to most of them in accordance with the following scale:

Consider the learning of algebra alone, and rate the term by this scale.

10. Essential for economical and efficient learning.
8. Very useful.
6. Probably worth learning.
4. Probably not worth learning.
2. The time is much better spent otherwise.
0. Of no value at all for the learning of algebra.

Use 1 when in doubt whether to use 0 or 2;

Use 3 when in doubt whether to use 2 or 4, etc.

Then consider the learning of elementary algebra, plane geometry and trigonometry, and rate each term in column 2 using the same scale.[1]

Each rating is an average of the averages of three groups, composed as follows:

(a) Six mathematicians and authors of textbooks on algebra of excellent repute

(b) Six psychologists with mathematical training and in all cases but one experience in teaching algebra, and

(c) Twelve graduate students of the teaching of mathematics in secondary schools.

When no rating is given, but only a − or +, the term was evaluated as almost certainly not worth learning or worth learning, respectively, by a preliminary rating by five competent judges.

These ratings, of course, represent thought influenced by custom. The psychologists are the most ready to criticize existing customary terms, but they on the whole tend to vote that whatever is, is right. For example, all but one of

[1] These second ratings are not reported here. Terms which have notable value apart from algebra will be as a rule easily distinguished.

them rated *trinomial* as essential or very useful. Yet it would seem that if *trinomial* were essential, *quadrinomial* ought to be fairly useful. Like *three-place-number* in arithmetic, *trinomial* in algebra is at times a convenient designation, but the writer cannot see wherein it facilitates the learning of algebra notably. With *monomial* and *polynomial* the matter seems very different, since these terms help to fix a distinction which is fundamental with respect to many operations. Yet these are rated actually a little lower than *trinomial.*

The ratings by the consensus represent a mixture of much sagacity and sound criticism and new ideas acting on a body of customary opinion — producing a reputable conservative total judgment, with the advantages and disadvantages which appertain to reputable conservatism. The teacher may feel safe from present criticism in omitting terms below 4.0 in credit[1] and in including terms above 6.0 in credit. Valid experimental investigations, however, may well prove that some of the 6.0's should have been 4.'0s and even that some of the 8.0's should have been 2.0's.

TABLE 27

RATINGS OF TERMS: FOR THE LEARNING OF ALGEBRA ALONE

Term	–		+	Term	–		+
Abacus	–			Affected quadratic			
Abscissa			+	equation	–		
Absolute term		4.2		Aggregation	–		
Absolute value			+	Ahmes	–		
Accumulating errors	–			Algebraic addition			+
Acute angle		4.5		Algebraic expression			+
Addend	–			Algebraic number		5.7	
Adjacent	–			Algebraic product			+
Adjacent angles	–			Algebraic solution			+
Aeq.	–			Algebraic subtraction			+
Aequalis	–			Al-jebr	–		
Aequatur	–			Alternate interior	–		

[1] So far as the teaching of algebra alone is concerned. Certain terms below 4.0 in credit are of notable value for geometry, physics, statistics, etc.

TABLE 27—Continued

Alternate exterior	–		
Alternation	–		
Altitude (*e. g.* of a mountain)		4.7	
Altitude of a triangle		4.2	
Angle			+
Angle of depression	–		
Angle of elevation	–		
Antecedent	–		
Antilogarithm		4.7	
Applying a formula			+
Approximation			+
Arc	–		
Arabic	–		
Area			+
Arithmetic average (mean)			+
Arithmetical numbers			+
Arrangement			+
Ascending powers			+
Assign		5.1	
Associative Law	–		
Average			+
Axes			+
Axioms			+
Bacon	–		
Bar diagram		5.7	
Bar graph			+
Base (of a triangle)			+
Base (distinct from power)		5.1	
Base angles	–		
Bearing	–		
Bhaskara	–		
Binomial			+
Binomial theorem			+
Bisect	–		
Bisection	–		
Bisector	–		
Brace		5.7	
Bracket			+
Broken Line	–		
Calculus	–		
Cancel			+
Cancellation			+
Cardan	–		
Carry	–		
Cartogram	–		
Cast out (nines, etc.)	–		
Center			+
Centigrade		5.7	
Central Angle	–		
Central tendency	–		
Chain (surveyor's)	–		
Characteristic		5.0	
Check			+
Circle			+
Circumference			+
Class interval	–		
Class limits	–		
Clearing of fractions			+
Coefficient			+
Coincide	–		
Collect		4.4	
Cologarithm			+
Combine			+
Common difference		5.9	
Common system of logs.		5.6	
Commutative law	–		
Compensating errors	–		
Complement	–		
Complementary angles	–		
Complete quadratic equation			+
Completing the square			+
Complex fractions			+
Complex numbers			+
Composition	–		
Conditional equation	–		
Conditions of the problem			+
Cone	–		
Conic sections	–		
Conjugate complex numbers	–		
Conjugate surds		4.2	
Consecutive			+
Consequent	–		

TABLE 27—Continued

Constant			+
Construct	–		
Convergent series		5.2	
Coördinates			+
Correct to n decimal places			+
Correct to n significant figures			+
Corresponding angles	–		
Corresponding parts	–		
Corresponding sides	–		
Cosine	–		
Cross products		5.9	
Cube			+
Cube root			+
Cubic (al)		5.1	
Cubus	–		
Cumulative errors	–		
Curve		5.9	
Curve of normal distribution		4.6	
Curve tracing		5.2	
Cylinder	–		
Data			+
Decagon	–		
Decimal			+
Degree of angle		4.3	
Degree of arc	–		
Degree of equations			+
Degree of latitude			+
Degree of longitude			+
Degree of a number		4.5	
Degree of a polynomial			+
Degree of a term			+
Denominator			+
Dependence			+
Dependent equations	–		
Dependent variable			+
Descartes		5.2	
Descending powers			+
Describe (an arc)	–		
Detached coefficients	–		
Determinant		5.9	
Determinant of n order	–		
Determined by (is)		4.8	
Diagonal	–		
Diameter		4.1	
Digit			+
Diophantus	–		
Direct variation			+
Directed number		5.4	
Discriminate	–		
Dissimilar terms		5.8	
Distributive law	–		
Divergent series		4.6	
Dividend			+
Division (in proportion)		4.5	
Divisor			+
Elimination			+
Ellipse		4.0	
English system	–		
Equal			+
Equality			+
Equate			+
Equations of n degree		5.5	
Equation of condition			+
Equations, identical			+
Equations in "n" unknowns			+
Equiangular	–		
Equidistant	–		
Equilateral			+
Equivalent equations		5.4	
Equivalent systems of equations	–		
Euclid	–		
Evaluation			+
Evolution		4.1	
Expand			+
Exponent			+
Exponential equations		5.0	
Expression			+
Exterior angles	–		
Extract a root			+
Extraneous roots			+
Extremes		5.1	
Factor			+

TABLE 27—Continued

Term			
Factor, common			+
Factor theorem	–		
Factorial			+
Fahrenheit		5.2	
Fallacious	–		
Ferrari	–		
Finite Series		5.8	
Force	–		
Formula			+
Fourth proportional		5.1	
Fractional equation			+
Fractions			+
Frequencies			+
Frequency table			+
Frustum of pyramid	–		
Fulcrum		4.2	
Function			+
Functions, trigonometric	–		
Fundamental operation			+
Gauss	–		
Geometric mean		5.7	
Geometry		4.2	
Given (equation)			+
Gives			+
Graph			+
Graphic			+
Graphical			+
Grouping			+
Hamilton	–		
Harriot	–		
Hexagon	–		
Highest common factor		5.9	
Hindu	–		
Hipparchus	–		
Homogeneous equation	–		
Homogeneous polynomial	–		
Horizontal axis			+
Hyperbola		4.2	
Hypotenuse			+
Identity			+
Imaginary numbers			+
Imaginary unit		4.8	
Included (side or angle)	–		
Incompatible equations	–		
Inconsistent equations			+
Independent equations			+
Independent variable			+
Indeterminate equations			+
Indeterminate forms		5.1	
Index (of root)			+
Indirect measurement		4.3	
Inequality			+
Inertia of large numbers	–		
Infinite series			+
Infinity			+
Initial side of angle	–		
Integer			+
Integral			+
Integral equation		5.2	
Intercept	–		
Interior angles	–		
Interior angles on same side of transversal	–		
In terms of			+
Interpolation		5.2	
Intersect	–		
Inverse variation			+
Inversion		5.1	
Invert			+
Inverted slide rule	–		
Involution	–		
Involving		5.7	
Irrational number			+
Isosceles	–		
Joint variation		5.6	
Kronecker	–		
Latitude		4.3	
Least common denominator (lowest)			+
Least common multiple (lowest)			+
Left side of an angle	–		

TABLE 27—Continued

Term	–		+
Legs	–		
Lever arm	–		
Like terms			+
Limit			+
Line			+
Line of sight	–		
Line segment	–		
Linear equation			+
Linear function			+
Literal equations			+
Literal number			+
Locus		4.4	
Logarithm			+
Logarithmic equations		5.0	
Longitude		4.2	
Mannheim	–		
Mantissa		5.2	
Maxima	–		
Mean proportional			+
Means		5.3	
Measurement of angle	–		
Measurement of arc	–		
Measurement of length	–		
Median			+
Median of triangle	–		
Members of equation			+
Meridian	–		
Metric system			+
Minima	–		
Minuend			+
Minutes of angle	–		
Mixed expression		4.2	
Mixed surd	–		
Mode	–		
Monomial			+
Moor	–		
Multiple			+
Multiplicand			+
Multiplier			+
Napier	–		
Nappes	–		
Negative angle	–		
Negative number			+
Newton		4.4	
N-gon	–		
Normal distribution		4.6	
Notation			+
Numerator			+
Numerical value			+
Oblique parallelopiped	–		
Obtain (an equation)		5.8	
Obtuse angle	–		
Octagon		5.8	
Operation			+
Opposite angles	–		
Opposite quantities		4.4	
Order of determinant	–		
Order of powers	–		
Order of radical		5.1	
Order of terms			+
Origin			+
Oughtred	–		
Parabola		4.7	
Parallel			+
Parallel right to right and left to left	–		
Parallel left to right and right to left	–		
Parallelopiped	–		
Parenthesis			+
Partial products		5.5	
Pentagon	–		
Perfect cube		5.9	
Perfect trinomial square			+
Perform			+
Perigon	–		
Perimeter		4.4	
Period (in square root of number)		4.8	
Perpendicular		4.6	
Perpendicular bisector	–		
Pictogram		4.2	
Plotting			+
Point (on graph)			+
Polygon		4.7	
Polynomial			+
Positive angle	–		

TABLE 27—Continued

Term	–	Value	+
Positive numbers			+
Power			+
Prime			+
Principia	–		
Probability		4.5	
Problems			+
Product			+
Progressions, arith			+
Progressions, geom			+
Progressions, inf. geom.			+
Proportion			+
Proportional			+
Protractor		4.3	
Pure quadratic equations		5.3	
Pyramid			
Pythagoras	–		
Quadrant		4.7	
Quadratic equation			+
Quadratic function			+
Quadratics			+
Quadratic surd		4.9	
Quadratic trinomial			+
Quadratic trinomial square	–		
Quadrilateral	–		
Quantity			+
Quotients			+
Radical			+
Radicand	–		
Radius		4.3	
Radix	–		
Raise to a power			+
Raleigh	–		
Random sampling		4.3	
Rate			+
Ratio			+
Rational			+
Rationalize			+
Real number			+
Recasting	–		
Reciprocal			+
Recorde	–		
Rectangle		4.9	

Term	–	Value	+
Rectangular parallelopiped	–		
Reduce (fractions)			+
Reflex	–		
Regiomontanus	–		
Remainder theorem		5.6	
Remainder			+
Rhombus	–		
Right angle		5.3	
Right parallelopiped	–		
Right side of angle	–		
Right triangle		5.1	
Root			+
Satisfy			+
Scale drawing			+
Scalene triangle	–		
Seconds of angle	–		
Sector	–		
Segment, line	–		
Segment, unit	–		
Semicircle	–		
Series			+
Series continuous	–		
Series discrete	–		
Series, infinite geom		5.3	
Side (of polygon)		4.3	
Signed quantity		5.3	
Significant			+
Significant figure			+
Signs			+
Similar terms			+
Similar polygons	–		
Similar triangles	–		
Simple equations			+
Simplify			+
Simultaneous equations			+
Sine	–		
Skewness	–		
Slide rule		4.6	
Slide rule runner		4.3	
Solid	–		
Solution			+
Solving equations			+

TABLE 27—Continued

Term	−		+
Solving triangles	−		
Specific gravity	−		
Sphere	−		
Square			+
Square root			+
Standard unit	−		
Statistical regularity	−		
Statistics			+
Stifel	−		
Straight angle	−		
Subscript			+
Substitution (simultaneous equations)			+
Substituting			+
Subtrahend			+
Sum			+
Supplement	−		
Supplementary	−		
Supplementary adj. angles	−		
Surd			+
Surface	−		
Symmetric equations	−		
Symmetry	−		
Synthetic division	−		
Synthetic multiplication	−		
System of equations			+
Tabulate			+
Tangent of an angle	−		
Terminal side of an angle	−		
Terms			+
Tetrahedron	−		
Theon	−		
Theorem		4.4	
Third proportional		5.2	
Transit	−		
Transposition			+
Transversal	−		
Trapezoid	−		
Triangle		5.3	
Triangular pyramid	−		
Trigonometric	−		
Trigonometry	−		
Trinomial			+
Turning tendency	−		
Unequal segments	−		
Unit of area	−		
Unit of segment	−		
Unit of volume	−		
Unknown			+
Unlike terms			+
Vanish			+
Variable			+
Variation			+
Velocity	−		
Verbal problems		5.4	
Verify			+
Vertex	−		
Vertical angles	−		
Vertical axis			+
Vieta	−		
Vinculum	−		
Volume		5.4	
Wallis	−		
Whole (number)			+

A certain interest attaches to the question of which terms the group of psychologists rates much lower than does the group of mathematicians and writers of textbooks. The former rates those of list A two points or more lower, and those of list B two points higher, than the latter. List A is, in our opinion, worth considering for suggestions for eliminating terms, beyond what reputable conservatism already advocates. List B bears witness to the influence of Rugg and Clark and Nunn upon the psychologists and to their interest in statistical work and algebra as a preparation for it.

A. *Psychologists lower by* 2.0

Absolute term
Absolute value
Algebraic number
Algebraic solution
Altitude
Altitude of a triangle
Arithmetical numbers
Arrangement
Ascending powers
Base (distinct from power)
Brace
Bracket
Complementary angles
Complex fractions
Complex numbers
Degree of equations
Degree of a polynomial
Degree of a term
Descartes
Descending powers
Digit
Dissimilar terms
Elimination
Ellipse
Equate
Equation of condition
Equations, identical
Equivalent equations
Extraneous roots
Factorial
Fundamental operations
Imaginary numbers
Imaginary unit
Inconsistent equations
Integer
Locus
Members of equations
Minuend
Newton
Notation
Operation
Polynomial
Pure quadratic equations
Pythagoras
Quadratic surd
Quadratic trinomial
Real number
Segment, unit
Signed quantity
Similar terms
Simple equation
Specific gravity
Subtrahend
Supplement
Supplementary angles
Supplementary adjacent angles
Surd
Synthetic division
Systems of equations
Trinomial
Unlike terms
Verify

B. *Mathematicians lower by* 2.0

Angle of depression
Angle of elevation
Applying a formula
Arithmetic average
Bar diagram
Central tendency
Class interval
Compensating errors
Construct

Cosine
Cross products
Curve of normal distribution
Degree of a number
Dependent variable
Diameter
Direct variation
Functions, trigonometric
Inertia of large numbers
Infinity
Interpolation
Inverse variation
Joint variation
Linear function
Logarithm
Mode
Normal distribution
Opposite quantities
Seconds of angles
Series, discrete
Sine
Solving triangles
Tangent of an angle
Trigonometric
Trigonometry

Superfluous Bonds

One of the nearest approximations to an axiom in education that we have is "Do not form two or more habits where one will do as well." Yet the teaching of algebra goes against this axiom in several of its main tasks. Consider the solving of quadratic equations. It is almost universal to teach first the solution by factoring, and to teach pupils confronted by a quadratic to be solved to try first to factor it. Later the pupil is taught to use some form of $\frac{-b}{2a}\pm\sqrt{\frac{b^2-4ac}{4a^2}}$, usually its simplest form $\frac{-b\pm\sqrt{b^2-4ac}}{2a}$. "Completing the square" by a "cut and try" process more or less systematized is often taught as a third intermediate method.

For the purposes of real life the use of the formula serves not only as well as all three together but better. So large a percentage of the quadratics that life will offer are not factorable that it is a waste of time to try.

The reasons why pupils have been and are taught two or three ways of doing one thing are probably: first, the belief that the difficult procedure of solving a quadratic may be made easier to learn by teaching it first in the limited case of a factorable expression where computation and checking are easy; and second, the delight in neat and ingenious manipulations, requiring intelligence and alertness. Neither reason seems adequate. The former is, I think, based on a misunderstanding of the learner's state of mind. He is not dismayed or even much disturbed by the difficulty of believing that $\frac{-b \pm \sqrt{b^2-4ac}}{2a}$ will give the right answers or the difficulty of understanding why it will. What really bothers him is the difficulty of choosing the right numbers to substitute for a, b, and c in the formula. In $x^2+\frac{x}{2}-7=4$ for example, he is disturbed by having apparently nothing to put for a, and may put 2 for b, and not know what the 4 is for, or put 7 or 11 instead of -11 for c. These difficulties are nowise reduced by experience in factoring, with the resulting confidence that an answer can be obtained. On the contrary, the habit of expecting integral answers, and of expecting two answers to be got *via two expressions* may make him less at home with $\frac{-b \pm \sqrt{b^2-4ac}}{2a}$ than he would have been if taught it at the outset. Intelligent confidence in a general formula and understanding of its derivation are hard, doubtless, but no harder without than with the previous exercises in factoring expressions set equal to 0. Checking reduces the former difficulty; and graded training with abstract formulas, the latter.

The teacher's delight in ingenious manipulations requiring intellect and alertness is pardonable. The brighter pupils share it, and may perhaps well be shown how to search for short cuts by factoring after thay have learned the general treatment. As it actually functions in the classroom, the solution of quadratics by factoring is not much nore intelligent or ingenious than substituting in the formıla. The pupils are taught to find the factors by routine ıs far as is possible.

Industry and patience mix with ingenuity and alertness n about the same proportions whether the pupils are trying o find what factors of the first term to try with what factors of the third, or are trying to decide how to put the equation nto the form $ax^2+bx+c=0$ and what in it will then correspond to a, b, and c respectively.

Consider next the treatment of numerical surds. We teach a pupil in some cases to rationalize the surd, in other cases to replace it by its numerical root, and in still others to leave it alone in hope that in the course of time it will be multiplied or divided by itself and cease to trouble us.

Consider replacing our present teaching by the rule, "In computing, replace $\sqrt{2}$, $\sqrt{3}$, $\sqrt{5}$, etc., by 1.414, 1.732, 2.236, etc., whenever you come to them, unless you see some easier way to get rid of them.[1]" This seems barbarous, but after all, if a pupil lacks the wit to see the easier ways of operating with surds, is it wise to try to teach him when to use them? Will not the actual cases of numerical surds that the student of algebra will meet outside of algebra be on the whole managed most successfully by those selfsame students by the one simple rule, even if many pupils always substitute the root? Is it not a gain rather than a loss to leave some leeway for initiative and adaptability here?

It is not so bad for pupils to lose time occasionally by using 1.414 where $\sqrt{2}$ would be better, as for them to be confused and half paralyzed mentally when they see, say, $\frac{1.6\sqrt{2.94}}{6+\sqrt{2}}$, tending to multiply by $\sqrt{2}$, or to multiply by

[1] It is assumed here that the pupil has learned that $\sqrt{a}\times\sqrt{b}=\sqrt{ab}$, that $\frac{\sqrt{a}}{\sqrt{b}}=\sqrt{\frac{a}{b}}$, and that like powers of equals are equal.

$6-\sqrt{2}$, or to multiply by $6+\sqrt{2}$, or to give it up, because they forget what one is supposed to do in such cases.[1]

Short Cuts

When two methods are taught one of which is of general applicability, but somewhat laborious, whereas the other is of limited applicability, but quick and easy within that field, we have the question of standard versus short-cut methods. Ability with the standard method plus ability with the short cuts is presumably better than the standard method alone. The question is whether ability with the short cut is of enough added value to justify the time spent upon it.

Much of the work that used to be done and some of the work that is still done by pupils in algebra consists of learning to make short cuts. Special products, factorizations, and a large part of the work with surds are essentially short cuts. $(a+b)^2=a^2+2ab+b^2$, $(a-b)^2=a^2-2ab+b^2$ and the binomial theorem have notable values over and above their service as short cut multiplications. The best of these other short cuts is that used in $(a+b)(a-b)=a^2-b^2$; $a^2-b^2=(a+b)(a-b)$ and the rationalization of denominators of the form $(\ )^{\frac{1}{2}}-(\ )^{\frac{1}{2}}$ or $(\ )^{\frac{1}{2}}+(\ )^{\frac{1}{2}}$. This is easy to learn, and has a substantial chance of sometime being useful in mathematical, scientific, and statistical work. The next best is probably the factorization of a trinomial of the form $c_1x^2+c_2x+c_3$, when it is the product of two binomials c_4x+c_5 and c_6x+c_7. This saves much time in adding or subtracting with fractions having such a trinomial as denominator to fractions having its binomial factors as denominators, and in solving

[1] A more fundamental improvement is possible by teaching fractional and negative exponents in place of much of the present treatment of radicals, as was suggested in Chapter VII.

quadratic equations where 0 equals such a trinomial. It is also an engaging task for youthful ingenuity to find the c_4, c_5, c_6, and c_7 which will fill the requirements.

We may then take this as a sample of the learning of short cuts that is rather a favored case. Even this case hardly justifies itself to the psychologist. In the actual work of science we meet fractions with trinomial denominators very rarely. A million high-school pupils living to an average age of 50 will probably not meet a hundred thousand of them. Of these hundred thousand probably not a thousand will be susceptible to the factorization in question. Of these thousand, probably not twenty will occur in a computational task having as another denominator, the c_4x+c_5 or c_6x+c_7 that fits. The time that the short cut saves in twenty cases would probably be less than one millionth of the time spent by the pupils in learning it. In the same way it appears that the time saved in life by the ability to solve certain quadratics by this factorization rather than by applying the general formula will be a very, very small fraction of the time spent in learning it.

Against the teaching of the short cut and its use in fractions, fractional equations and the solution of quadratics stands the fact that it forms or strengthens the pernicious habit of expecting algebraic results to "come out even," and of using the attainment of integral numbers and simple literal expressions as a partial check. In order to preserve insight and analysis and practice in using principles from being swamped and thwarted by computational difficulties, it is desirable and even necessary to have much of the pupil's algebraic work be with whole numbers, even multiples, perfect squares, and the like. We must not waste these by using them elsewhere. We should be very cautious in teaching short cuts which work only with certain special combinations of whole numbers, even multiples, and the like.

There is the further psychological objection that learning these short cuts weakens the pupil's sense that one chief purpose of literal numbers is to tell the story of certain real quantities which stand in certain real relations. For convenience and economy of time in learning to compute with literal numbers, much work with a's and b's taken abstractly without reference to any real situation or relation, must be done. Is it wise to add hundreds of exercises made up to secure facility in a short cut?

Finally, as is so often the case, those who introduced the learning of this particular short cut into first-year algebra and those who retain it, probably confuse value as a symptom with value as training. The pupil who excels others in general intelligence and mathematical capacity probably excels others more in finding the factors of a factorable $c_1x^2+c_2x+c_3$ than he does in straightforward computation or uniform application of a formula like $\frac{-b\pm\sqrt{b^2-4ac}}{2a}$. But this does not imply that learning to find such factors is a better training for him than learning the straightforward computation. Still less does it imply that it is better training for the group as a whole.

The only unequivocal merit of this short cut seems to be that it gives a certain extension and added stability to the insight that a or b or c can represent *any* number. One of the main trunk lines of alegbraic ability starts with such simple formulas about "any number" as $a\times a=a^2$, $a\times a\times a=a^3$, $3a+2a=(3+2)a$, continuing on through a long series to such as $a^{-m}=\frac{1}{a^m}$, or b^2-4ac being >0 the roots are real and unequal, or the binomial theorem. The short-cut formulas in general enrich this series. That $c_1a^2+c_2a+c_3=(c_4a+c_5)(c_6a+c_7)$ if $c_4c_6=c_1$, if $c_5c_7=c_3$ and if $c_7c_4+c_6c_5=c_2$, doubtless helps a little.

If the arguments presented hold in the case of finding the factors of a product of $ax+b$ and $cx+d$, they will hold *a fortiori* for such cases as a^3+b^3, a^3-b^3, $a^3+3a^2b+3ab^2+b^3$, $a^3-3ab^2-b^3$, and for $a^2+2ab+b^2$ and $a^2-2ab+b^2$.

A further reason why it is wise to keep the minimum standard course free from these short-cut factorizations is that teachers are tempted, if they teach them, to spend much further time in having attractive puzzles solved by their aid, such as:

1. Simplify
$$2a^2+3b^2-\frac{4a^4+9b^4-x^4-49+12a^2b^2+14x^2}{(2a^2+3b^2+3)-(x^2-4)}$$
2. Factor $6a^2-9ay-15by-20bx-12ax+10ab$
3. Factor $64a^3-125b^3-c^3-48a^2+12a-75b^2c-15bc^2-1$

If they are taught only to the abler pupils and if no exercises involving their use are given save genuine computations actually made in the sciences or in the derivations and proofs of later mathematics where their use is demonstrably profitable, they may be allowed in the elementary course.

L. C. M. and H. C. F.

The computation of a least common denominator as a step in adding or subtracting fractions or clearing an equation of fractions is a short cut that has been elevated into the position of the standard procedure. Its value is problematical. Suppose that one thousand pupils were taught to find the least common denominator, whereas another thousand were taught, in adding and subtracting, to reduce to *any* common denominator that suggested itself to them, and in clearing of fractions to multiply both sides by any denominator, and if the equation was not then in shape for solution, to multiply by any of the denominators remaining, and so on, equal times being spent by the two thousands in learning.

It is a fair question which thousand would succeed best with the fractions they would encounter outside of school. So far as disciplinary value goes the second thousand would seem to have a slight advantage; for the methods taught them emphasize more fundamental principles and leave more room for intelligent choice. Operating by finding the least common denominator, as with the first thousand pupils, is a little more likely to become a mystic rite.

The computation of highest common factor may be considered here though it is not a short, but a very long, cut for the computations that pupils will encounter. Pupils should be taught *not* to find it, but to reduce fractions by canceling out any common factors as fast as they can locate them.

Algebra and Arithmetic

Other things being equal, bonds already formed in arithmetic should be dealt with in the light of that fact. Doubtless many pupils have forgotten, and some have never learned, arithmetic, but it is bad policy to discount completely the work of the elementary school. It is better psychology to take the attitude that the pupils do know arithmetic, and to refer them to the proper parts of their textbooks in arithmetic to review the matter when they do not, and only as a last resort to teach the arithmetic from the beginning. Yet the facts of similar triangles, the hypotenuse rules, and sometimes other matters, are often taught in algebra as if the pupils had never had any previous experience with them. The pupil who naturally expects to learn something new in algebra, may be actually confused and perturbed by the very ease of the learning. What is a review of arithmetic should be clearly stated to be such.

Other things being equal, bonds should not be formed between certain situations and algebraic treatment thereof, when the arithmetical treatment which the pupil already knows is superior. It degrades algebra to invoke it to do what the pupil can do better without it. Yet just this is the stock method of introducing the equation used by almost all textbooks and teachers. For example,

> Six times a certain number equals 18. Find the number.
> Let $x =$ the number.
> Then $6x =$ six times the number
> Since $18 =$ six times the number
> $6x = 18$
> Dividing by 6, $x = 3$
> Therefore the number is 3.

The excuse offered for thus making algebra seem a long, dull way to do things will be, of course, that we need to have very simple cases as introductions, so that the algebraic fact or principle may stand out clearly, undisturbed by any other difficulties. We do need such simplicity, but we should try to secure it with a task which also illustrates the need for and profit from algebraic treatment. Indeed we may say that the difficulty of not seeing why anybody should "Let x," etc., etc., instead of just dividing 18 by 3 is one of the chief difficulties to be avoided. A better problem for the purpose would be:

> Six times a certain number equals $a+2b$. What is the number when $a=10$ and $b=4$?

In the first steps with equations, in the use of mensuration and interest formulas, in ratio and proportion, and in variation, the teaching of today often teaches pupils that algebra is arithmetic done in a harder way, or at least comes dangerously near to such teaching.

Integral Answers as Checks

The connection between the attainment of a simple answer, such as an integral number or a familiar fraction or mixed number (like $\frac{1}{2}$, $\frac{1}{3}$, $2\frac{1}{2}$ or $12\frac{1}{2}$), or a or $\frac{b}{c}$ or $m+n$, and confidence that the answer is right, and the connection between decimal and other non-simple answers and re-computing, should not be formed. So far as numerical results go, this use of "coming out even" as a check would be folly in real life. With literal answers it may not be quite so bad, but is still mischievous.

It is desirable in algebraic work done without pencil and in work made easy in all respects save the principle or process that is being taught, to have very easy computations. Consequently a certain percentage of answers will "come out even." This desirable percentage is however far exceeded by most textbooks, and probably still more by the tasks added by teachers. The facts for a number of texts are given below.

When for emphasis on some principle or process, the computation is deliberately made easy, it is probably wise to announce frankly that "The answers to problems .... to .... are all integers" or the like, so that the pupil will use the check then and not at other times.

TABLE 28

Relative Frequencies in Percents of Numerical Answers[1] which are Integers, Common Fractions, Decimal Fractions, Surds and Imaginaries in Nine Standard Textbooks.

Book	A	B	C	D	E	F	G	H	I	All books combined
Integers....	71.7	71.7	63.7	62.6	61.1	59.1	58.1	49.4	43.0	59.4
Fractions...	23.4	10.6	19.7	15.7	15.6	14.2	15.1	14.6	13.4	15.4
Decimals...	4.1	6.7	9.1	20.7	17.6	17.1	9.0	18.6	40.4	15.8
Surds.......	0.8	9.1	7.5	1.1	5.7	7.7	15.3	13.2	3.2	8.1
Imaginaries.	0.0	1.8	0.0	0.0	0.0	1.9	2.4	4.2	0.0	1.4

[1] The answers are counted from the keys and so exclude many or all of those obtained in work to be done without pencil.

CHAPTER X

New Types of Exercises In Algebra

We very much need experiments on the feasibility of teaching algebraic symbolism and computational technique in subordination to the study of formulas or equations that state general relations. It is desirable that pupils should gain very early a clear and emphatic sense that a, b, c, x, y, z, etc., mean real numbers and quantities. At the present time many of them do not.

It is also desirable that, in the words of the National Committee:

> The primary and underlying principle of the course, particularly in connection with algebra and trigonometry, should be the notion of relationship between variables, including the methods of determining and expressing such relationship. The notion of relationship is fundamental both in algebra and in geometry. The teacher should have it constantly in mind, and the pupil's advancement should be consciously directed along the lines which will present first one and then another of the details upon which finally the formation of the general concept of functionality depends.

Special practice is required for this. As Hedrick says [1921, p. 193]:

> The acquisition of such ideas is a very slow process. It must be begun early in a very simple manner; it must be presented first only by individual instances of a simple and numerical character; it must be fixed in the mind by repe-

tition after repetition, and by instance after instance, until thinking in such terms becomes habitual with the individual. Only in this way can an individual acquire what has been called the "habit of functional thinking."

We also need exercises planned more to stimulate pupils to learn algebra by what they themselves do and to guide their action, and less to test their understanding of matters which they have been told.

Miss Woodyard and Mr. Orleans have worked out in part a series of exercises which are designed to further these two purposes. Some of these are presented here.

B. FORMULA

I

$N=nr$ is the formula for the number (N) of trees in an orchard having r rows with n trees in each row.

In the formula thus written nr means that n is to be multiplied by r.

Thus if $n=7$ and $r=6$, then $N=6\times7=42$; while if $n=10$ and $r=8$, then $N=80$.

Fill in the missing values N for the given values of n and r in Table I.

N	n	r
42	6	7
80	10	8
54	6	9
	5	8
	10	13
	6	12
	14	20
	20	15
	14	12

TABLE I

II

A church is divided by a middle aisle into 2 sections, each having p pews, seating n persons in each pew. The total number of persons (N) that the church will seat is found from the formula $N=2pn$.

a. Explain what $2pn$ means.

b. When $p=25$ and $N=8$, then $N=2\times25\times8=400$. Fill in the missing values of N for the given values of p and n in Table II.

N	p	n
400	25	8
360	20	9
	25	15
	15	10
	24	12
	16	8
	22	8
	24	12

TABLE II

(Exercises III, IV, and V continue the treatment.)

VI

The velocity in feet per second (V) with which a wave is moving is equal to the length of the wave in feet (L) divided by the time (T) of one oscillation (rise and fall) of the wave in seconds. This is expressed by the formula:

$$V=\frac{L}{T}$$

In the formula thus written $\frac{L}{T}$ means "L divided by T." Thus if $L=16$ and $T=8$ then $V=\frac{16}{8}=2$.

Fill in the missing values of V for the given values of L and T in Table III.

V	L	T
2	16	8
	20	12
	20	14
	16	10
	15	10
	24	18
	28	18
	18	10

TABLE III

(Exercise VII is similar to VI.)

VIII

a. The density of a substance is equal to the mass of the substance in grams divided by its volume in cubic centimeters. Write this as a formula using D for density, M for mass, and V for volume.

b. Make a table using the following values for M and V and evaluate D (find what D equals) for these values.

1. $M=258$ $V=100$
2. $M=170$ $V=20$
3. $M=97.9$ $V=11$
4. $M=1.2$ $V=5$
5. $M=31.2$ $V=12$
6. $M=489.5$ $V=55$
7. $M=757.1$ $V=67$
8. $M=1145.9$ $V=53.5$
9. $M=64.2$ $V=4.2$
10. $M=41.89$ $V=5.6$

c. **1.** Does D increase as M increases?
2. Does D increase as V increases?
3. Does D decrease as V decreases?
4. Does D decrease as M decreases?
5. Does D decrease as V increases?
6. Does D increase as M decreases?

(Exercise IX extends this work using $E=\frac{L_1}{L_0t}$.)

X

a. The velocity of the recoil of a gun is found by multiplying the weight of the projectile (in pounds) by the velocity (in feet a second) of the ball at the muzzle of the gun, and dividing the product by the weight of the gun in pounds. Write this law as a formula using

V for the velocity of the recoil of the gun
w for the weight of the projectile
v for the velocity of the ball at the muzzle
W for the weight of the gun

b. Make a table using the following values for w, v, and W and calculate the corresponding values of V.

1. $v = 80$ $w = 1000$ $W = 10{,}000$
2. $v = 400$ $w = 1600$ $W = 44{,}000$
3. $v = 350$ $w = 1400$ $W = 30{,}300$
4. $v = 360$ $w = 1500$ $W = 45{,}000$
5. $v = 125$ $w = 1000$ $W = 11{,}000$
6. $v = 135$ $w = 1200$ $W = 25{,}000$
7. $v = 245$ $w = 1250$ $W = 30{,}000$
8. $v = 180$ $w = 1150$ $W = 22{,}500$

c. **1.** If w increases, does V increase?
2. If v decreases, does V decrease?
3. If W decreases, does V increase?
4. If W increases, does V increase?
5. If W increases, does V decrease?

XI

a. In making a weak solution from a strong one a hospital nurse uses the formula

$$N = \frac{S}{W} - 1$$

in which N means the number of gallons of water to be added for each gallon of the strong solution,

S means the strength of the strong solution
W means the strength of the weak solution

$$\text{If } S = .08 \text{ and } W = .005 \text{ then } N = \frac{.08}{.005} - 1 = 16 - 1 = 15$$

b. Evaluate N for the following table of values for S and W:

N	S	W
	.08	.005
	.05	.02
	.10	.01
	.20	.05
	.40	.18
	.14	.009

TABLE VI

c. **1.** If S increases does N increase?
2. If S decreases does N decrease?
3. If W increases does N increase?
4. If W decreases does N increase?
5. If W increases does N decrease?

(Exercises XII, XIII, and XIV are omitted here.)

XV

a. A brick wall is to be whitewashed. It is L feet long, H feet high and contains 3 windows l feet long and h feet high. Write the formula for the number of square feet (A) in the area to be whitewashed.

b. Evaluate this formula when

A	L	H	l	h
	40	8	6	$3\frac{1}{2}$
	30	$6\frac{1}{2}$	6	3
	25	7	$5\frac{1}{2}$	$2\frac{1}{2}$
	50	10	$6\frac{1}{2}$	4
	45	8	6	$3\frac{1}{2}$

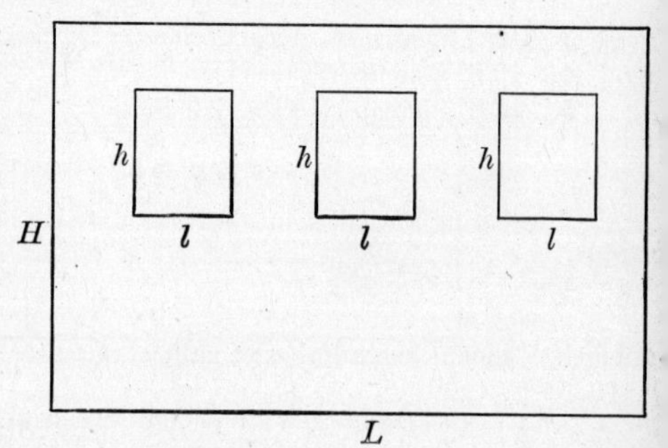

c. If it is desired to express A in square yards when L, H, l and h are given in feet, what change must be made in the formula?

d. Evaluate A in square yards in each of the above examples. Express results correct to 2 decimal places.

e. If L, H, l, and h were given in yards in what units would A be expressed?

f. What is the name given to 3, in this formula?

g. If the wall contained only one window, the formula would be

$$A = LH - lh$$

When the coefficient is 1 it need not be expressed, because any number multiplied by 1 equals the number itself; e. g., b and $1b$ are the same. What is the coefficient of LH? of lh?

C. VARIATION

II

(Exercise I is still simpler than Exercise II.)

$N = nr$ is the formula for the number (N) of trees in an orchard having r rows with n trees in each row. Thus when $n = 10$ and $r = 7$, $N = 10 \times 7 = 70$
when $n = 10$ and $r = 14$, $N = 10 \times 14 = 140$

a. In Table E evaluate N for the given values of n and r.

$n =$	10	10	15	16	20	30	32	15	30	30	15	32
$r =$	7	14	6	11	14	6	22	12	7	14	18	11
$N =$	70	140										

TABLE E

b. In Tables F, G, and H, fill in the missing values of N for the given values of n and r.

$n =$	15	30	16	32	10	10	15	15
$r =$	6	6	11	11	7	14	6	12
$N =$								

TABLE F

c. Study Tables F, G, and H and then write the correct word or words on the dotted lines.

$n=$	10	30	15	15
$r=$	7	7	6	18
$N=$				

TABLE G

$n=$	10	20	16	32
$r=$	7	14	11	22
$N=$				

TABLE H

1. If n becomes twice as large and r remains the same, N becomes........ ....as large.
2. If n remains the same, and r becomes twice as large, N becomes........ ....as large.
3. If n remains the same, and r becomes half as large, N becomes........ ....as large.
4. If n becomes half as large, and r remains the same, N becomes........ as large.
5. If n remains the same, and r becomes three times as large, N..........
6. If n becomes a third as large, and r remains the same, N..............
7. If n becomes three times as large, and r remains the same, N..........
8. If n remains the same, and r becomes a third as large, N..............
9. If n becomes twice as large, and r becomes twice as large, N..........
10. If n becomes twice as large, and r becomes three times as large, N......
11. If n becomes twice as large, and r becomes half as large, N...........
12. If n becomes half as large, and r becomes twice as large, N...........

(Exercises III and IV lead up to Exercises V, VI, and VII.)

V

$D=\frac{M}{V}$ is the formula for the density of a substance in which

D means the density of the substance
M means the number of grams in a given mass of the substance
V means the number of cubic centimeters in that mass.

a. Make a table using the following values of M and V. Evaluate D for these values.

1. $M=258$ $V=100$
2. $M=129$ $V=50$
3. $M=129$ $V=100$
4. $M=258$ $V=25$
5. $M=64.5$ $V=100$
6. $M=43$ $V=50$
7. $M=86$ $V=100$
8. $M=86$ $V=150$
9. $M=172$ $V=150$
10. $M=43$ $V=25$

b. Write the correct word or words on the dotted lines.

1. If M becomes twice as large, and V remains the same, D. . . . .
2. If M becomes twice as large, and V becomes twice as large, D. . . . .
3. If M becomes three times as large, and V becomes three times as large, D. . . . .
4. If M becomes half as large, and V becomes half as large, D. . . . .
5. If M remains the same, and V becomes twice as large, D. . . . .
6. If M remains the same, and V becoms three times as large, D. . . . .
7. If M remains the same, and V becomes half as large, D . . . .
8. If M becomes half as large, and V remains the same, D. . . . .
9. If M becomes three times as large, and V remains the same, D. . . . .
10. If M becomes half as large, and V becomes twice as large, D. . . . .
11. If M becomes twice as large, and V becomes half as large, D. . . . .
12. D becomes twice as large if . . . becomes twice as large and . . . remains the same.
13. D becomes half as large if . . . . becomes half as large and . . . . remains the same.
14. D becomes three times as large if . . . . becomes . . . . as large and. . . . . .
15. D becomes twice as large if . . . . becomes . . . . as large and. . . . .
16. D becomes four times as large if . . . . becomes twice as large and . . . . becomes half as large.

VI

$W=\frac{g}{ab}$ is a formula for the weight (W) of a square inch of metal, if g is the weight of a rectangular piece of the metal a inches long and b inches wide.

a. Make a table using the following values of g, a, and b, and evaluate W for these values. Thus if $g=42$, $a=3$, $b=7$, $W=\frac{42}{3\times 7}=2$.

1. $g=42$ $a=3$ $b=7$

2. $g=84$ $a=12$ $b=7$

3. $g=84$ $a=16$ $b=9$

4. $g=28$ $a=4$ $b=4$

5. $g=35$ $a=5$ $b=2$

6. $g=27$ $a=9$ $b=1$

7. $g=72$ $a=15$ $b=11$

8. $g=56$ $a=10$ $b=4$

b. Write the correct word or words on the dotted lines.

1. If g becomes twice as large, and a and b remain the same, W..........

2. If g and a each become twice as large, and b remains the same, W......

3. If g and b each become twice as large, and a remains the same, W......

4. If g, a, and b each become twice as large, W..............................

5. If g remains the same, and a and b each become twice as large, W......

6. If g and a remain the same, and b becomes twice as large, W..........

7. If g and b remain the same, and a becomes twice as large, W..........

8. If g becomes half as large, and a and b remain the same, W...........

9. If g remains the same, and a and b each become half as large, W.......

10. If g and a each become half as large, and b remains the same, W.......

11. If g and b each become half as large, and a remains the same, W.......

12. If g and a each become half as large, and b becomes twice as large, W............

13. If g remains the same, a becomes twice as large, and b becomes half as large, W............

14. If g and a each become three times as large, and b remains the same, W............

15. If g becomes three times as large, and a and b each become twice as large, W............

16. If g increases while a and b remain the same, W......................

17. If g and a remain the same, while b increases, W......................

18. If g decreases while a and b remain the same, W......................

19. If g and b remain the same while a decreases, W......................

20. If g increases while a and b both decrease, W........................

VII

If W = the weight of a bag of marbles
b = the weight of the bag alone
n = the number of marbles in the bag
m = the weight of one marble

then the formula for the weight of the bag of marbles is: $W = b + nm$.

a. Make a table using the given values of b, n and m and evaluate W. Thus if $b=10$, $n=44$, $m=1\frac{1}{2}$, then $W=10+44\times\frac{3}{2}=76$.

1. $b=10$ $n=44$ $m=1\frac{1}{2}$
2. $b=20$ $n=44$ $m=1\frac{1}{2}$
3. $b=20$ $n=22$ $m=1\frac{1}{2}$
4. $b=10$ $n=44$ $m=\frac{3}{4}$
5. $b=20$ $n=44$ $m=\frac{3}{4}$
6. $b=30$ $n=66$ $m=1\frac{1}{2}$
7. $b=20$ $n=66$ $m=\frac{3}{4}$
8. $b=10$ $n=22$ $m=\frac{3}{4}$
9. $b=20$ $n=33$ $m=1\frac{1}{2}$
10. $b=10$ $n=33$ $m=\frac{3}{4}$

b. Write the correct word or words on the dotted lines.

1. If b increases while n and m remain the same, W....................
2. If b and m remain the same, while n increases, W....................
3. If b decreases while n and m remain the same, W....................
4. If b and n remain the same while m increases, W....................
5. If b and m remain the same while n decreases, W....................
6. If b and n each become twice as large while m remains the same, W............
7. If b and n remain the same while m decreases, W....................
8. If b and m each become twice as large while n remains the same, W...... ....
9. W will become twice as large if .... become twice as large and .... remains the same.
10. If b and n each become half as large while m remains the same, W......
11. If b and m each become half as large while n remains the same, W......
12. If b and n each become 3 times as large while m remains the same, W............
13. W will become half as large if .... become .... as large, and .... remain the same.
14. W will become a third as large if .... become .... as large, and .... remains the same.
15. W will become twice as large if .... become twice as large, and .... remain the same.

(Exercise VIII is similar to VII, but with $L=4j+3g$.)

IX

$A=CD-cd$ is the formula for the area of a stone wall to be whitewashed, the length of the wall being C feet and its height D feet, and there being in the wall a window whose length is c feet and whose height is d feet.

Thus if $C=60$, $D=48$, $c=4$, $d=3$, then $A=60\times48-4\times3=2868$
if $C=60$, $D=24$, $c=4$, $d=3$, then $A=60\times24-4\times3=1428$

a. Evaluate A for the given values of C, D, c, and d in Table P.

$C=$	60	60	30	30	30	60	60	30	30
$D=$	48	24	24	24	24	24	24	48	24
$c=$	4	4	4	4	2	2	4	2	2
$d=$	3	3	3	$1\frac{1}{2}$	$1\frac{1}{2}$	3	$1\frac{1}{2}$	$1\frac{1}{2}$	3
$A=$	2868	1428							

TABLE P

b. Evaluate A for the given values of C, D, c, and d in Tables Q, R, S, T, U, and V.

$C=$	30	60	30	60
$D=$	24	24	24	24
$c=$	4	4	4	4
$d=$	3	3	$1\frac{1}{2}$	$1\frac{1}{2}$
$A=$				

TABLE Q

$C=$	60	60	30	30
$D=$	24	48	24	48
$c=$	4	4	2	2
$d=$	3	3	$1\frac{1}{2}$	$1\frac{1}{2}$
$A=$				

TABLE R

$C=$	30	30	60	60
$D=$	24	24	24	24
$c=$	2	4	2	4
$d=$	$1\frac{1}{2}$	$1\frac{1}{2}$	3	3
$A=$				

TABLE S

$C=$	30	30	60	60
$D=$	24	24	24	24
$c=$	4	4	4	4
$d=$	$1\frac{1}{2}$	$1\frac{1}{2}$	3	3
$A=$				

TABLE T

$C=$	30	60	30	60
$D=$	24	24	24	24
$c=$	2	4	2	4
$d=$	$1\frac{1}{2}$	$1\frac{1}{2}$	3	3
$A=$				

TABLE U

$C=$	60	60	30	30
$D=$	24	48	24	48
$c=$	4	4	2	2
$d=$	$1\frac{1}{2}$	3	$1\frac{1}{2}$	3
$A=$				

TABLE V

c. Study Tables Q, R, S, T, U, and V, and then write the correct word or words on the dotted lines.

1. If C increases and D, c and d remain the same, A....................

2. If D increases and C, c and d remain the same, A....................

3. If c increases and C, D and d remain the same, A....................

4. If d increases and C, D and c remain the same, A....................

5. If C decreases and D, c and d remain the same, A....................

6. If D decreases and C, c and d remain the same, A....................

7. If c decreases and C, D and d remain the same, A....................

8. If d decreases and C, D and c remain the same, A....................

9. If C and c both become twice as large and D and d remain the same, A............

10. If D and d both become twice as large and C and c remain the same, A.............

11. If C and d both become twice as large and D and c remain the same, A............

12. If D and c both become twice as large and C and d remain the same, A............

13. If C, D, c and d all become twice as large, A........................

14. If C and d both become half as large, and D and c remain the same, A............

d. **1.** If C becomes twice as large, and D, c, and d remain the same, does A become twice as large? If not, how does A change?

2. If d becomes twice as large and C, D, and c remain the same, does A become twice as large? If not, how does A change?

3. If D becomes half as large and C, c, and d remain the same, does A become half as large? If not, how does A change?

4. If c becomes half as large, and C, D, and d remain the same, does A become half as large? If not, how does A change?

E. FORMULAS

I

a. If a rectangular sheet of metal c inches long and d inches wide weighs g ounces, what is its weight (w) to the square inch?

b. **1.** What will be the weight (W) of a piece m inches long and k inches wide?

2. What will be the weight in grams to the square inch of this same kind of metal?

3. What will be, the weight in grams (Wg) to the square centimeter?

4. What will be the weight in grams (W) of a piece e centimeters long and f centimeters wide?

(See Tables for equivalent English and metric units.)

c. Evaluate W from the following table of values:

W	w	c	d
oz.	10 oz. per sq. in.	24 in.	12 in.
lb.	8.3 oz. per sq. in.	20 in.	18 in.
gm.	132 gm. per sq. cm.	16 in.	12 in.
gm.	160 gm. per sq. cm.	42 cm.	25 cm.
lb.	12.6 oz. per sq. in.	75 cm.	45 cm.
oz.	12 oz. per sq. in.	45 cm.	30 cm.
oz.	150 gm. per sq. cm.	2 ft.	$1\frac{1}{2}$ ft.

II

If there are s subscribers in a telephone system, the number of connections (N) a central office can make is $\frac{s(s-1)}{2}$.

a. Write the formula as a statement.

b. Evaluate when $s=100$; 250; 500; 1000; 480; 291; 100,000.

N	s

c. Make a table to show changes in N when $s=0$, 10, 20, 30, 40, 50, 60, 70, 80, 90, 100; thus:

s	0	10	20	30	40	50	60				
N	0	45	190	435							

Increase in N	45	145	245								

d. Make another table to show the changes in N when $s=100, 200, 300, 400, 500$; thus:

s					
N					

Increase in N				

III

When piles are driven into the ground by the ram of a piledriver, the weight each pile will bear without sinking or breaking, is given by the formula

$$L=\frac{Wh}{d(W-P)}$$

where L = the greatest load or weight in tons that the pile will bear
W = the weight of the ram in cwt.
h = the height in feet through which the ram falls in driving the pile
d = the distance in inches the pile was driven in by the last blow of the ram
P = the weight of the pile in cwt.

Evaluate L when

L	P	h	d	W
	6	4	$1\frac{1}{2}$	15
	7	$5\frac{1}{2}$	2	10
	5	$4\frac{1}{2}$	$2\frac{1}{2}$	$7\frac{1}{2}$
	6	5	2	8
	8	$4\frac{1}{2}$	$1\frac{1}{2}$	10

IV

When a building is heated it loses heat through the walls to the colder outside air. The formula used to compute the loss of heat from this cause is

$$H = t\left(\frac{nc}{55} + g + \frac{W}{4}\right)$$

in which H = the loss of heat per hour in British Thermal Units (B.T.U.)
t = the difference between temperature inside and outside in degrees Fahrenheit
n = the number of times the air is changed in an hour
c = contents of the building in cubic feet
g = the area in square feet of the glass in the outside walls
W = the area in square feet of exposed wall surface.

a. In a 10-story building 60 ft. × 100 ft. in size and 150 ft. high with 40 windows on each floor, each 4 ft. × 8 ft. in size, the air is changed once every hour. On a winter day the inside temperature is kept at 65°, while the outside temperature is 20°. What is the loss of heat per hour in B.T.U.?

b. In a 22-story building 80 ft. × 120 ft. × 325 ft. high with 50 windows on each floor, each 4 ft. × $7\frac{1}{2}$ ft. in size, the air is changed once every hour and a half. What is the loss of heat per hour in B.T.U. on a day when the inside temperature is kept at 65°, while the outside temperature is 15°?

V

The horsepower used by a steamer depends on the speed of the steamer, the area of the cross section of the ship, and its wetted surface. The formula used to compute it is

$$H.P. = .0088s^3(.05A + .005S)$$

in which

$H.P.$ = the horsepower used
s = the speed in knots per hour
S = the wetted surface in square feet
A = the cross section of the ship in square feet.

a. Find $H.P.$ when

$H.P.$	s	S	A
	10	200	1800
	12	240	2000
	9	185	2500
	8	150	1500
	16	150	1500

b. Show that the formula

$$H.P. = .00044s^3\left(A+\frac{S}{10}\right)$$

is the same as the formula just given.

c. **1.** What effect on $H.P.$ does doubling s have while A and S remain the same?

2. Would doubling S have the same effect on $H.P.$ as doubling s?

3. Would doubling A have the same effect on $H.P.$ as doubling s?

4. What must be done to S and A to have the same effect on $H.P.$ as results from doubling s?

VI

The horsepower of an automobile engine is given by the formula

$$H.P. = KND(D-1)\left(\frac{S}{D}+2\right)$$

in which

$H.P.$ = the horsepower of the engine
K = a constant depending on the kind of car (for touring cars $K = .197$)
N = the number of cylinders
D = the diameter of the cylinders in inches
S = the length of the stroke in inches.

a. What is the horsepower of an engine in a touring car if

$H.P.$	N	D	S
	4	4	5
	6	$4\frac{1}{2}$	6
	8	4	$5\frac{1}{2}$
	8	$4\frac{1}{2}$	6

b. Show that the formula

$$H.P. = KN(D-1)(S+2D)$$

is the same as the above formula.

c. **1.** What is the effect on $H.P.$ of doubling N?

2. What other letter or letters if doubled will have the same effect on $H.P.$ as doubling N has?

VII

The pressure carrying the draft up a chimney is computed by the formula

$$P=H\left(\frac{7.6}{460+T}-\frac{7.9}{460+t}\right)$$

in which P = the pressure (measured by the height in inches of a column of water which it will support in a tube)
H = the height of the chimney (in feet)
T = the temperature of the outside air (Fahrenheit)
t = the temperature of the inside air (Fahrenheit).

a. Find P when

P	H	T	t
	150	60	600
	200	32	650
	125	20	550
	250	0	1000
	225	10	800
	200	−10	600

b. Show that the given formula is the same as

$$P=7.6H\left(\frac{1}{460+T}-\frac{1.039}{460+t}\right)$$

or

$$P=7.9H\left(\frac{.962}{460+T}-\frac{1}{460+t}\right)$$

c. **1.** If H is doubled, what happens to P (T and t remaining unchanged)?

2. If H becomes half as large, what happens to P (T and t remaining unchanged)?

3. Does a change in the value of T make a correspondingly large change in the value of P?

4. Does a change in the value of t make a correspondingly large change in the value of P?

5. If T becomes larger, what is the effect on P?

6. If t becomes larger, what is the effect on P?

7. If t becomes smaller, what is the effect on P?

8. If it is important to increase P what are the ways of doing so? Which would be easiest to do?

9. If it is important to decrease P what are the ways of doing so? Which would be easiest to do?

VIII

a. Door plates are to be cut from a sheet of metal weighing c ounces to the square inch. The plates are to be rectangular, 3 inches by 5 inches in size. They are to have a hole in each corner $\frac{5}{16}$ of an inch square.

1. What is the weight (w) of a single plate?

2. What is the weight by the dozen?

3. What is the weight by the hundred in pounds?

(Give answers correct to 2 decimal places.)

b. If the plates are to be r inches by s inches and the holes are to be $\frac{1}{6}$ of an inch square, what is the weight in pounds by the hundred?

c. **1.** If the plates are to be square, p inches on a side, and the holes are to be circles, t tenths of an inch in diameter, what is the weight (w) in ounces by the single plate?

2. In pounds by the hundred?

IX

In a business, the "turnover" of the business is found from the formula

$$T=\frac{C+\dfrac{C}{n}}{t}$$

in which T = the turnover

C = the capital invested

$\frac{1}{n}$ = the fraction of a dollar that can be borrowed on each dollar of capital

t = the time in months required for turnover.

a. Find T when

T	C	$\frac{1}{n}$	t
	50,000	$\frac{2}{3}$	2
	250,000	$\frac{9}{10}$	6
	100,000	$\frac{3}{4}$	3
	10,000	$\frac{2}{3}$	2

b. **1.** If t increases while C and n remain the same, does T increase or decrease?

2. If C increases while n and t are constant, does T increase or decrease?

3. If n increases while C and t are constant, does T increase or decrease?

4. By what changes in C, n, and t can T be made half as large?

(Exercise X is of the same sort as XI, but with $C=\dfrac{E}{\frac{r}{n}+R}$.)

XI

The pressure of water on a pipe that will not burst the pipe is called the safe working pressure. In cast iron pipes, such as carry water in a city water system, the safe working pressure is given by the formula

$$P=\frac{7200}{D}\left[T-\frac{1-\frac{D}{100}}{3}\right]-100$$

in which P = the pressure per square inch
D = the inside diameter of the pipe in inches
T = the thickness of the iron shell in inches.

a. Find P when

P	D	T
	6	$\frac{3}{4}$
	10	$\frac{11}{16}$
	48	$2\frac{1}{8}$
	24	1
	30	$1\frac{1}{8}$
	36	$1\frac{1}{2}$

b. Show that the given formula is the same as

1. $P=\dfrac{7200}{D}\left[T-\dfrac{1}{3}+\dfrac{D}{300}\right]-100$

or

2. $P=\dfrac{2400}{D}\left[3T-1\right]-76$

c. **1.** In what ways can the pressure be decreased?

2. In what ways can the pressure be increased?

CHAPTER XI

THE ARRANGEMENT OF TOPICS IN ALGEBRA

The bonds to be formed having been chosen, we inquire concerning the best order in which to form them — the order providing the maximum of facilitation and the minimum of interference, and the best opportunity for the development of an effective algebraic ability.

Certain principles of order are axiomatic, as when one set of bonds demands another as a prerequisite. For example, the solution of a set of linear equations must follow the solution of a single linear equation. Certain principles are almost certain, as when the teaching of literal exponents is delayed until after a wide experience with literal symbolism and algebraic generalizations has been given. With these axioms and high probabilities, already understood and used by competent teachers, we need not deal, but may proceed at once to certain principles suggested by psychology and classroom experiments which seem likely to improve teaching.

Teachers tend to favor systematic arrangements whereby a subject of study is divided up into topics, and these into subtopics, to be taught one at a time and in a certain order. They are intellectual, and intellectual persons favor such systems; also it is an advantage to them to think of algebra in that way. An outline like those following enables them to know quickly what their task is and how far they have gone with it.

I

The Formula: General Number.
The Equation.
Positive and Negative Numbers.
Addition and Subtraction of Literal Expressions.
Multiplication and Division of Literal Expressions.
Linear Equations: Problems.
Special Products and Quotients.
Factors. Multiples. Equations Solved by Factoring.
Fractions.
Fractional Equations. Problems. Formulas.
Proportion. Variables.
Systems of Linear Equations.
Square Root. Quadratic Surds.
Quadratic Equations.
Systems Involving Quadratic Equations.
Exponents.

II

Negative Numbers.
Algebraic Expressions.
Addition.
Subtraction.
Multiplication.
Division.
Simple Equations with One Unknown Quantity.
Special Products and Quotients.
Factors.
Fractions.
Simple Fractional Equations.
Ratio, Proportion, and Variation.
Simultaneous Simple Equations.
Graphs.
Powers and Roots.
Quadrative Equations.
Simultaneous Quadratic Equations.
Powers and Roots Completed.
Progressions.
The Binomial Theorem.
Logarithms.

An order which is excellent as a means of arranging algebraic abilities for contemplation, or for keeping track of whether they are learned, may not be a good order in which to acquire them. The outline which we can survey in a few minutes is spread out over a year for the learner. We who know what is to come can use a system which is valueless to him who has yet to learn what is to come. What he most needs is a system and order that is good to learn by not to look at.

To us who know algebra, it is natural to think of monomial as a name for expressions like $3ab^2$ that should be learned along with the expressions, but to the pupil monomial is probably largely an extra burden until he comes to compute and work with polynomials.

The same can be said for practically all definitions. The usual practice is to begin a chapter containing a new topic with complete logical statements of the meaning of all the new concepts to be dealt with therein. For instance, when parentheses first occur, usually after subtraction, definitions are given for parenthesis, brace, bracket, vinculum, and symbols of aggregation, with perhaps an historical note concerning their origin. The first time that equations appear in the older texts they are introduced by definitions of some or all of the following: (a) equation, (b) identity, (c) equation of condition, (d) unknown quantity, (e) roots of an equation, (f) right and left members, (g) solution of an equation, (h) axiom, (i) formal statement of the first four axiomatic operations upon equations (however, the axioms of raising to powers and taking roots often are never stated in the book). The chapter on radicals is likely to define (a) real number, (b) imaginary number, (c) rational number, (d) irrational number, (e) radical, (f) surd, (g) mixed surd, (h) entire surd, (i) index of radical, (j) order of radical, (k) radicand, (l) principal root, (m) fractional exponent, (n) simplification of radicals, (o) like radicals, (p) unlike radicals, (q) radicals of same order, and (r) rationalization, before any manipulation is given. The pupil's mind is thus overloaded with new concepts which he does not really understand until he has succeeded in performing most of the examples under the topic.

To acquire meanings through use, then to have the definitions as models of accurate, economical statement of these meanings would help the learning of algebra and incidentally be of advantage in the mastery of English expression.

After a person knows algebra it is natural to think of a before $4a$ or $6a$, and of $a+a=2a$ before $3a+5a=8a$, and to think of a as simpler than $4a$ or $6a$. This is, however, because the person knows so well that a means $1a$. To the learner $4a$ or $6a$ is easier to understand than a, and $3a+5a=8a$ is a better first problem in addition than $a+a$.

It is an easier thing for the beginner to see meaning and value in 2 times $3a$ than in 2 plus $3a$. Certainly a times a is harder than a^3 times a^4 and $3a$ plus $4b$ is harder than $3a$ times $4b$. It is an even chance that $3b$ times $4b$ is easier than $3b$ plus $4b$. The child sees that *something* happens to the b's as well as to the 3 and 4 in the former, while in the latter he must think them separately in some such fashion as "something to do with 3 and 4 but nothing to do with the b's except annex one of them to the answer — where the other one goes I don't know." It is amazingly difficult for the person skilled in algebra to understand the interference these opposing ideas set up in the ordinary child's mind; even few teachers remember to clear up his wonder as to what happens to the other b.

Certain of the items which are recognized as difficult places for the average pupil, where his wrong learning must be undone (if it ever is) only at the price of much toil for the teacher and discomfort for himself, become much easier with a changed order of learning. To select a case, the unhappy tendency of the learner to cancel terms rather than factors in the reduction of fractions to lowest terms (as when he reduces $\frac{x^2+8x+15}{x^2+4x+3}$ thus, $\frac{\overset{2}{\cancel{x^2}+\cancel{8x}}+\overset{5}{\cancel{15}}}{\cancel{x^2}+\cancel{4x}+\cancel{3}}$, and arrives at $2+5$ or 7 for answer) is not the outcropping of original sin, but a highly natural application of the carefully learned habit of reducing fractions with monomial terms. (In such an example as $\frac{35a^4x^2y^3}{21a^2x^5y^4}$ he is taught to get his answer thus $\frac{\overset{5a^2}{\cancel{35}\cancel{a^4}\cancel{x^2}\cancel{y^3}}}{\underset{3\ \ x^3y}{\cancel{21}\cancel{a^2}\cancel{x^5}\cancel{y^4}}}=\frac{5a^2}{3x^3y}$.) There is no necessity in the nature of the material why reduction of fractions with polynomial terms should not be taught first and reduction of fractions with monomial terms be made subordinate to it. So far as our experimentation goes, the results confirm the implications of this statement.

An equally prolific source of error for many pupils lies in the minus sign preceding a fraction in an equation. So long as the minus parenthesis is made a subhead under subtraction great care will be required to prevent and correct such errors. Numerous experimenters besides ourselves have found that making the general topic of removing parentheses subordinate to multiplication rather than to addition and subtraction decreases the difficulty.

A certain courage in considering and experimenting with arrangements which seem likely to aid learning is thus desirable even if they seem "scrappy" and "confused" to the person who already knows algebra.

There are doubtless many orders or arrangements, all of which are tolerable, and perhaps nearly equally effective. The order which seems to us the most promising on psychological grounds is shown on the pages following.

DIVISION A: THE FORMULA

(With such concepts and computation and manipulation as are needed.)

1. Easy formulas.
   a. To understand.
   b. To evaluate.
   c. To make.
2. Formulas with parentheses and complex fractions. (As above.)
3. Formulas containing radicals. (As above.)
4. Formulas with abstract quantities. (As above.)
5. Formulas with negative numbers. (As above.)
6. Formulas: to "change the subject."

DIVISION B: THE PROBLEM

Any quantitative issue can be solved by framing a suitable special formula or set of formulas and solving.

7. Genuine problems naturally treated as having one unknown.

8. Genuine problems naturally treated as having two or more unknowns.
   a. Solved by substitution.
   b. Solved by addition or subtraction (treated as a short cut.)

DIVISION C: RELATED VARIABLES

9. Statistical graphs.

a. To understand and interpret graphs on concrete real issues, e. g., cost of living, growth of population, practice curves, in $++$ quadrant only.

b. To make such graphs.

10. Mathematical graphs.

a. The coördinate system, with extension to $-+$, $--$ and $+-$ quadrants.

b. Important curves, e.g., $y=Kx$, $y=K$ per cent of x, $y=x^2$, $y=\sqrt{x}$, $y=x^3$, $y=\sqrt[3]{x}$.

11. Direct and inverse proportion.

a. Equation $y=kx$ and graph.

b. Equation $y=k/x$ and graph.

12. The general linear graph.

a. Equation $y=kx+b$.

13. Simultaneous linear equations.

a. To find constants for the general linear equation $y=kx+b$, given two sets of values of x and y.

14. Square root.

a. Of numbers.

b. Of trinomials.

15. Trigonometric ratios.

a. Tangent, sine, cosine of angles in 1st quadrant.

16. Logarithms.

a. For ease of computation.

b. Slight theory of logs. based upon graph of $y=10^x$.

17. Quadratic equations in 1 unknown.

a. By graph.

b. By completing the square.

c. By formula.

18. The general notion of variation: summary and systematization.

DIVISION D: ABSTRACT FORMULAS

19. The series.
    a. Arithmetic.
    b. Geometric.
20. Fractional and negative exponents.
21. The binomial theorem.
22. Abstract formulas.

This order of presentation differs from the customary arrangements chiefly as follows:

Symbolism and computation are made instrumental to the understanding, evaluation, framing, and transforming of formulas. Fractions and radicals are introduced early, evaluation of formulas takes the place of solving many numerical equations and changing the subject of the formula replaces the usual work in abstract literal equations. The straight-line graph follows a number of curvilinear graphs. The chief use made of simultaneous first degree equations is for determining the constants in the linear equation of two variables, $y=ax+b$, two sets of values for x and y being known. Similarly the chief use of the quadratic expression ax^2+bx+c is not to solve the equation $ax^2+bx+c=0$, but to determine the constants when $y=ax^2+bx+c$, three paired values of x and y being given. The solution of simultaneous first degree equations in three unknowns thus becomes important.

The early use of simplification of radicals in the two useful cases (e.g., $\sqrt{18}$, $\sqrt{\frac{2}{3}}$) is in line with the attempt to attain early mastery of the most widely useful parts of algebra. These operations with easy radicals profitably can replace the customary manipulation of complicated polynomials in the early study of algebra. Such an ordering permits genuine applications of algebraic technique in work dealing with diagonals of squares and rectangles, and the altitudes of triangles, and facilitates computation of trigonometric ratios for given angles.

It seems probable that much of the practice given in the past in the solution of quadratic equations was misplaced. The field of usefulness (outside the algebra text) of the quadratic equation is much more to be found in graphic representation and in evaluation of $y=ax^2+bx+c$ than in the solution of the special case $ax^2+bx+c=0$.

We shall not present here the psychological facts which have led us to approve this order. In many cases, they involve a rather elaborate balancing of goods and ills. One of us is engaged in discovering its merits and defects by actual experimentation. We shall, however, consider these facts in the case of two of its features, one where it departs widely from universal custom, and one where it adheres to the older custom against the example of both Rugg and Clark and Nunn.

Problem Solving as a Unit versus Problem Solving as a Series of Applications of all the Techniques Learned

The customary order makes use of problems to apply and give drill in the technique learned in each new topic. Instead, we have grouped them as a whole between the group on the formula and that on the relationship of variables. Certain considerations have led to this decision.

First, we wish to make sure that the primary educational value of the verbal problem seeking a single answer rather than a general rule is realized. This value we take to be the assured knowledge that any quantitative issue solvable from the data given can be solved by putting the data in a suitable equation or set of equations, and performing the computations needed to solve them; and further the habit of trying to frame such an equation or equations when a simpler means of guiding computation aright does not suggest itself. This value seems much more likely to be attained by concentrating most of the practice under the topic of problem solving than by scattering it under linear equations, fractional linear equations, simultaneous linear equations, quadratic equations, equations involving radicals, and so on.

Second, the use of problems after each technique as an application of it is likely to give so much direction in respect to the sort of equation to be framed as seriously to impair the value of the problems. The pupil working a certain group of problems almost knows that the equation must be a quadratic, or that an arithmetical series is to be used, or that he is expected to make two equations.

Third, organization by the situations involved is generally recognized as superior to organization by the character of the equations framed. Treating problem solving in a division by itself permits one to follow this principle without the customary restraints.

Fourth, we have a higher regard for the making of *general* equations or formulas or relation lines and a somewhat lower regard for the making of *special* equations to solve the typical verbal problems, than is usual. We would secure significance for computations and practice with computations by such general equations and graphs more often and by the typical special verbal problems less often. "Thinking outcomes" from the study of algebra are highly desirable, consequently such material as is real and applicable outside school should be sought and utilized. Such matter is to be found not only in problems leading to particular equations, but in data from which to make and handle general formulas and graphs. That these two abilities will yield as great disciplinary returns on the investment of effort seems likely. Ability to interpret a graph correctly and ability to use a formula with precision are as valuable and as indicative of the scientific attitude of thought as ability in the customary problem solving.

Fifth, there are not enough genuine, comprehensible, and serviceable problems to parallel the techniques of intricate fractional equations, quadratic equations, equations with radicals, and the progressions, by series of problems applying them. Of the five hundred or so verbal problems given in a standard manual, not over a hundred are such as will ever occur and require an algebraic solution in real life. Genuine problems demanding algebraic treatment do occur in statistics, economics, physics, chemistry, mechanics, surveying, engineering, and the like; but these are seldom comprehensible to first-year pupils. All the genuine problems concerning states of affairs which the pupils can comprehend can be included in the course outlined here.

The concentration of most of the problem solving of the customary sort in one division will not, of course, prevent the use of problems as introductions, explanations, evidences of utility, and the like, later, wherever they seem profitable for learning. We are not eliminating problems from other divisions but are eliminating their systematic insertion in the customary routine: technique A, technique A applied to problems; technique B, technique B applied to problems; and so on.

The Placement of Negative Numbers

The proper place for the introduction of the negative number has been the subject of some experimentation.

Nunn [1914, p. 53] makes a case for delaying this extension of the number series until after the following topics have been presented: making, interpretation and evaluation of formulas; factorizations (common monomial and difference of 2 squares) and the converse expansions; easy fractions; changing the subject of a formula; functionality as found in direct and inverse proportions; using the trigonometric ratios especially as applied to navigation questions; graphic representation and interpretation; square root; radicals (such as $\sqrt{8}, \frac{1}{\sqrt{2}}$); approximation formulas for the squares and cubes of the sum and difference of two numbers and the converse square and cube roots; calculation of mean, median, and quartile in statistical data.[1]

Rugg and Clark [1918, p. ix], after experimentation in classes, delay the introduction of the negative number until the second semester's work, presenting previously these topics: literal notation as an aid to problem solving; easy equations in one unknown and problems leading to such equations; use and evaluation of algebraic expressions and formulas; making and reading scale drawings using ruler and protractor; solving for unknown lengths proportions derived from similar triangles; the tangent and cosine ratios used with the Pythagorean theorem for solving triangles; reading and making statistical tables and graphs with computation of mode, median, and mean; and the functional linear graph with its equation or formula, $y=ax+b$.

The evidence as presented for this delay in introducing negative numbers seems inadequate. So many other factors

[1] In connection with Nunn's location of the negative number it should be remembered that a child in England using his text would begin the study of algebra at eleven or twelve years of age and carry it on in two or three lessons a week parallel with the study of geometry up to the age of sixteen, whereas our children begin algebra at about fourteen and study it for a year five lessons per week.

enter into any experiment of this type that, until the partial correlations are published, the completeness and trustworthiness of the proof will be a matter of doubt.

The following reasons lead us to prefer a location of this topic more in line with the traditional position. First, there are many generalizations and statements made into formulas which are only half truths until the comprehension of the negative number permits them to run the entire gamut of values. The painstaking care that must be exercised to make evaluation possible when arranging numbers for substitution in a formula in which a minus sign occurs tends to rob such expressions of reality. To illustrate: "i. Write a formula for the distance apart after a given time of 2 cars which start from the same point and travel opposite ways at a given speed (d, t, s_1, s_2); ii. The same, the cars being already a certain distance apart to begin with; iii. The same, the cars starting from the same point but in the same direction; iv. The same, the cars going in the same direction and the faster being a given distance ahead of the slower at the beginning; v. The same, the slower car being a given distance ahead of the faster at the beginning." [Nunn, 1914 b, p. 19.] The last result has fuller meaning in case the faster overtakes the slower and passes it, if the child understands the negative number.

Second, the failure to introduce the negative number reasonably early often makes necessary the writing of two formulas where one should suffice. Numerous instances to support this contention are available. To choose an easy one from Nunn: "i. A boy sells a model flying machine and a number of white mice and spends most but not all of the proceeds in buying rabbits. Write a formula for the money left over, given the sum he receives for the flying machine, the prices of a mouse and a rabbit, and the numbers sold and bought (R, f, m, r, n_1, n_2). ii. Give a formula for the money he still requires if the cost of the rabbits is greater than the proceeds of his sale." To make two formulas grow where but one is needed does not benefit the pupil.

In the third place, an interference factor is to be reckoned with when a child taught to make and use numerous algebraic expressions on the assumption of positive numbers only is required to readjust to the concept and use of negative numbers in the same or similar expressions. In case he has so learned his algebra that it yields him satisfaction as a tool of operation, his resentment at the new upsetting doctrine will be pronounced — and possibly the keener the more diligently he has studied the earlier algebra. To have mastered the method of algebraic addition, multiplication, factorization, and treatment of fractions, on a basis of positive numbers, renders relearning the same topics with negative numbers a doubly disagreeable task. Even to the teacher it will tend to be boring to supply sufficient drill matter to habituate the pupil in these processes using negative numbers if the processes have been well taught earlier in the course with positive numbers.

Again, there is on the part of most pupils a readiness to acquire new points of view at the beginning of the course in algebra of which it is worth while to take advantage. By most American children a new subject, a new teacher, a new classroom, a new textbook, a new group of fellow students are met in one happy readjustment. All surroundings thus conspire to amenability to revolutionary doctrine concerning what has seemed one of the fixed items of the universe of abstract truths, namely that numbers start at 1 (for a few children, at zero) and progress by unit increase to a very large sum. To take advantage of this flood tide in the affairs of pupils seems sound psychology.

For these reasons in our judgment the fundamental concept of the negative number should be taught as soon as the usage of literal notation is mastered — probably about the sixth week of the first semester.

The chief reasons for delaying the treatment as Nunn and Rugg and Clark do are: (1) that two great contributions of algebraic learning — symbolism and the relation of one variable to another — can be taught without the use of negative numbers; (2) that these ideas are more likely to be mastered if the pupil is without an added burden and possible source of mystification; and (3) that both these ideas, and the extension of them and of all computational techniques to include negative numbers, will profit by double treatment of much of algebra, first with unsigned numbers, and later in a more general manner with signed or "directed" numbers.

We appreciate the weight of these reasons, and would at once admit that an excellent brief course could be given without generalizing to signed numbers at all, with all graphs and equations in the + + quadrant. Of the three things — symbolism, the treatment of related variables and the extension of computation to include signed numbers — we also would rate the third as the least valuable. If it were really too great a burden and confusion to begin work with all three in the first three months of the course (as given in American schools) so that one must be delayed, we also should delay the third. It seems to us, however, that sufficient care and ingenuity in a graded treatment will reduce the burden and confusion to amounts so small that the gain from the greater generality of treatment will outweigh them. In particular, the burden will be lightened if proofs that are beyond his comprehension are replaced by the principle and experience that the laws of signs are right because they always give the right results [as in (100+2) (100−2), checked by (102×98)]; and if only genuine, needed uses are brought forward; and if exercises based on credits and penalties in school tests and on deviations plus and minus from standards for ages, grades, and the like in school tests are used in illustration and application. Some of the difficulty with signed numbers is due to our explanations, and some of the confusion is due to our requiring the pupils to use them in cases where he would naturally and properly use plain arithmetic.

CHAPTER XII

The Strength of Algebraic Connections

As things are now, pupils lack mastery of the elements of algebra. The extent to which this is the case can be understood and appreciated best by the consideration of actual test results. The tasks shown in Table 29 were the first twenty-eight of forty making a test for which from ninety to one hundred minutes was allowed and which could be done by first rate algebraists in twenty-five minutes without errors save an occasional lapse.[1] Few or no complaints were made about insufficient time, and almost all the pupils attempted all of these twenty-eight tasks, and others beyond them. The schools were either private schools with excellent facilities, or public high schools in cities which rank much above the average of the country in their provision for education. In both cases the pupils would, beyond question, be superior to the average of second-year high-school pupils in general intellect and capacity for mathematics. All the pupils had studied algebra for at least one year. Most of them were continuing their study of it at the time the test was given (in October and November and December, 1921).

It does not seem an exaggeration to say that, on the whole, these students of algebra had mastery of nothing

[1] Four forms of the test were used, different in the concrete details of each task, but constructed on the same plan and of almost exactly equal difficulty. The results where forms B, C, and D were used may therefore be safely used to make the percentages more reliable; and this has been done.

TABLE 29

PERCENTAGES OF WRONG RESPONSES AND FAILURES TO RESPOND TO TWENTY-EIGHT TASKS IN TEN SCHOOLS

Task	School									
	A	B	C	D	E	F	G	H	I	J
1. $(5a^2+4-3a^2)+(a^2-2-7a^2-5)$	11	4	10	7	23	18	28	54	46	47
2. $(-2a^3-10a-4a^3)+(5a+3a^3-4a)$	14	19	4	7	20	25	35	45	44	51
3. From $3a+4b$ subtract $5a-9b-3c$	6	9	8	7	17	17	35	35	52	65
4. From $5a-b-2c$ subtract $3c-3a$	4	6	7	10	9	18	51	32	59	65
5. $8a+8b-(3a+6b)$	4	9	7	12	11	17	39	38	53	63
6. $(5d-e)-(7e+2f)$	14	11	12	21	20	28	42	41	66	61
7. $7d \times 2de^2$	7	2	1	0	6	11	26	30	28	26
8. $de^2 \times d^2e$	9	9	8	2	8	24	28	38	48	37
9. $8-5\ (d+2)$	10	6	24	21	20	34	29	61	80	63
10. $4e^2+e\ (-4e-3)$	10	6	17	14	14	30	43	60	73	70
11. $5np-3p\ (4n+3p)$	14	9	21	14	15	28	36	58	74	61
12. $m+\frac{8\,m^2n^2}{mn}$	39	11	31	26	20	44	72	71	75	81
13. $4m+\frac{2m^2n\,p^3}{mp}$	39	26	36	45	23	40	78	80	88	84
14. $\frac{m^2np}{mn}-\frac{m^2n-mp}{m}$	47	30	44	45	42	62	87	84	94	95
15. $(m^5n)\ (m^2n^3)$	3	2	9	5	5	23	39	58	52	35
16. $(2a-7)^2$	4	6	7	14	6	13	26	34	46	49
17. If $a=2$, and $b=3$, what does $5a^2-2ab$ equal?	11	0	8	5	15	18	32	44	62	51
18. If $a=.7$, and $b=1.2$, what does $2a^2-5ab$ equal?	57	45	46	50	63	64	80	78	91	91
19. If $a=1$, $b=2$, $c=.4$, and $d=100$, what does a^2-b+cd equal?	44	26	36	36	32	54	71	69	90	84
20. If $a=12$, $b=6$, $c=5$, $d=3$, and $e=1$, what does $\frac{22}{7}[ab+c(d-e)]$ equal?	59	30	42	33	37	63	71	74	93	84
21. If $d=2$, $e=3$, $f=4$, what does $\frac{\frac{df}{d+e}}{ef}$ equal?	16	9	17	19	22	28	48	63	81	77

TABLE 29—Continued

Task	School									
	A	B	C	D	E	F	G	H	I	J
22. $d=\frac{ef}{g}$. What does f equal?	14	15	13	21	31	35	58	70	91	86
23. $\frac{e}{W}=\frac{r}{R}$. What does W equal?	31	17	9	21	35	28	67	80	95	91
24. $\frac{PV}{T}=\frac{P_1V_1}{T_1}$. What does V equal?	40	30	12	21	49	44	77	82	96	91
25. $4q=7q+5$. What does q equal?	17	19	9	17	26	33	46	63	71	63
26. $15=7w-4$. What does w equal?	13	9	7	10	25	26	33	55	58	49
27. $\frac{V}{4}=a-2$. What does V equal?	10	2	5	5	22	19	49	62	74	74
28. $\frac{6}{2-u}-\frac{11}{3-u}=0$. What does u equal?	36	21	28	50	32	54	62	89	95	98

whatsoever. There was literally nothing in the test that they could do with anything like 100% efficiency. If they had been asked to add $3a$ to $7a$, or to multiply $3b$ by $2b$, we might have had nearly perfect records, but that would not have meant mastery of $3a+4a$ or $3b\times2b$. Complicate the situation slightly, as in $7d\times2de^2$ or $de^2\times d^2e$ or $4e^2+e(-4e-3)$, which are Nos. 7, 8, and 10 of the test, and the pupils fail.

These results are supported by the findings in all tests of algebraic abilities that have been published. They have not stood out in such clear relief before, since in some of the previously given tests the pupils have been urged especially to speed, and in others the tasks have been more elaborate, complex, and difficult than those which we used. Since the communities and schools which share voluntarily in educational tests and experiments tend to be far above the average in intellectual abilities and in educational wisdom and devotion, it is safe to assume that the results from Monroe, Rugg and Clark, Hotz, and Douglas represent the work of superior pupils taught by superior teachers. Our results surely do.

We quote in Table 30 the facts found by Hotz [1918] for pupils who had studied algebra nine months. Hotz reports [1918, p. 4] that "Evidence collected by a system of checks . . . seems to indicate that the time allotment was ample. . . . In ninety-six out of two hundred of the tests submitted to nine months' students, exercises No. 19 and No. 25 in the equation and formula test were interchanged. They were then submitted to classes on a "fifty-fifty" basis. The results showed that exercise No. 25 was solved correctly about three times as often when it came last in the list, while No. 19, on the other hand, was solved more frequently when it came nineteenth on the list. Similar checks were employed in each of the other tests, with the exception of the graph test, and similar results were obtained." Since we quote the results for the first [1] and easiest sixteen out of twenty-four tasks in the addition and subtraction test, for the first and easiest sixteen tasks out of twenty-four in the multiplication and division test, and for the first [2] and easiest seventeen tasks out of twenty-five in the equation and formula test, there is still less danger of improper influence of insufficient time in the case of the tasks quoted here than for the tests in general.

[1] Not absolutely so; one was 17th in order.

[2] Not absolutely so; one was 18th in order.

TABLE 30

PERCENTAGES OF WRONG RESPONSES AND FAILURES TO RESPOND IN THE CASE OF PUPILS WHO HAVE STUDIED ALGEBRA NINE MONTHS. AFTER HOTZ

Addition and Subtraction		Multiplication and Division		Equation and Formula	
$4r+3r+2r=$	1	$3\cdot 7y=$	0	$2x=4$	0
$2x+3x=$	1	$\frac{12n}{4}=$	1	$7m=3m+12$	2
$12b+6b-3b=$	2	$2a\cdot 4ab^2=$	3	$3x+3=9$	4
$2c+\frac{1}{2}c=$	5	$6c^3\div 2c^2=$	4	$5a+5=61-3a$	7
$7x-x+6-4=$	7	$\frac{2}{3}$ of $9m=$	5	$7n-12-3n+4=0$	7
$3a-4b+5a-2b=$	9	$\frac{-8a^2b}{4a^2}=$	8	$10-11z=4-8z$	10
$5m+(-4m)=$	13	$4x\cdot(-3xy^3)=$	11	$\frac{2}{3}z=6$	10
$20x-(10x+5x)=$	14	$a^3\cdot(-3a)\cdot(-2a)=$	12	$c-2\ (3-4c)=12$	10
$(4r-5t)+(s-3r)=$	17	$\frac{18m^2n-27mn^2}{9mn}=$	20	$\frac{1}{2}x+\frac{1}{4}x=3$	22
$8c-(-6+3c)=$	24	$\frac{4x^4}{5}\div 2x^2=$	15	$\frac{2x}{3}=\frac{5}{8}$	20
$3a^2-3b-(2a^2+3b-4)=$	25	$(2a^2+7a-9)(5a-1)=$	30	The area of a triangle $=\frac{1}{2}bh$, in which $b=$ length of the base, and $h=$ height of the triangle. How many square feet are there in a triangle whose base is 10 feet, and whose height is 8 feet?	31
$5x-[4x-(3x-1)]=$	36	$\frac{n^4+7n^2-30}{n^2-3}=$	24	$\frac{y}{3}=\frac{5}{2}-\frac{y}{4}$	30
$\frac{3c}{4}-\frac{3c}{8}=$	36	$\frac{7a}{15}\div\frac{7a^2}{20}=$	16	$\frac{1}{4}\ (x+5)=5$	36
$\frac{3x-2}{3}+\frac{x+4}{6}=$	45	$\frac{-12x^2y^2\cdot(x-2)}{-3x^2y^2}=$	37	$3m+7n=34$, $7m+8n=46$	30
$\frac{1}{a-x}-\frac{3x}{a^2-x^2}=$	51	$\frac{m+n}{a}\cdot\frac{b}{m^2-n^2}=$	22	$\frac{4}{3-x}=\frac{2}{1+x}$	31
$\frac{r}{r+z}+\frac{r}{r-z}=$	54	$(-3xy^3)^4=$	38	The area of a circle $=\pi r^2$ in which $r=$ radius of the circle and $\pi=3\frac{1}{7}$. Find the area in square feet of a circle whose radius is 7 feet.	52
				In the formula $RM=EL$, find the value of M	49

We present in Table 31 the facts found by Childs [1917, p. 171 ff.] for the median of ten schools at the end of a year's study of algebra. Having many specimens of one kind of task in each test may suggest to pupils that they work more rapidly and less accurately than they ordinarily would; on the other hand it should aid precision by securing a "set" or adaptation to the task.

TABLE 31

	Median Number of "Attempts" per Minute	Median Percent of Correct Answers
Test I. Subtract. Time, 2 minutes. e.g. $\begin{array}{r} 10a-17b+240 \\ \underline{-3a+\ 2b-100} \end{array}$	3.6	49
Test II. Multiply and remove parenthesis. Time, 1 minute. e.g. $\pm 4\ (\pm 3x \pm 7)$	7.65	91
Test III. Solve for x. Time, 1 minute e.g. $\pm 15x = \pm 4$	9.25	63
Test IV. Divide. Time, 6 minutes e.g. $3x-2\sqrt{-3x^2+17x-10}$	.81	52
Test V. Transpose terms. Time, 1 minute e.g. $-4x+5=3x-9$	5.35	82
Test VI. Collect terms. Time, 2 minutes e.g. $-53x-10-115-40x=$	3.88	69
Test VII. State as an equation. Time, 4 minutes e.g. The sum of two numbers is 160. The greater is four times the less. Find each number.	1.59	73

The data of Monroe [1915] and Rugg [1917] represent substantially the same status.

There is then, obviously, need for considering the psychology of the strength of the mental connections or bonds required in algebra.

We may first consider two basic propositions:

A. These pupils could have gained the abilities needed to enable them to do these simple tasks with far fewer errors. The tasks are not beyond the intellects of most of them; the trouble is not that tasks 1 to 28 contain subtleties which they cannot comprehend.

B. Many of them would profit educationally if some of the time and thought that they and their teachers have spent in other ways had been spent in enabling them to subtract $3c-3a$ from $5a-b-2c$; to multiply $2de^2$ by $7d$, to find V when $\frac{PV}{T}=\frac{P_1V_1}{T_1}$, and the like. It would be better if they thoroughly knew what is required to master half or two-thirds of this work rather than knew a little about it all.

Probably very, very few experts in mathematics, teaching, or psychology, could be found to dispute either of these propositions. The truth of A, for all save the few pupils who are either below Stanford Mental Age 13.5 or Army Alpha score 56 (first trial), or suffer from a special mathematical disability, could probably be deduced from psychological facts. It is also demonstrated *a posteriori* by the fact that certain schools, not superior in their student personnel, do secure substantial efficiency at such tasks.

The truth of B is argued as follows:

The disciplinary value of algebra shrinks toward zero when pupils operate it so as to fail with one out of four simple tasks. The lessons of logic, precision, and economy cannot well be transferred if they have not been learned for algebra itself. The value of algebra as a tool may fall below zero when pupils are so insecure in its technique. It may be actually better for them to earn money to hire somebody to do their algebra for them than to trust their own work. The value of algebra as an inspiration and enrichment becomes very dubious. One fears that children who are so much at a loss in operating with symbols and equations lack any very beneficial ideas about symbolism or the equation.

It will be understood that we are not upholding B for all algebraic knowledge, but only, for the present, for such fundamental connections or bonds as are needed for such tasks as Nos. 1 to 28 of Table 29. Indeed, one of the most promising ways to secure something like 100% efficiency with certain bonds is to sacrifice others. For example, a rather long list can be made of mnemonic bonds now often formed at considerable time cost, all of which might perhaps be replaced by "Copy these formulas carefully on a card and put such a card in a pocket of every suit of clothes (dress) you own." Another long list could be made of bonds between various disguises of a^2-b^2, $a^2+2ab+b^2$, etc., and their factors where time might be saved, these tasks being left to be done, if at all, as "originals." Other cases where bonds may be formed to only slight or even zero strength will suggest themselves.

Two objections will be made to emphasis on the strengthening of bonds by thinkers who, while admitting the validity of propositions A and B, deprecate any tendency that may sacrifice the applications of algebra and its study of relations to formal work with symbols. They will object that the formal work has already more than its fair share of attention and that we should not be interested in creating skillful, rapid algebraic computers.

We may sympathize with these objections without abandoning the view that certain bonds need to be far stronger than they now are. We could, in fact, reduce the relative amount of formal work enormously and still give more practice to the fundamental bonds than they now receive. For example, the elimination of all work with polynomial denominators, division by a polynomial, and square root and cube root of polynomials, would leave much time free for strengthening basic bonds. Moreover, it may be that interesting applications of algebra are the very best means of strengthening them.

As to creating computers, the objection states a true and important fact, but it is not an objection. We should not care much about training algebraic computers; the "practical" utility of even the simplest algebraic computations, as such, is not widespread, as is the utility of simple arithmetical computation. Algebraic computation is, however, much more than a practical tool. It is also an evidence of understanding of the algebraic principles learned and an aid in learning others. Unless the pupil has mastery for such tasks as Nos. 1 to 28, he can hardly have any real appreciation of the nature of algebraic symbolism, negative numbers, exponents, equations, or the axioms used in solving them. Nor is he probably fit to follow the derivations and proofs of formulas, or to select the formulas which fit given problems in applied algebra, or to apply them properly when selected. More attention to the fundamental bonds will probably be profitable, entirely apart from the improvement in computation for computation's sake.

Without further debate about the importance of strengthening these fundamental bonds,[1] let us consider promising means of doing it.

[1] By an unfortunate choice of words, it is customary to say that the basic mental connections involved in the use of the axioms, the laws of signs, exponents in multiplication and division, removal of parentheses, and the like, should be *automatic*. Automatic is used by many psychologists to mean "*unconscious*," "*without awareness*"! We do *not* wish them to be automatic in the sense of *without awareness*. On the contrary, it is rather an advantage for a pupil to be fully aware that in $\frac{a^4x^2}{a^2bx^3}$ he is cancelling, that he leaves a^2 above as the balance from a^4 and a^2, and leaves x below as the balance from x^2 and x^3 and that b must stay in, too. It is surety and readiness to act, not the absence of awareness or consciousness, that we desire. *Strong, perfect, errorless, habitual, fluent,* would perhaps be better adjectives than *automatic.*

The first is a better and earlier understanding of the essential fact that letters represent numbers. A pupil may think that $ab \times b = ab^2$ with the attitude "The product of two numbers times one of them = what?" and think, when he obtains the ab^2, "This is a rule that will be true of the product of any two numbers by one of them," and half-think "$cd \times d$ would be cd^2, $xy \times y$ would be xy^2." $ab \times b = ab^2$ is to him a meaningful series like "A dog has four legs." He may, on the other hand, see or hear $ab \times b = ab^2$ without any "set" of his mind toward "generalized arithmetic," and without thinking of numbers, or even, in any proper sense, of anything. $ab \times b = ab^2$ is then a nonsense series like "rig fan tu lo." It will then be hard to learn and to remember, and will be a dead item of memory unrelated to $cd \times d$, $xy \times y$, and hardly differentiated from $\frac{ab}{b}$ or $ab + b$. If the "set" or attitude of the mind *toward the first hundred or so operations with literal numbers is permitted* to become that of learning a queer game, where you pretend to add, subtract, multiply and divide letters, there is certainty that these bonds themselves will be weak, and probability that all later practice will be much less effective than it should be. As a result of their experiments in teaching, Rugg and Clark were led to provide painstakingly for full and repeated attention to the fact that $a, b, c \ldots\ldots x, y$ mean real numbers of some real objects or quantities. Work in evaluation is of great merit in this respect, as they found.

The second means of strengthening the fundamental algebraic bonds is to form and justify the habit of expecting the operations to give a trustworthy, useful result. If, by keeping the tasks within the pupil's powers, by providing them with keys and checks for use when needed, and by other means, we give them cause to trust their algebraic results as they trust their addition of 2 and 2 or their multiplication of 10 by 10, the connections will be made with distinctness, emphasis, and satisfaction. Consequently, they will grow strong rapidly. If, on the other hand, the pupil thinks $ax \times 3ax^2 = 3a^2x^3$ with no sense of security, the gain from the practice will be slight. If he feels much the same when he calls $ax \times 3ax^2$ "$3ax^3$" as when he calls it "$3a^2x^3$", we cannot expect rapid strengthening of the latter. Unless we are skillful, some of the pupils' practice will be practice in error and much of it will be practice in insecurity.

Some of the devices which have been found helpful in arithmetic deserve trial in algebra. Such, for example, are keyed exercises wherein the pupil can learn at once whether his response is right or wrong; and practice drills wherein he acquires a specified mastery of certain bonds before proceeding to form others. Consider material like that shown on page 331 for early work in multiplication. The pupil covers the answers with a card and later looks at them to verify his answers. In early stages he may verify each answer as he obtains it. Later he may write some or all before verifying any.

Consider material like that shown on page 332[1], which the pupil uses with the directions: "Practice with these until you can write the right answers in 15 minutes." This material also may be keyed, the keys being planned for convenient use with both oral and written practice, and used so as to economize the pupil's time and encourage him to do without the key as soon as is wise.

If the attention of the class is held, rapid oral exercises are useful in algebra as in arithmetic, having the merit that a wrong response meets immediate correction.

[1] Further illustrations of this sort of work will be found in recent textbooks, for example, on pages 47, 176, 206, 254, 269, 279, 280, 305 and 309 of *Fundamentals of High School Mathematics*, by Rugg and Clark.

Exercise 1

$3\times5a$	$15a$	$3\times a$	$3a$
$4\times7b$	$28b$	$4\times b$	$4b$
$5\times6n$	$30n$	$9\times k$	$9k$
$7\times8p$	$56p$	$7\times t$	$7t$
$8\times10q$	$80q$	$6\times y$	$6y$
$2\times6a^2$	$12a^2$	$5\times c^2$	$5c^2$
$3\times4c^2$	$12c^2$	$7\times d^3$	$7d^3$
$5\times9d^2$	$45d^2$	$6\times x^2$	$6x^2$
$7\times6f^2$	$42f^2$	$4\times ay$	$4ay$
$10\times2m^2$	$20m^2$	$8\times bm$	$8bm$
$5\times4p^3$	$20p^3$	$2\times cq^2$	$2cq^2$
$6\times3q^3$	$18q^3$	$5\times ep^3$	$5ep^3$
$7\times5x^4$	$35x^4$	$9\times d^2z$	$9d^2z$
$8\times2y^4$	$16y^4$	$2\times adx$	$2adx$
$9\times3q^5$	$27q^5$	$3\times cy^2z^3$	$3cy^2z^3$
$2\times4ac$	$8ac$	$2a\times3a$	$6a^2$
$3\times5bd$	$15bd$	$5b\times9b$	$45b^2$
$4\times7ex$	$28ex$	$6c\times4c$	$24c^2$
$5\times8mt$	$40mt$	$7x\times2x$	$14x^2$
$6\times2xy$	$12xy$	$9y\times8y$	$72y^2$
$7\times2ac^2$	$14ac^2$	$3c\times4ab$	$12abc$
$8\times3c^2k$	$24c^2k$	$8c\times9cg$	$72c^2g$
$9\times2ef^2$	$18ef^2$	$5x\times6xy$	$30x^2y$
$8\times4h^2m$	$32h^2m$	$3p\times2pv$	$6p^2v$
$6\times5mn^2$	$30mn^2$	$4m\times7mp$	$28m^2p$
$4\times3ab^2d$	$12ab^2d$	$8d\times2ad$	$16ad^2$
$5\times5de^2f^2$	$25de^2f^2$	$3b\times8ab$	$24ab^2$
$7\times3p^2q^2t$	$21p^2q^2t$	$2k\times7dk$	$14dk^2$
$8\times2x^2yz^3$	$16x^2yz^3$	$5x\times9cx$	$45cx^2$
$9\times3w^3y^2z$	$27w^3y^2z$	$4y\times5my$	$20my^2$
$2\times7x$	$14x$	$3ab\times2ax$	$6a^2bx$
$3\times5u^2t$	$15u^2t$	$8ck\times4cy$	$32c^2ky$
$4\times4a^3$	$16a^3$	$7mx\times6my$	$42m^2xy$
$5\times4mn^2p$	$20mn^2p$	$8ep\times5kp$	$40ekp^2$
$6\times9axy$	$54axy$	$9mp\times3ap$	$27amp^2$

EXERCISE 2

$6 \times 9m$	$2d\ (a-6cd^2)$
$4 \times 7axy$	$-am\ (mn-a)$
$3ab \times 6ax$	$p^2\ (5px+s)$
$d \times d^3$	$-cx\ (x-5y)$
$cm \times 1.6h$	$acy\ (2y-c)$
$2x \times 1.4$	$3y\ (n-asy)$
$x \times 8$	$bfk\ (b+4k)$
$7 \times 5x^4$	$-bx\ (y-b)$
$an \times np^2$	$3b\ (6-bc)$
$ad \times 4md$	$-dv^2\ (3d-8v)$
$ay \times 3x$	$-b^2\ (-b+y)$
$4ch \times cfh^2$	$4an^2\ (cn+2a^2)$
$5df \times 3np$	$e\ (p-8bc)$
$2ady \times 4nd^2y$	$-d^2x\ (11-2ax)$
$8 \times 2x^2y^2$	$4y\ (x+c)$
$9 \times 3ef^2$	$5m\ (x-am^2)$
$x \times x^3$	$-bdy\ (y-3)$
$2cv \times 3v$	$ax^2\ (dx+12)$
$ab \times cx$	$f\ (x-16)$
$7 \times a^2z$	$-cp\ (y-ap)$
$5 \times 4p^3$	$-4p\ (p+4)$
$3 \times m$	$kms\ (4-2p)$
$2y^2 \times y$	$-mn\ (pm-n)$
$4dw \times 5d$	$9n\ (-a-4n)$
$-3n \times 5p$	$ay^2\ (r+5)$
$5b \times 6b$	$2q^2\ (3q-5)$
$3x \times 8cx$	$-a^2d^2\ (a+ab)$
$4b \times 7ab$	$m^2p\ (mp-8)$
$7d \times 2de^2$	$-8n^3\ (b-an)$
$-pw^2 \times p^2w$	$e^2y\ (y-3)$
$5 \times 4mn^2p$	$5x\ (7+akx)$
$6bkx \times 3ky$	$7n^2\ (3d-2b)$
$2dy \times 4ady$	$-x^2y^2\ (x-3dy)$
$b \times x$	$6a\ (9+b)$
$p \times y^2$	$-ps\ (p-8s)$

The third means is by infusing the process of learning with interest, so that the pupils care about obtaining right answers. Drills can probably be devised that will be as suitable in the bonds formed and much more attractive than those on pages 331 and 332.

Group competition and competition by individuals each with his own past record will be found useful. The teaching of algebraic computation as a means of solving for any one of the elements of a formula and deriving new formulas from those already known will show the utility of the computations, and may thereby increase interest. Nunn's treatment should be studied from this point of view, since he has used brilliant ingenuity and much care in introducing computing as a tool for "changing the subject" of a formula.[1]

A fourth means is the provision of aids to bridge the transition from learning A and B and C and D to learning to operate A and B together, and C and D together, and later A, B, C, and D all together. Thus a pupil learns to find any product of the form $a \times bx$, and any product[2] of the form $x \times y$, and any product of the form $x^a \times x^b$, and learns that $+\times+$ gives $+$, $-\times-$ gives $+$, and $+\times-$ or $-\times+$ gives $-$.

To multiply $3p \times 4qr$, he has to use the first two in coöperation; to multiply $(p^2)\ (-p^3)$ he has to use the last two in coöperation; to multiply $(2^c mx) \times (-.03m^2px^3)$ he has to use all four (and in fact certain other bonds as well) in the right coöperative arrangement.

[1] It is, however, a question whether the formulas of science and engineering are very much more interesting to pupils than the a's and b's and x's, and whether changing the subject of a formula and deriving new formulas from a given formula are *much* more real issues to them than finding sums, differences, products, and quotients. Nunn's procedure is correct and means some gain, but we should not expect too much from it.

[2] Letting a, b, and c represent any numerals, and letting x, y, and z represent any literal factors expressed by single letters.

The organization and coöperative use of habits needs guidance as truly as their separate formation. Graduation of the tasks and keyed exercises will help to prevent practice in error and blundering. Surveying the results with respect to signs, coefficients, letters, and exponents may help. The more elaborate the selection, arrangement and relations of the habits are, the more profit there will be from checking. Pupils who are confused and react in a hit or miss way may be aided by being led to state just what they plan to do and why they plan to do it.

The most obvious means of improvement we have not yet mentioned — namely, a general increase in the amount of practice on computation. We have not mentioned it because it is doubtful whether a general increase in the kind of practice now given is an economical means of securing mastery. We do not, for example, know that the use of textbooks in which the general computation is reduced enormously, results in weaker fundamental bonds. The quality of the practice is certainly the thing for science to improve. Anybody can increase its general amount. Certain inequalities and special insufficiencies should, however, receive attention. They will be listed in chapter XIII.

Finally it is obvious that any improvements made in the conditions and methods of learning will tend to secure greater strength of these bonds, other things being equal. There is a positive correlation among schools between mastery of them and ability with more elaborate calculations and with problem solving.

So much for the bonds that need to be made stronger than they are made now. Consider bonds that may well be left weaker than they are now. We have first any bonds that are useful only for abilities which have been recommended for discard. They need only zero strength. Next we have such specific memory bonds as those for the formulas of arithmetic and geometric progressions and the binomial theorem. Pupils might perhaps gain by being permitted to look these up in the book or by being given time to derive them instead of being required to learn them as now. They are evidently hard to remember; for it is a regular procedure for pupils who take college entrance examinations to study the formulas just before the examination, and write them out on the question paper as soon as they receive it, before even looking to see which it calls for! Their teachers train them to do this. If a pupil really understands them, however, it would seem that he ought to be able to remember or re-derive them after a reasonable amount of practice in applying them.

In general, in a course in algebra such as would embody the recommendations so far made in this volume, there are not many bonds formed that are not worth forming to a strength of, say, "Right 49 times out of 50," when operating along with other bonds in the ordinary applications of algebra.

One very special case remains — that of crutches, or connections which are formed for temporary use only, to give way later to others. Such are: writing 1 as coefficient, writing a parenthesis around a polynomial which is a numerator or denominator or under a radical sign, and writing 1 as exponent. Such crutches are very rarely advocated by authors of textbooks or courses of study and are not much used by teachers of algebra.

The general principle is to avoid them except for reasons of weight, in accord with the general psychological maxim, "Other things being equal form a connection in ways in which it is to be used." When there do seem to be reasons of weight the bad consequences of the use of crutches can be reduced by attaching the standard procedure to one mental set or attitude, and the provisional "crutch" procedure to a clearly differentiated set. For example, the pupils may be given work in this form:

In this column you may change a to a^1, b to b^1, c to c^1, etc., to help you to remember that when no exponent is printed, the exponent 1 is understood.

$a^{\frac{1}{2}}\ (a^{\frac{1}{2}}+a)$
$b^{\frac{1}{2}}\ (a+a^{\frac{3}{2}}+b^{\frac{3}{2}})$
$c^2\ (c^{\frac{1}{2}}-c+c^{\frac{3}{2}})$
$d(\sqrt{d}-\sqrt{cd})$
$(e+e^{\frac{1}{2}})\ (e-e^{\frac{1}{2}})$
etc.

In doing the work of this column, remember that when no exponent is printed, the exponent 1 is understood.

$a\ (a^{\frac{3}{2}}-a)$
$b^{\frac{1}{2}}\ (a^{\frac{1}{2}}b+ab^{\frac{1}{2}}+b)$
$(c^{\frac{1}{2}}+d)\ (c+d^{\frac{1}{2}})$
$e^{\frac{2}{3}}\ (e^{\frac{1}{3}}-e^{\frac{1}{2}}+e)$
etc.

In some cases where a certain procedure eventually gives way to another the former should still be maintained at a substantial strength, because of its value as a part of the pupil's total system of algebraic abilities and as an insurance against rote learning and other calamities. Such, for example, are the first applications of the axioms in the arrangement of equations for solving. Adding .... to both sides and subtracting .... from both sides do give way to "transposing" but they should not be permitted to perish for lack of exercise thereafter. It is true that we do not wish a pupil, after attaining $2p+4=p+6$ or $10p=100$, to think laboriously, "I will subtract p from both sides and 4 from both sides," and "I will divide both by 10." On the other hand we do wish him to retain the axiom bonds strong for use when needed. The use of equations which result in $14.5p=92.6$ and the like will serve this and other useful ends.

There is some evidence that teachers of algebra let the axiom bonds weaken too much and too soon from disuse. For example, many pupils have no clear and sure ideas of why the common denominators vanish when an equation is "cleared of fractions" and do not vanish when fractions are added or subtracted. In fact if, after the training in clearing of fractions, tasks in adding fractions are assigned, a considerable percentage of pupils discard the denominators.

In many respects it would be profitable to teach pupils at the beginning to clear equations of fractions gradually by multiplying by the "largest" denominator first and then by the largest that remained and so on. As a general procedure for after-life, this is perhaps better than finding the least common denominator, reducing all terms to it, and then letting all denominators disappear. It is easier to remember, and nearly or quite as economical of time for the sorts of operations that life offers. As a procedure for school use it has the merit of reinforcing the fundamental knowledge of the equation and the use of the axioms. If pupils later learn to obtain the least common denominator and operate accordingly they will be less likely to learn it as an unreasoning routine.

Teachers sometimes treat what we have called the regular procedure almost as a crutch assisting the pupil to mastery of a short cut which replaces it. For example, a pupil would probably be scorned for multiplying out $(2m+7)$ $(2m+7)$ instead of applying the $a^2+2ab+b^2$ formula, or for writing $(2p-q)-(3p+q)$ in column form and subtracting instead of changing signs and collecting terms, if he did either after the short cut had been learned. He might be scorned for dividing $m^6n^3-27p^3$ by m^2n-3p instead of writing the result directly with the aid of $a^3-b^3=(a-b)$ (a^2+ab+b^2), or for not transposing two terms from each side in one step.

In view of the very low degree of strength of the fundamental bonds, it seems unwise to abandon them so soon. Agility with algebraic manipulations is of value chiefly as a symptom of understanding of literal and negative numbers, formulas, equations, and the laws of generalized arithmetic. Mastery of the regular operations usually teaches these lessons better than facility with short cuts, and the short cuts themselves are most instructive when based on mastery of the regular operations.

CHAPTER XIII

THE PSYCHOLOGY OF DRILL IN ALGEBRA: THE AMOUNT OF PRACTICE

TEACHERS' ESTIMATES OF AMOUNTS OF PRACTICE

Teachers of algebra have, in general, vague and erroneous ideas concerning the amount of practice that they give on the various features of algebraic learning. The reader may convince himself of this by having a score of teachers of mathematics make independent estimates as directed below, either for an average textbook, as stated, or for any given textbook with which they are familiar. It will be found that they differ very greatly one from another and from the real facts.

We report here some of the results found in the estimates of sixty-eight teachers of mathematics. These teachers were probably mostly from the top fifth of teachers of mathematics in respect to ability to make such estimates. Two-thirds of them were members in attendance upon a meeting of an important association of teachers of mathematics, and one-third were graduate students in an advanced course on the teaching of mathematics at a large university. They used the following Estimate Record:

If a pupil does all the work given in an average algebra textbook for Grade 9, that is, for a one-year course in algebra, how many times will he do each of the following? (Write estimates on the dotted lines):

.... **1.** Represent a number by a letter.
.... **2.** Form an equation.
.... **3.** Translate an algebraic statement into words.

.... **4.** Transform a formula.
.... **5.** Add with unlike signs.[1]
.... **6.** Add with like signs.[1]
.... **7.** Remove a negative parenthesis.
.... **8.** Use the fact that $+ \div -$ gives $-$ and $- \div +$ gives $-$.
.... **9.** Add or subtract with polynomials.
.... **10.** Use the fact that like powers of equals are equal.
.... **11.** Use the fact that like roots of equals are equal.
.... **12.** Factor x^2+bx+c.
.... **13.** Factor ax^2+bx+c.
.... **14.** Solve simultaneous equations by addition or subtraction.
.... **15.** Solve simultaneous equations by substitution.
.... **16.** Express a ratio.
.... **17.** Change the sign of one term of a fraction, and before the fraction.
.... **18.** Factor the difference of two perfect squares.
.... **19.** Cancel in fractions.
.... **20.** Divide by a fraction.

In what follows a, b, c, mean any numerals; x, y, z, mean any letters.

.... **21.** $ax \div x$
.... **22.** $x \div ax$
.... **23.** $x^2 \div x$
.... **24.** $x \div x^2$
.... **25.** $a \div bx$
.... **26.** Add or subtract fractions.
.... **27.** $x^2 \div x^3$
.... **28.** $x^3 \div x^2$
.... **29.** $\frac{\sqrt{a}}{\sqrt{b}}=\sqrt{\frac{a}{b}}$ and $\frac{\sqrt{x}}{\sqrt{y}}=\sqrt{\frac{x}{y}}$
.... **30.** $\sqrt{\frac{a}{b}}=\frac{\sqrt{a}}{\sqrt{b}}$ and $\sqrt{\frac{x}{y}}=\frac{\sqrt{x}}{\sqrt{y}}$

Table 32 shows the estimates for the first twenty items. Table 33 shows, for comparison, the facts actually found in counts of the operations performed by a pupil in doing all the work assigned for the first year in four textbooks.[2] It

[1] Of the sixty-eight teachers, twenty-three were directed not to include estimates on No. 5 and No. 6. The estimates upon these two items are, none the less, more variable than those upon any other items in the list. The forty-five teachers varied from one hundred to one million in their estimates of No. 5 and No. 6! See Table 32.

[2] The method of making these counts is described on pages 348 and 349. One of the four counts was of the work assigned for two years in a textbook in which algebra, geometry, and some trigonometry are taught as a combined course.

will be observed that the highest estimate of the sixty-eight is almost always over one hundred times the lowest, and that to include even only half of the estimates, we require differences of four to one, five to one, eight to one, and the like.

TABLE 32

ESTIMATES, BY 68 TEACHERS OF MATHEMATICS, OF THE AMOUNT OF PRACTICE ON VARIOUS ABILITIES IN THE AVERAGE FIRST-YEAR ALGEBRA TEXTBOOK

	Represent a Number by a Letter	Form an Equation	Add with Unlike Signs	Add with Like Signs	Use the Facts that $\frac{+}{-} = -$ and $\frac{-}{+} = -$	Add or Subtract with Polynomials
	1	2	5[1]	6[1]	8	9
0 to 99	3	6		1	8	18
100 to 199	5	15	6	3	12	14
200 to 299	14	12	1	2	7	4
300 to 399	4	3	5	4	5	5
400 to 499	3	4	3	2	5	2
500 to 599	4	11	5	5	10	5
600 to 699	1	1		2	3	1
700 to 799		2	1	1		1
800 to 899	4	2	2	1	1	3
900 to 999			1	2		
1000 to 1099	8	4	8	8	5	3
1100 to 1199						
1200 to 1299	2	3				1
1300 to 1399						
1400 to 1499	1					
	also 1,500 2,000(3) 2,500 3,000(2) 5,000(2) 6,000(2) 7,500 8,000 10,000(4) 20,000 50,000	also 2 at 2,000 2,200 3,000	also 1,800 2,000 (3) 2,500 (3) 3,000 10,000 (3) 20,000 1,000,000	also 2,000 (5) 2,500 3,000 3,500 4,000 5,000 10,000 (2) 20,000 1,000,000	also 2,000(6) 2,500 5,000(4) 50,000	also 1,700 2,000(4) 2,500 3,000 5,000(2) 8.000 One teacher failed to estimate 9

[1] Only 45 teachers estimated for 5 and 6.

TABLE 32 (Continued)

	Translate an Algebraic Statement into Words 3	Transform a Formula 4	Remove a Negative Parenthesis 7	Use the Fact that Like Powers of Equals are Equal 10	Use the Fact that Like Roots of Equals are Equal 11	Factor x^2+bx+c (x = Any Letter, b and c Any Numeral) 12	Factor ax^2+bx+c (x = Any Letter, a, b and c Any Numeral) 13
0 to 9	2			2	2	1	2
10 to 19	1	5	1	7	10		
20 to 29	8	9	5	8	10	4	11
30 to 39	2	1		2		3	2
40 to 49		1	4		1	2	3
50 to 59	7	17	7	18	12	8	10
60 to 69		2	1	1	1	1	2
70 to 79	3	3	3	4	5	7	6
80 to 89	1			2	1		
90 to 99	1		1		1		1
100 to 109	13	7	9	7	9	9	6
110 to 119							
120 to 129						1	
130 to 139							
140 to 149			1				
150 to 159	3	3	1	3	4	2	4
160 to 169						1	
170 to 179	1		2	1			
180 to 189			1				
190 to 199		1					
200 to 209	8	10	4	3	3	9	7
	also	also	also	also	also	also	also
	250(3)	250	250(2)	300	250	250(2)	300(4)
	300(4)	300	300(4)	400(3)	300	300(4)	350
	400	350	400(4)	500(4)	400(3)	350	400(2)
	500(4)	400	500(10)	2000	500(3)	500(6)	500(2)
	600	500(3)	600	10000	6000	600(2)	600
	800	1000	700			625	625
	1500	3000	900			800(2)	900(2)
	2000(2)		1000(3)			1000(2)	1000
	4000		2000(2)				

TABLE 32 (Continued)

	Solve Simultaneous Equations by Addition or Subtraction	Solve Simultaneous Equations by Substitution	Express a Ratio	Change the Sign of One Term of a Fraction, and Before the Fraction	Factor the Difference of Two Perfect Squares	Cancel in Fractions	Divide by a Fraction
	14	**15**	**16**	**17**	**18**	**19**	**20**
0 to 9		4	6	1			
10 to 19	2	6	14	7	1		5
20 to 29	4	13	11	18	1	4	9
30 to 39	3	4	1	7	5	1	3
40 to 49	3	4	5	3	3	2	3
50 to 59	17	13	11	9	5	4	10
60 to 69	2	1		1		3	
70 to 79	2	3			5	2	2
80 to 89	1			1	2		
90 to 99	1	1					1
100 to 109	14	7	8	8	7	15	6
110 to 119							
120 to 129					1		3
130 to 139							
140 to 149							
150 to 159	3	1	2		4	2	6
160 to 169							
170 to 179	1	1	1			1	2
180 to 189							1
190 to 199							
200 to 209	6	3	2	2	9	6	
	also 250(2) 300 500 600 800 1000 1500 1650	also 250 300(2) 500 750 800	also 300 1000(2) 2500	also 250 300(3) 400(3) 1000 1500 2500	also 250(5) 300(7) 350(2) 400(3) 500 550 600 1000(2) 1250 1500 2500	also 250(5) 300(5) 350 400 500(6) 600 800(2) 1000(5) 1500 3000	also 250(5) 275 300(3) 500(3) 600 800 900 1000(2)

TABLE 33

		Book A	Book B	Book C	Book D
....	**1.** Represent a number by a letter	427	559	367	507
....	**2.** Form an equation	476	536	429	627
....	**3.** Translate an algebraic statement into words	49	30	52	72
....	**4.** Transform a formula	43	25	54	43
....	**7.** Remove a negative parenthesis	200	232	41	121
....	**8.** Use the fact that $+\div-$ gives $-$ and $-\div+$ gives $-$	412	813	$87+9k$	138
....	**9.** Add or subtract with polynomials	38 258	39 49	1 3	39 38
....	**10.** Use the fact that like powers of equals are equal	38	30	1	6
....	**11.** Use the fact that like roots of equals are equal	107	98	102	48
....	**12.** Factor x^2+bx+c	215	208	$131+18k$	61
....	**13.** Factor ax^2+bx+c	164	123	—	66
....	**14.** Solve simultaneous equations by addition or subtraction	131	132	60	80
....	**15.** Solve simultaneous equations by substitution	47	78	50	48
....	**16.** Express a ratio	14	13	30	49
....	**17.** Change the sign of one term of a fraction, and before the fraction	30	43	8	14
....	**18.** Factor the difference of two perfect squares	186	280	—	139
....	**19.** Cancel in fractions. Mono[1]	278	144	196	178
	Cancel in fractions. Poly.	266	181	77	58
....	**20.** Divide by a fraction	60	62	30	38

In what follows a, b, c, mean any numerals; x, y, z mean any letters.

		Book A	Book B	Book C	Book D
....	**21.** $ax\div x$	124	141	5	18
....	**22.** $x\div ax$	0	3	5	3
....	**23.** $x^2\div x$	185	107	88	9
....	**24.** $x\div x^2$	2	10	0	1
....	**25.** $a\div bx$	4	0	1	0
....	**26.** Add or subtract fractions	231	129	90	102
....	**27.** $x^2\div x^3$	0	0	0	0
....	**28.** $x^3\div x^2$	8	16	2	12
....	**29.** $\frac{\sqrt{a}}{\sqrt{b}}=\sqrt{\frac{a}{b}}$ and $\frac{\sqrt{x}}{\sqrt{y}}=\sqrt{\frac{x}{y}}$ **30.** $\sqrt{\frac{a}{b}}=\frac{\sqrt{a}}{\sqrt{b}}$ and $\sqrt{\frac{x}{y}}=\frac{\sqrt{x}}{\sqrt{y}}$	109	112	106	1

[1] Factors just alike.

Table 34 shows the estimates for all twenty-eight items summed together,[1] giving a rough index of the teachers' notions of how much practice in general a standard instrument of instruction gives. There is, here, a range from under one thousand to about one hundred thousand!

TABLE 34

FREQUENCIES OF DIFFERENT ESTIMATES FOR ALL TWENTY-EIGHT ITEMS TOGETHER

SUM OF ESTIMATES FOR 28 ITEMS	NUMBER OF TEACHERS
0 to 999	1
1000 to 1999	12
2000 to 2999	5
3000 to 3999	7
4000 to 4999	7
5000 to 5999	1
6000 to 6999	2
7000 to 7999	7
8000 to 8999	
9000 to 9999	2
10000 to 14999	9
15000 to 19999	5
	also 21000+
	22000+
	27000+
	30000+
	34000+
	44000+
	50000+
	53000+
	66000+
	94000+

By dividing a teacher's estimate for each of the twenty-eight estimates by the sum of the twenty-eight, we have measures of his opinion concerning the relative amount of practice given to each item, which are comparable to similar measures for other teachers. Table 35 exhibits these opinions concerning relative amounts of practice, with the effect of general over- or under-estimation of practice eliminated.

[1] That is, the sum of the twenty-eight estimates was obtained for each teacher. Items 5 and 6 were not used.

There is still an enormous variation one from another and from the truth.

TABLE 35

ESTIMATES OF RELATIVE AMOUNTS OF PRACTICE: ITEMS 1, 2, 8, 9, 19, 21, 22, 23, 24, 26

ITEM ÷ SUM OF 28	1	2	8	9[1]	19	21	22	23	24	26[2]
.000 to .009	1	2	7	17	11	7	23	6	19	8
.010 to .019	2	2	6	17	13	21	18	18	17	18
.02 to .029	3	11	6	6	11	9	7	8	8	11
.03 to .039	2	8	2	4	8	8	4	13	3	8
.04 to .049	3	9	4	4	8	6	7	4	6	9
.05 to .059	7	7	6	3	2	4	4	6	6	7
.06 to .069	3	9	7	2	5	2	1	3		2
.07 to .079	3	4	3	3	3	4	2	3	3	
.08 to .089	5	1	1		2			1		
.09 to .099	3	3	2	1	1		1	1	1	1
.10 to .109	3	2	4	6	1	2	1	2	1	
.11 to .119	2	2	1	1				1		
.12 to .129	4		1	2						1
.13 to .139	1	3	3		1	1			1	1
.14 to .149	1	1		1						
.15 to .159			1	2	2	3				
.16 to .169	2	1		1					2	
.17 to .179	1	1	1							
.18 to .189	2	1						1		
.19 to .199			3	2						
.20 to .249	2	1	4	1		1		1	1	
.25 to .299	8		2	4						
.30 to .349	1		2							
.35 to .399			1							
.40 to .449	1			1						
.45 to .459	1									
.50 or over	7		1							

In the items credited in general with the larger amounts of practice the range of the estimates is usually from about 1 per cent to over 20 per cent; in the items credited in general with smaller amounts of practice the range is usually from under one-half of one per cent to twenty times as much.

[1] +1 omitted.
[2] +2 omitted.

TABLE 35 (Continued)

ESTIMATES OF RELATIVE AMOUNTS OF PRACTICE:
ITEMS 3, 4, 7, 10, 11, 12, 13, 14, 15, 16, 17, 18, 20

ITEM ÷ SUM OF 28	**3**	**4**	**7**	**10**	**11**	**12**	**13**	**14**	**15**[1]	**16**[2]	**17**[3]	**18**	**20**
.000 to .004	10	8	5	13	10	5	8	8	15	23	22	5	6
.005 to .009	14	19	6	17	23	6	15	11	20	20	12	6	18
.010 to .014	3	13	9	12	9	14	16	12	12	12	12	8	13
.015 to .019	3	5	3	9	9	10	3	9	7	5	6	6	7
.020 to .024	5	5	2	2	3	4	2	9	3		2	6	9
.025 to .029	7	8	9	5	4	8	8	6	4	1	4	8	5
.030 to .034	4	2	3	1	2	1	5	3	1		2	7	5
.035 to .039	6	1	4	4	2	3	3	3	2	1		4	1
.040 to .044	3	2	6	1	2	3	2	4	1	1	1	2	2
.045 to .049	2		7			2	2	2	1			2	
.050 to .054	2	2	3	1		1	1				1	4	1
.055 to .059	2		2				1			1	1	1	
.060 to .064	2					1	1				2	2	
.065 to .069		2	3		2	2						1	
.070 to .074	1					1							
.075 to .079				1	1	3						2	
.080 to .084	1				1	3							1
.085 to .089													
.090 to .094	1												
.095 to .099						1			1			2	
	2 at .10	1 at .10	1 at .10	1 at .14			1 at .12	1 at .12		1 at .21	2 at .10	1 at .12	
			1 at .12	1 at .20								1 at .20	
			3 at .13										
			1 at .14										

The variation remains very great if we compute the ratios of the estimates for two items made by each individual. For example, the opinions of the sixty-eight teachers in answer to the question "How many times as often will the pupil 'Form an equation' as he will 'Translate an algebraic statement into words' " were as follows, for twenty

[1] +1 omitted.
[2] +3 omitted.
[3] +1 omitted.

teachers, taken at random: $\frac{2}{3}$, $\frac{9}{10}$, 1, $1\frac{1}{2}$, $1\frac{2}{3}$, 1.7, 2, 2, 2, $2\frac{1}{2}$, 3, 4, 5, 6, $6\frac{1}{2}$, $6\frac{2}{3}$, 7, 8, 10, 15.

The ratios for (8) "Use the fact that $+ \div -$ gives $-$ and $- \div +$ gives $-$" and (14) "Solve simultaneous equations by addition or subtraction" were, for twenty teachers taken at random: $\frac{1}{5}$, $\frac{1}{2}$, 1, 2, 2, 3, 3, 4, 5, 5+, 6, 10, 10, 10, 12−, $12\frac{1}{2}$, 16+, 20, 15, 83.

Some of the variations with some of the items are doubtless due to different interpretations of "Represent a number by a letter," "Form an equation," and the rest. This cannot, however, be a large factor, since the variation is nearly or quite as large for items so clear as "Remove a negative parenthesis," or "$ax \div x$." Some of the variations with some of the items are doubltess due to familiarity with different textbooks, which assign different amounts of work with, say, ratio, factoring, and radicals for the first-year course. This too is not a very large factor; for the variation is nearly or quite as large for items like "Use the fact that $+ \div -$ gives $-$" which everybody knows are in all first-year courses, as for the items involving roots and powers.

The plain fact is that nobody — not even the author of a textbook — has exact knowledge of the amount of practice it contains unless he actually makes the count. Such counts are excessively laborious and, so far as we can learn, the ones reported here are the first that have ever been made. Still less has anybody exact knowledge of the amount of practice in the classroom apart from the textbook.

MEASUREMENTS OF AMOUNTS OF PRACTICE

The only entirely satisfactory way to measure the amount of practice which any group of pupils have in the case of any feature of algebraic learning would be to observe the work of each pupil in class and out. This would, of course, be enormously expensive in time and is out of the question for us. Our best means of securing something approximating the knowledge which is needed is to make inventories of just what practice a pupil will have if he does the work of a standard textbook in the way in which it is ordinarily done.

We have done this, using three standard texts for first-year algebra and the first two books of a combined course where algebra and geometry are taught together during two years. The method is as follows: The person making the inventory does the work of the book, page by page, as directed by the book. He records each bit of his activity, using a list like that shown (in part) below.

1. Representing a number by a letter.
2. Representing several numbers in terms of one.
3. Forming an equation.
11. Adding like signs.
12. Adding unlike signs.
13. $a+x$, $a-x$, $x+a$, $x-a$, $ax+b$, $ax-b$[1] (includes knowledge that these cannot be added to make one term).
14. $x+bx$, $x-bx$.
15. $ax+x$, $ax-x$.
36. $a \cdot x$.
37. $a \cdot x^2$, $a \cdot x^3$, and ax^3.
94. Special product. Product of sum and difference of two numbers.
240. Using simultaneous equations in solving problems.

The inventory is kept separate for each successive ten pages of the text, so that we can see how much practice is given, when it is given, and what it is related to. For example, in doing the work of a certain text, the act of removing a negative parenthesis occurs 200 times, these occur-

[1] a, b, c, etc., are used to mean ordinary numbers; x, y, z, etc., are used to mean letters.

rences being distributed in successive ten-page sections as follows:

0, 0, 0, 0, 0, 58, 17, 20, 0, 14, 0, 0, 0, 0, 0

0, 0, 10, 15, 3, 10, 0, 23, 4, 2, 4, 1, 19

The person making the inventory is unlike the pupil in that he does not misunderstand the directions, or make mistakes, or give up when a task is hard, or use awkward, unnecessary procedures. He does, however, avoid taking short cuts which ordinary pupils would not think of, or using a procedure like transposing or canceling or clearing of fractions at an earlier stage in the course than the ordinary pupil would use it. Where there is a choice of correct ways he takes the way which he thinks most pupils would take.

The resulting inventories do not, probably, measure what is actually being done in any class. Few teachers assign all the work that a textbook contains. Still fewer assign nothing beyond what it contains. Few pupils do correctly all that they are told to do, in such a way as to obtain full practice value from it. The inventories do, probably, measure the relative amounts of practice given to various algebraic abilities rather well. The authors represent educational leadership in algebra; teachers in general will proportion their additional assignments somewhat as the text does, the number who deliberately set out to improve on the author's plan of work being small.

In any case, they are our only present accessible store of facts about the amount of practice. Let us therefore examine them, letting further cautions and reservations wait until they are needed.

Such inventories as these are not entirely objective or absolutely exact. They are not entirely objective because again and again there are tasks which a pupil studying the book in question might do in one of two or more ways. The investigator has to judge which way would probably be taken. For example, after a time the pupil ceases to think "add $2x$ to both sides" and "subtract 3 from both sides," thinking rather "transpose $2x$ and 3, changing their signs." In some simultaneous linear equations it is hard to decide whether the ordinary pupil would substitute or multiply and add or subtract. Two investigators counting the processes for the same ten pages will then differ, and even the same investigator making a recount would not duplicate his first results. These differences may be large in some cases. For example, one psychologist may consider that the pupil will use the principle of least common multiple in clearing of fractions an equation like $\frac{x}{3}+\frac{x}{6}=6$; another may think this principle will not be used unless the denominators are much more varied, and much harder to handle by mere inspection.

The inventories are not absolutely exact, because nobody, no matter how skillful and painstaking, can keep perfect account of just what he does and just which of the hundreds of mental connections play a part therein, and enter the record for each without error.

We have checked three of the four inventories from which quotations are made here by having parts of each book (120, 120 and 200 pages respectively) inventoried independently by other observers. Wherever the counts reported have a probable error that would impair the certainty of the theoretical or practical inference drawn here, the fact is noted in connection with the table. The single entries in these tables should not be used for other purposes than the ostensible ones, since any single entry may have a large probable error. For example, if two counts quoted show 200 for X and 20 for Y, and if the check gives 250 for X and 10 for Y, it is still safe to infer that X has much more practice than Y, though both estimates are highly unreliable, but the comparison of Y, as 20, with a Z which is 5 or 10 would not be allowable.

The first fact shown by the inventories is that there is no accepted standard of how much practice any given algebraic ability shall receive. Books A and B are of the type that was almost universal until five years ago. Book C is a book that subordinates manipulation to the study of relations. Book D is one wherein algebra is taught along with geometry. Characteristic resulting differences in the amount of practice are shown in Table 36. Elaborate addition, subtraction, multiplication, and division are given ten times as much practice in A or in B as in C. Factorizations of even the commoner sorts are given three times as much. Work in understanding, framing, and transforming formulas, on

TABLE 36

VARIATIONS IN INSTRUMENTS OF INSTRUCTION IN RESPECT TO THE AMOUNT OF PRACTICE WITH ELABORATE COMPUTATIONS, FACTORIZATIONS, AND UNDERSTANDING FORMULAS

	A	B	C	D
Polynomial × polynomial (exclusive of special products and binomial × binomial)	28	61	0	106
Polynomial ÷ polynomial (exclusive of special factorizations)	73	71	21	33
Factor trinomials, difference or sum of two cubes and difference of two squares	790	865	$226+21K$	432
Making and transforming formulas	54	25	119	65

the other hand, is given only one-half as much practice in A, as in C, only one-fifth as much in B as in C.

In general, Books A and B provide much more practice in computational abilities of all sorts than Book C. This is shown by Table 37, which reports the counts for a dozen items which everybody, conservative or radical, includes in algebra, and which indicate fairly the general amount of practice in algebraic computation.[1] If Books A and B give

[1] The count for Book D was not complete in respect to items 1, 2, and 11.

approximately the right amount, Book C can hardly be thought to give enough, and vice versa.[1]

TABLE 37

THE AMOUNT OF PRACTICE IN EACH OF FOUR STANDARD INSTRUMENTS OF INSTRUCTION FOR FIRST-YEAR ALGEBRA, IN THE CASE OF REPRESENTATIVE OPERATIONS

	A	B	C	D
1. Adding like signs 2. Adding unlike signs 3. Changing sign of subtrahend	3371	4193	1564+256K	?
4. $a \cdot x$ 5. $a \cdot x^2$ or $a \cdot x^3$ or $a \cdot x^c$ 6. $a \cdot b \cdot x$ or $a \cdot bx$ or $a \cdot b \cdot x^c$ 7. $x \cdot x^2$ or $x^2 \cdot x$ or $x \cdot x^3$ or $x^3 \cdot x$ or $x^2 \cdot x^3$ or $x^3 \cdot x^2$ or $x^2 \cdot x^2$ or $x^c x^d$	3059	3240	1290	1542
8. Multiplication of any monomial by any monomial, save those already listed in 4 to 7, and excluding $x \cdot x$, $(x)^2$, $(ax)^2$, $(x^2)^2$ and other cases treated as raised to powers	1736	1596	356+13K	1471
9. $x \cdot x$ 10. Monomial squared	905	848	249+16K	440
11. + divided by − gives −; − divided by + gives −	412	813	87+13K	?
12. The use of equals + equals, equals − equals, transposing, equals ÷ equals, equals × equals, upon equations or formulas	2954	3048	2321+223K	2520
Total for all 12 items [2]	12437	13738	5867+521K	9200[3]

[1] It should be noted that in Book C some of the practice is in the excellent form, "Practice with these until you can, etc." Since we could not estimate how much practice that would be we have entered the practices which are assigned in this form once in the general total and again as so many k. In Table 36, for example, 226+21k means that the pupil had 226 practices plus 21 times the number of repetitions of the exercises of the "Practice until" type. If he did these exercises twice his total would be 247. If he did them three times, it would be 268.

[2] The probable errors for the totals are A 9%, B 1%, and C 3%. For the separate components they are higher.

[3] Estimated.

There is evidence that teachers and authors of textbooks conduct their work without any accurate knowledge of how much practice they do provide. The teachers fill as many class periods as is customary and the authors fill as many pages as is customary without any clear awareness of how much they put in. Evidence was given at the beginning of this chapter. Further evidence is of the type presented in Table 38. If the gifted authors of Book A had known that they had provided only one-twentieth as many insertions of negative parentheses as of positive, only a hundred and fiftieth as many divisions like $\frac{x}{4x}$ and $\frac{x^2}{x^3}$ as divisions like $\frac{4x}{x}$ and $\frac{x^3}{x^2}$, only a third as many uses of the axiom for powers as of the axiom for roots, and nearly three times as much practice on a^3-b^3 as on a^3+b^3,—if they had known that these things were so, they would not have left them so. One can provide this or that defense for one or another of these, but no competent person would deliberately assign amounts of practice in these proportions to these five pairs.

Sixty-nine[1] experienced teachers of algebra were asked to estimate as follows:

How many times as much practice would you give in first-year algebra to A as to B in the case of each of the following pairs? Write your estimate on the dotted lines in each case

$\frac{A}{B}$ = ... A. Inserting positive parentheses.
B. Inserting negative parentheses.

$\frac{A}{B}$ = ... A. $ax \div x$
B. $x \div ax$

$\frac{A}{B}$ = ... A. $x^2 \div x$
B. $x \div x^2$

$\frac{A}{B}$ = ... A. Like powers of equals are equal.
B. Like roots of equals are equal.

$\frac{A}{B}$ = ... A. Finding factors for the sum of two perfect cubes.
B. Finding factors for the difference of two perfect cubes.

[1] Five more gave incomplete records, which were not used.

Their median estimates were 0.5, 1.0, 1.0, 1.0 and 1.0.

The facts of Table 38 are chosen as being helpful by illustrating inequitable assignments of practice of which all or nearly all of us are guilty, by reason of a nearly universal ignorance of what we have assigned. The ignorance of an individual author concerning his own particular assignments is much greater.

TABLE 38

UNWISE ALLOTMENTS OF PRACTICE. THE A/B RATIOS ARE IN GENERAL TOO HIGH

		A	B	C	D
A.	Inserting positive parenthesis	1308	1361	393	35
B.	Inserting negative parenthesis	60	67	0	0
A.	$ax \div x$	124	141	5	18
B.	$x \div ax$	0	3	5	3
A.	$x^2 \div x$	185	107	88	9
B.	$x \div x^2$	2	10	0	1
A.	Like roots of equals are equal	107	98	102	48
B.	Like powers of equals are equal	38	30	1	6
A.	Factor difference of two perfect cubes	59	54	—	13
B.	Factor sum of two perfect cubes	22	39	—	18

The unreliability of some of the gross counts of this table is considerable. It is, for example, possible that the 1361 and 1308 would in an average of many counts turn out to be as low as 1000 or as high as 1600; or the 141 and 124 might change places. The general fact about the A/B ratios is, however, entirely sound.

Table 39 also presents what seem to a psychologist facts of misplaced emphasis, like those of Table 38, but all characterized by the neglect of the more fundamental and general in favor of a derived or special case. There is room for argument here, and the authors' procedure may have been deliberate. The derived or special case is easier for pupils to manage, and it may be argued that we need much easy work to accustom the pupil to factoring trinomials, eliminating, and solving a quadratic, before we teach him the really trustworthy general procedures. The psychology of these three cases is worth considering.

TABLE 39

MISPLACED EMPHASIS IN ALLOTMENTS OF PRACTICE

TOO LITTLE PRACTICE WITH THE FUNDAMENTAL AND GENERAL PROCEDURE IN COMPARISON WITH THE AMOUNT OF PRACTICE WITH A DERIVED OR PARTIAL PROCEDURE

The counts have been checked only for the second pair. The result there is satisfactory, and the other pairs concern such definite procedures that they are almost certainly sufficiently precise.

			A	B	C	D
1.	A.	Factor x^2+bx+c	215	208	—[1]	61
	B.	Factor ax^2+bx+c	164	123	—[1]	66
2.	A.	Simultaneous equations solved by adding or subtracting	131	132	60	80
	B.	Simultaneous equations solved by substitution	47	78	50	48
3.	A.	Quadratic equations solved by factoring	111	224	30	46
	B.	Quadratic equations solved by formula	50[2]	29[2]	—[3]	29

Factoring x^2+bx+c and ax^2+bx+c is learned chiefly as a neat and attractive way of solving certain quadratics. If a pupil has mastery of solving by the formula this need disappears. Leading up to the formula by much factoring of ax^2+bx+c, and leading up to factoring ax^2+bx+c by still more factoring of x^2+bx+c seems a curious approach. It is a question whether factoring ax^2+bx+c provides as much aid as interference with learning to understand and use the formula. It is a question whether factoring x^2+bx+c gives much more aid than interference with learning to factor ax^2+bx+c. The pupil who does two hundred examples of x^2+bx+c forms rather firmly the habit of paying no attention to x^2 or its coefficient, and of choosing d and e so that their product equals c and their sum equals b. The procedure of finding factors for ax^2+bx+c is not a very hard matter to understand or do, and does not require elaborate introduction of any sort, if it is taught frankly as the matter

[1] A is taught as a sub-case of B.

[2] Counts of the work for the second year in algebra would probably make the amounts for B higher proportions of those for A.

[3] Method B is not given during the first year at all by Book C.

of intelligent guessing which it really is. Some of the fact-torizations may take a *long time* for a dull pupil to work out, but length is not synonymous with difficulty.

In the case of eliminating in simultaneous equations, substitution is a universally valid method; and one method is enough for the pupil to learn for any of his theoretical or practical purposes. Instead of learning another, and inspecting equations to see if he can save a little time by adding or subtracting, he should substitute forthwith. The only aid to learning the technique of substitution which practice in eliminating by adding or subtracting gives is in the form of a confidence that the answers are obtainable. This is not needed, and except for this there is interference. It is, however, true that solving simultaneous equations by adding or subtracting is very well liked by pupils.

The case of quadratics solved by factoring and by the formula has already been discussed.

UNDERLEARNING

Our inventories show certain cases where the practice seems insufficient to guarantee learning, even with superior zeal and ability. Table 40 reports some such cases, especially those where all of the books, or all but one, agree in the insufficiency.

TABLE 40

UNDERLEARNING

CASES WHERE THE AMOUNT OF PRACTICE IN TWO OR MORE OF THE BOOKS SEEMS INSUFFICIENT TO GUARANTEE MASTERY

	A	B	C	D
Writing formulas	11	0	65	21
$a \div bx$	4	0	1	0
$x \div ax$	0	3	5	3
$x \div x^2$	2	10	0	1
$x \div x^3$	0	5	0	0
$ax \div bx^2$	0	11	0	0
$x^2 \div x^3$	0	0	0	0
Change in sign of one term of a fraction with change of the sign before the fraction	30	43	8	14
Change in sign of both terms of a fraction	0	16	0	4

OVERLEARNING

More practice than is needed for mastery may be wasteful both because of the time it uses up and because of its tediousness and consequent bad effect upon interest in algebraic work. In certain cases, however, overlearning is necessary and beneficial.

First, a bond may be formed strongly enough for mastery in certain simple situations, but require more practice for mastery in changed and more complex situations. Thus "To subtract a number change its sign and add it," may be mastered for cases like *Subtract* 11 *from* -2, or *Subtract* $-3x$ *from* $4x$, by a hundred or so practices, but needs added practice when the tasks are to *Solve* $n+11=-2$ *by subtracting* 11 *from both sides*, and $4x=14-3x$ *by subtracting* $-3x$ *from both sides*. This added practice may involve overlearning of the bond in its first applications.

Second, we often wish to introduce or illustrate some new principle or technique by material which is perfectly mastered. Thus in teaching the extraction of roots as the reverse of raising to powers, we might use:

$$\begin{array}{ll} a \cdot a = a^2 & (a^2)^{\frac{1}{2}} = a \\ b \cdot b = b^2 & (b^2)^{\frac{1}{2}} = b \\ c \cdot c = c^2 & (c^2)^{\frac{1}{2}} = c \\ a^2 \cdot a^2 = a^4 & (a^4)^{\frac{1}{2}} = a^2 \end{array}$$

just because $a \cdot a = a^2$, etc., were already perfectly mastered, so that attention and thought could be free to a maximum degree to study the new material.

Third, overlearning is relative. The pupil needs to know that $(\sqrt{a}+\sqrt{b})\ (\sqrt{a}-\sqrt{b})=a-b$ well enough to be reminded of it upon consideration of an expression which can profitably be rationalized by its use, but it is of little consequence whether it takes him one second or ten seconds or twenty seconds to think of it then. He needs to know that $\frac{ax}{x}$ gives a well enough to call it to mind on suitable occasions in much less than twenty seconds. Practice beyond what is needed for surety may still be profitable by increasing speed in the case of those bonds whose use is of great importance.

Moreover, very little time is required for much practice, once mastery is attained. Sometimes almost none is required, the bond acting along with others without appreciably slowing them up, as when the pupil practices the law of signs in multiplying -3 by $+4$ in $(x-3)(x+4)$. Even when the bond acts in isolation it is often a matter of from less than a second to two seconds, so that a thousand practices can be had in less than half an hour.

We have measured the time required for a first-rate student who has mastery of algebra to do all the work of a first-year course. Mr. A did all the work in Hawkes, Louby and Touton, *First Course in Algebra*, edition of 1917, in 25 hours. Miss B did all the work on pages 1–272 and 386–397 of Wells and Hart, *New High School Algebra*, edition of 1912, in 24.5 hours. The time spent, exclusive of the verbal problems, was about 19 hours in the former case and 17 hours in the latter. The time cost of overlearning after mastery is thus a trivial matter.

The reduction of practice beyond mastery on things worth mastery is not so promising a means of improving instruction as the total elimination of things which are not worth mastery and, save for general mental exercise, are hardly worth learning at all.

We should, of course, avoid wasteful overlearning of even the best abilities, and Table 41 reports certain suspected cases found in our inventories. They are to be suspected either on general grounds or because the variation among the books is so great. For example, if Book C needs only 674 cases of transposing, does Book B need 1262? The reader will recall that Table 2 showed a very great excess in practice in Books A and B over that given in Books C and D. If children learn algebra as well from Books C and D as from A and B, the latter may be suspected of wasteful excess of practice.

TABLE 41

CASES OF POSSIBLE WASTEFUL EXCESS OF PRACTICE

	A	B	C	D
$(x)^2=x^2$	395	368	49	167
Transposing	909	1262	674	—[1]
Square root of a monomial perfect square	771	1016	231	358
Cube root of a monomial perfect cube	126	163	5	54
Factor difference of two perfect squares	186	280	—[2]	139

So far we have suggested improvements by adding practice where the present customary amounts seem insufficient, eliminating wasteful overlearning, and adjusting cases where the relative amounts of practice seem ill-judged. We have not definitely set any amounts of practice as suitable to give mastery in this, that, and the other ability. Nor does it seem wise to do so now. In arithmetic this can reasonably be attempted with such bonds as $2+3=5$, $3+2=5$, $2+4=6$, $4\times7=28$, which are specific, and with such bonds as "Divisor × quotient should equal dividend," or "Number of decimal places in the product should equal the sum of the number in the multiplier and the number in the multiplicand," which, though general, are definite and uniform in their action. In algebra there are few bonds of the former

[1] Counted under equals + equals and equals − equals.

[2] Treated as a special case under ax^2+bx+c.

sort ($\sqrt{}$ means $^{\frac{1}{2}}$, $\sqrt[3]{}$ means $^{\frac{1}{3}}$, are samples); few of the general bonds are as definite as those of arithmetic ("No numeral before a literal number means 1 times that number;" "No exponent after a letter means that 1 is the exponent;" "(any number)^2 means that number times that number," are samples of the most definite). Even so apparently definite a rule as "To subtract, change signs and add," means different things according as you subtract $+a$, a, $+\frac{a}{2}$, $-a+b+c$, $-\frac{a+b}{c-d}$, $-\{a-(b+c)\}$ and $a^{-\frac{2}{3}}$. With $+a$ you change from $+$ to $-$; with a you change from $+$ understood to $-$; with $+\frac{a}{2}$ you change to $-\frac{a}{2}$ but do not alter the sign of the a or the 2; with $-a+b+c$ you must change them all; with $\frac{3}{c-d}-\frac{a+b}{c-d}$ you change the a and the b both; with $-\{a-(b+c)\}$ you change the a once and the b and c both twice; with $a^{-\frac{2}{3}}$ you change the a but do nothing to the $^{-\frac{2}{3}}$.

The number of practices desirable for any algebraic bond also depends to a very great extent upon the support it is to receive from other bonds, and especially upon the extent to which all have been organized and integrated to produce what may be called an algebraic sense, or good judgment in operating with literal numbers. For example, there should be some experiences in operating with capital letters, but it does not greatly matter that these should appear in connection with every process; and the number needed will depend on the extent to which algebraic habits have been otherwise freed from subserviency to a, b, x, and y, by work with other letters, primes, subscripts, angles, etc.

There should be some practice in understanding and applying computational formulas, like:

$$ax+ay=a(x+y)$$[1]
$$ax\times by=(a\times b)xy$$
$$\frac{ax}{by}\times\frac{cw}{dz}=\frac{(a\times c)xw}{(b\times d)yz}$$
$$(x+y)(x-y)=x^2-y^2$$
$$x^a\times x^b=x^{a+b}$$

but the amount will depend upon the liveliness of the pupil's appreciation that letters stand for real numbers of real things, the amount of practice he has had with formulas other than computational, and the extent to which he has organized his experiences and activities into a sense that tells him what terms are "like," what to take as the "coefficient," and what to treat as "one number" for any given purpose.

The amount of practice needed for any bond thus depends upon the total teaching plan of which the formation of that bond is a part. It is better to set standards of achievement to be reached than of amounts of practice to be given.

In general, at the end of a year's study, the easier single bonds like "$3x+5x=8x$" or "$a^2\times a^3=a^5$" should operate infallibly except for occasional lapses (say 99 times out of 100) in $2\frac{1}{2}$ seconds or less. The harder single bonds like "$-(a\ .\ .\ .)$ equals $-a\ .\ .\ .$" or "12 can be factored into 3 and 4," or "12 can be factored into 2 and 6," or "$\sqrt[3]{a}$ means $a^{\frac{1}{3}}$" should operate correctly 99 out of 100 times in 4 seconds or less. Operations which involve reference of the case to a principle and a single clear application of the principle, such as "$\frac{ab}{a}=b$", "$\frac{a}{ab}=\frac{1}{b}$", "$x-2a=3a$ is reducible to $x=5a$", or "$(a^2)^3=a^6$" should operate correctly 99 out of 100 times in 6 or 7 seconds. Where four or five

[1] a, b, c, etc., here represent any numerals, and x, y, z, etc., represent any literals.

bonds of the same sort operate together as in $x^2yz^3 \times xyz^2$, there should be enough time saved because the mind has to be adjusted to the task only once to make up for the time spent in discovering that a single adjustment will serve and what the adjustment is.

Where two or more different adjustments have to be made and two or more elementary bonds chosen and operated with the right parts of the situation or in the right order, we may add a rather generous time and error allowance. It is not improper that the pupil should stop two or three seconds to think what he is to do in $8-7\ (d-e)$ or $5m+\frac{m^2np^3}{mp}$.

If we consider tasks 1 to 11 listed below as a sample, we find them to consist of the operation of about sixty bonds

1. $(5a^2+4-3a^2)+(a^2-2-7a^2-5)$
2. $(-2a^3-10a-4a^3)+(5a+3a^3-4a)$
3. From $3a+4b$ subtract $5a-9b-3c$
4. From $5a-b-2c$ subtract $3c-3a$
5. $8a+8b-(3a+6b)$
6. $(5d-e)-(7e+2f)$
7. $7d \times 2de^2$
8. $de^2 \times d^2e$
9. $8-5(d+2)$
10. $4e^2+e(-4e-3)$
11. $5np-3p(4n+3p)$
12. $m+\frac{8m^2n^2}{mn}$
13. $4m+\frac{2m^2np^3}{mp}$.
14. $\frac{m^2np}{mn}-\frac{m^2n-mp}{m}$
15. $(m^5n)\ (m^2n^3)$
16. $(2a-7)^2$
17. If $a=2$, and $b=3$, what does $5a^2-2ab$ equal?

(mostly easy ones), with rather simple adjustments and selections of what to do and when to do it. At an average of $2\frac{1}{2}$ seconds per bond and 10 failures of the bond to act per thousand, we have 150 seconds and 0.6 errors. Making an allowance of 4 seconds per task for adjustment and control we have 194 seconds. Making an allowance of one error per fifty bonds (nine of these tasks) for omitting some part of the process, or confusing letters, or misplacing letters, and for clerical errors in reading, copying, etc., we have 194 seconds and 1.8 errors as a reasonable allowance for the eleven tasks. If the pupil checks all his work until he obtains two agreeing answers for each task he will then require about 40 seconds per task in order to have substantially errorless work. If he works carefully checking piecemeal as he thinks is desirable, he should then do the tasks at an average of 30 seconds or less and have nine out of ten right.

Gifted adults who have mastered algebra do such tasks as these eleven at an average rate of 8 seconds per task and have nineteen right out of twenty. In a first experiment they made 14 errors in a total of 176 tasks; on being warned to be more careful in a second experiment, they made only 5, of which 2 were not algebraic errors in any sense, but misperceptions excusable by the way the tasks were written.

The gifted adults do such tasks as Nos. 12 to 17 above at an average rate of 11 seconds and have eleven right out of twelve. One checking with rechecking of discrepant answers would enable them to have practically errorless work at a rate of 17 seconds per task for Nos. 1 to 11 and 24 seconds per task for Nos. 12 to 17.

We therefore recommend that practice on such fundamentals as appear in Nos. 1 to 11 should be sufficient to secure the ability to do such tasks (without full checking)in 30 seconds or less with at least nine right answers out of ten;

with Nos. 12 to 17 the time may be increased to 40 seconds. The pupil should be able, if time for full checking is available, to turn in such work substantially errorless.

Our standard of precision may be thought to be too severe for ordinary attainment since these gifted minds do not far surpass it, except after warning to be careful. However, there should be no very great superiority of gifted over ordinary minds in the surety of these routine computations or in the clerical work of reading and copying. Having nearly four times as long, the school pupil ought to do nearly as well. Relaxation to seventeen right out of twenty is defensible, but not in our opinion desirable.

It is true that pupils may have an understanding of all the principles in question and yet make many slips in operating them and very many slips in reading, writing, copying, and doing all the different things that each task requires. Indeed a standard of one right out of two of these tasks could not be attained without substantial understanding of the principles. For the reasons stated in Chapter XII, however, it seems wise to give the pupil greater mastery of the basic mental connections than that which comes simply from understanding of the principles.

For operations that require more selective thinking (as in solving, where the axiom to be used must be chosen and what to add or multiply by, etc., must be chosen) setting standards of speed becomes less important and less desirable, because of the greater influence of differences in native capacity. Substantial infallibility (to be secured, where necessary, by checking) should be demanded. The gifted child will discern quickly what is the thing to do, the duller child will have to consider and perhaps try many possibilities.

As has been stated, it seems best to make the time spent in practice and the number of repetitions subordinate to the attainment of certain standards. Since, however, almost all of the textbooks now available lay out the work by repetitions, teachers may be helped by some approximate estimate of the number of repetitions desirable in representative cases. Consider, as representative cases, the twelve listed below:

About how many practices would you think it wise to have on each of the twelve procedures listed below, in a first-year course in algebra supposed to take one-fourth of the pupil's time? Each time that the pupil does the thing, whether by itself or as an incident in some other procedure, is to count one. You will, of course, consider the claims of the course as a whole, in making the estimates. Write your estimate on the dotted lines under the heading "Reasonable Allowance."

		REASONABLE ALLOWANCE
1.	Translate an algebraic expression or equation or formula into words.	 1
2.	Form an equation to express the facts of, and provide the solution for, a particular problem stated in words.	 2
3.	Write a formula to express some general rule or relation.	 3
4.	Transform a formula.	 4
5.	Remove a negative parenthesis.	 5
6.	Add or subtract fractions with numerals as denominators.	 6
7.	Add or subtract fractions with literals as denominators.	 7
8.	"Transpose" (counting each term moved as one).	 8
9.	Find the square root of a perfect monomial square.	 9
10.	Solve simultaneous linear equations (2 unknowns).	10
11.	Factor the difference of two perfect squares.	11
12.	Interpret a graph.	12

Five psychologists, of whom three were experienced teachers of algebra, assigned estimates in accordance with the directions quoted. The estimates of these psychologists varied rather widely as shown below (Table 42) and none of them felt able to defend his estimates save within rather wide limits. Seventy-four teachers of mathematics made similar estimates. They varied still more widely, and most of them expressed a similar uncertainty about their estimates. It is, however, the case that the variability is not significantly greater among those who felt that they were merely guessing than among those who felt sure that their estimates were in most cases somewhere rather near the truth.

TABLE 42

ESTIMATES MADE BY FIVE PSYCHOLOGISTS, A, B, C, D, AND E

	A	B	C	D	E
1. Translate an algebraic expression or equation or formula into words...	50	200	150	125	75
2. Form an equation to express the facts of, and provide the solution for, a particular problem stated in words.	300	300	350	250	350
3. Write a formula to express some general rule or relation..........	100	200	250	125	50
4. Transform a formula..............	70	200	400	200	150
5. Remove a negative parenthesis.....	50	100	75	100	600
6. Add or subtract fractions with numerals as denominators..........	50	150	125	75	300
7. Add or subtract fractions with literals as denominators............	50	150	60	150	100
8. "Transpose" (counting each term moved as one)..................	300	260	500	1000	2000
9. Find the square root of a perfect monomial square................	60	120	175	200	250
10. Solve simultaneous linear equations (2 unknowns)..................	50	120	35	75	75
11. Factor the difference of two perfect squares........................	50	80	50	75	100
12. Interpret a graph.................	100	120	40	125	50
Total........................	1230	2000	2210	2500	4050

One might dismiss these estimates as just a curious collection of opinions, but for three facts. First, if these teachers and psychologists do not know how many repetitions are desirable, who does? Second, there is a fair probability that the median judgment of each group is free from large and serious constant errors. There is no question of partisanship or general doctrines or personal fortune involved. Indeed, few of the teachers ever considered the issue before. The teachers are doubtless influenced by the textbooks which they are using and have used, with a resultant constant error toward emphasis where it has been put. This error cannot be of very great magnitude, however, since, as we have seen, teachers do not know how many repetitions there are in the textbooks, nor their relative frequency. The psychologists were probably influenced by a general preference for such work as 1, 2, 3, 4, and 12; and this constant error in part balances the constant error of the teachers. In the third place the medians for the two groups are remarkably close, as shown in Table 43. This could easily happen if both were wise opinions based on fact, but would be improbable if both were mere echos of traditions, since the traditions were different for the two groups.

TABLE 43

APPROXIMATE MEDIAN RATINGS

ABILITY	PSYCHOLOGISTS	TEACHERS	APPROXIMATE AVERAGE OF MEDIANS
1	125	125	125
2	300	215	260
3	125	100	110
4	200	110	155
5	100	100	100
6	125	100	110
7	100	100	100
8	500	500	500
9	175	100	140
10	75	100	90
11	75	85	80
12	100	75	90
Sum of Medians	2000	1710	

There is one suspicious feature in these medians. Three of one series and six of the other are at 100, and five more of the former and four more of the latter are within 25 of 100. A cynic might infer that the medians represented mainly persons who, not knowing what to think, put down "one hundred." This is, however, certainly not true of the psychologists, and any suspicions that the reader may have of the median teacher's approximations to 100 will, we think, be reduced rather than confirmed if he will experiment with the teaching of these twelve features of algebra learning. Until such experiments have been made we may consider the combined estimates of Table 43 as reasonable allowances for the number of repetitions for the abilities in question.

It is interesting to compare these combined estimates, and the separate estimates also, with the median number of repetitions arranged for in the four standard textbooks. This is done in Table 44.

TABLE 44

COMPARISON OF AMOUNTS OF PRACTICE

	Combined Estimate of Psychologists and Teachers	Median of Four Textbooks
1. Translate an algebraic expression or equation or formula into words	125	51
2. Form an equation to express the facts of, and provide the solution for, a particular problem stated in words	260	506
3. Write a formula to express some general rule or relation	110	16
4. Transform a formula	155	43
5. Remove a negative parenthesis	100	161
6. Add or subtract fractions with numerals as denominators	110	116 (6 and 7 together)
7. Add or subtract fractions with literals as denominators	100	
8. "Transpose" (counting each term moved as one)	500	909
9. Find the square root of a perfect monomial square	140	565
10. Solve simultaneous linear equations (2 unknowns)	90	156
11. Factor the difference of two perfect squares	80	163
12. Interpret a graph	90	20

CHAPTER XIV

The Psychology of Drill in Algebra: The Distribution of Practice

The same amount of practice may be distributed in many ways. If the total is 100 practices, the number of different distributions possible is, of course, enormous. A few are illustrated in Figs. 17 to 22. In these figures each tenth of

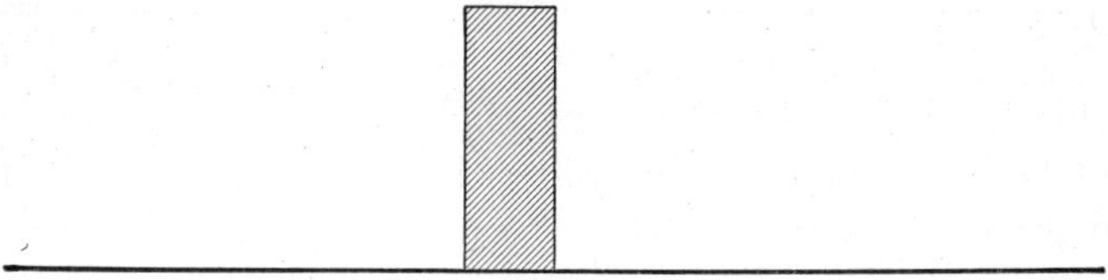

Fig. 17

of an inch along the base line equals one-thirty-sixth of the first year's work or one week.  Each hundredth of a square inch of shaded area equals four practices. Fig. 17 shows all the practice given in three successive weeks, equal amounts in each. Fig. 18 shows the practice spread out evenly over the whole year, no more being given during

Fig. 18

the week when the ability was first taken up than in any other. Fig. 19 shows the ability taken up for learning in the tenth week and given fifty practices then. The practice thereafter is haphazard. Fig. 20 shows a first learning with sixty-four practices in the sixth week of the year, a review with twenty-six a month later, and a review with ten a month later. Some such plan of reviews as this is commo in courses of study in grades 1 to 8 and to a less degree i grades 9 to 12. Fig. 21 shows a distribution of practice wit the first learning followed by reviews of decreasing amoun at increasing intervals.

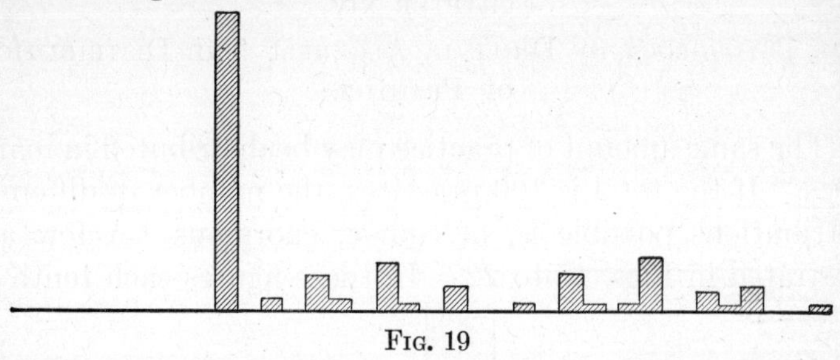

Fig. 19

Fig. 22 is like Fig. 21, save that some of the practice i interspersed irregularly instead of being concentrated in th reviews.

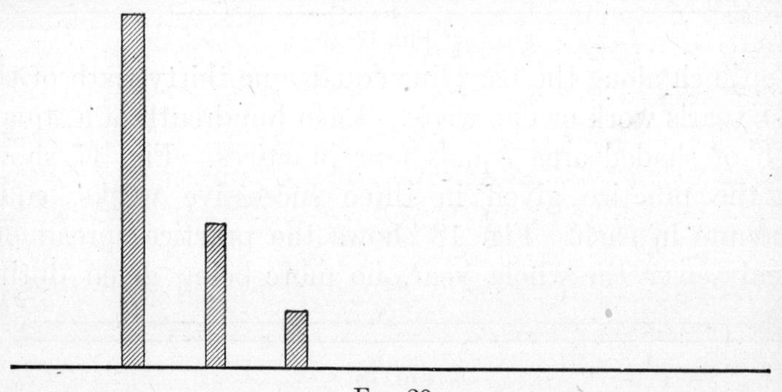

Fig. 20

Other things being equal, a distribution of the type o Fig. 21 or Fig. 22 is probably the most effective. There i enough practice at the time of first learning to give the abilit healthy birth; after a time it is examined and any weaknesse which have developed in it are cured; after a longer perio (with or without some casual practice) the ability is agai tested and treated; and so on until it can safely be left to be kept alive and well by such practice as the ordinary course of events offers. The arrangement of Fig. 17 is commonly undesirable, there being a likelihood that pupils will lose interest during the learning, and will make fewer connections between it and the rest of algebra than we desire,

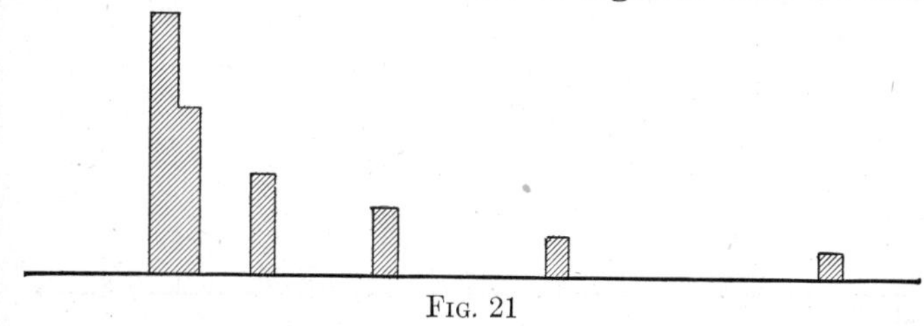

FIG. 21

and will develop errors or inabilities in respect to it. The arrangement of Fig. 18 is undesirable because if three practices are enough in the week of first learning, ninety-seven are too many for later reviews. If the total amount of practice is very large, as with $+ \times -$ and $- \times +$ give $-$, or equals added to equals give equals, such a distribution as that of Fig. 18 is not, *per se*, objectionable. It will of course arouse suspicion that the amount of practice is excessive.

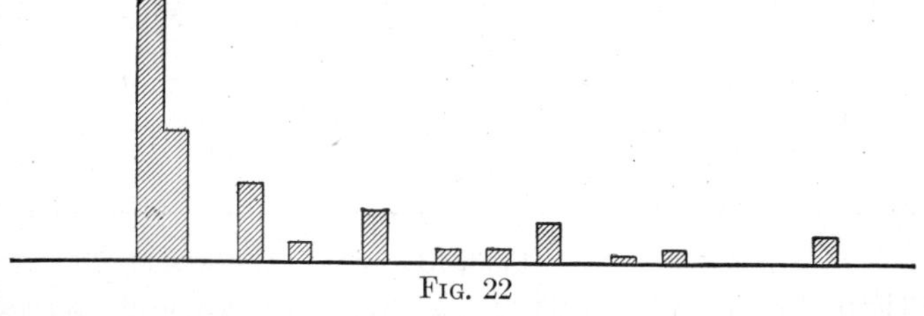

FIG. 22

The plan of Fig. 19, of establishing an ability, and then giving it practice from time to time, is defensible, but it is better to reduce the amount of practice per week as time goes on. The pupil can hardly need as much if our teaching has been adequate. Moreover, obviously, if all abilities learned are treated as in Fig. 19, he will soon have no time left to learn any new ones!

The plan of Fig. 20 is based on the same general principles as those of Fig. 21 and Fig. 22, but is cruder.

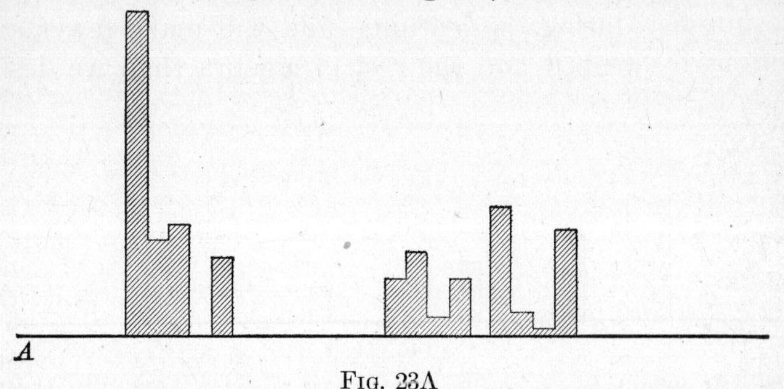

Fig. 23A

Consider now actual distributions of practice in algebra, as found in some of our best instruments of instruction.[1]

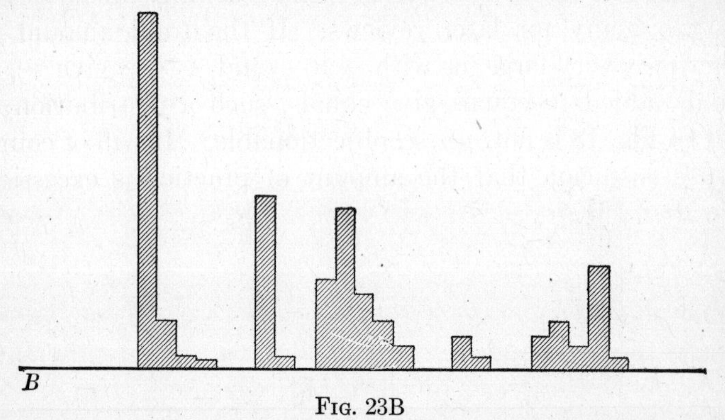

Fig. 23B

[1] The reader will remember the cautions concerning these textbook counts which were stated in the previous chapter. The distributions of practice shown here would not be identical with those determined by another person's count, and in any one case in any one particular, they are imperfect representations of the fact. The general impression which they give is, however, very closely similar to the general impression that would be given by the average of many independent counts. There is no reason to expect that such an average of many counts would show distributions much more or much less like the ideal.

Figs. 23A, 23B, and 23C show the number of times that a pupil will remove a negative parenthesis week after week if he does all the work contained in the assignment for the first year in Books A, B, and C. Each tenth of an inch

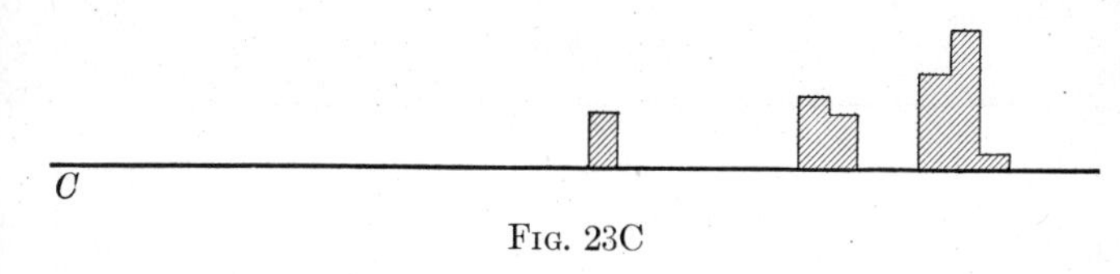

FIG. 23C

along the base line here equals ten successive pages of the textbook in question. This will be on the average about one week's work, though any one ten pages may be much more or much less. An area of one hundredth of a square inch equals four practices at removing a negative parenthesis. A column one-half inch high and 0.1 inch wide thus equals twenty practices; 0.1 square inch (a column 0.1 inch wide and one inch high) = forty practices. Fig. 23D is on the

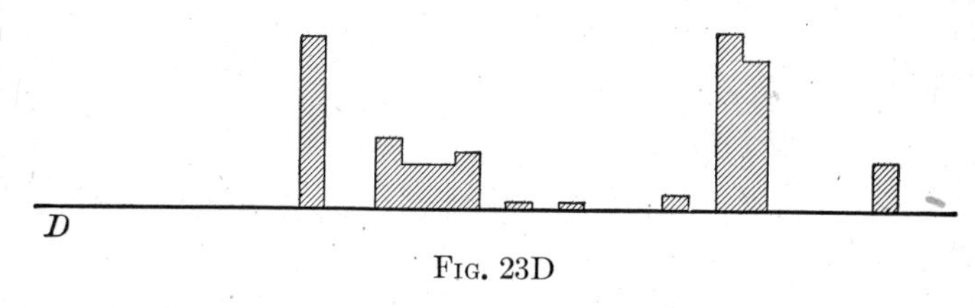

FIG. 23D

same plan except that it represents the assignment for two years, the algebra being combined with other mathematics in Book D, and each tenth of an inch along the base line equals twenty successive pages. An area of one hundredth of a square inch equals four practices, just as in the other diagrams. So Fig. 23D may be compared with the others if the reader will remember that the practices in 23D which seem to the eye as far apart as those in 23A or B or C are really twice as far apart.

Figs. 24A, 24B, and 24D represent practices on $(p+q)(p-q)=p^2-q^2$ in the same way, D differing from A and B as before.

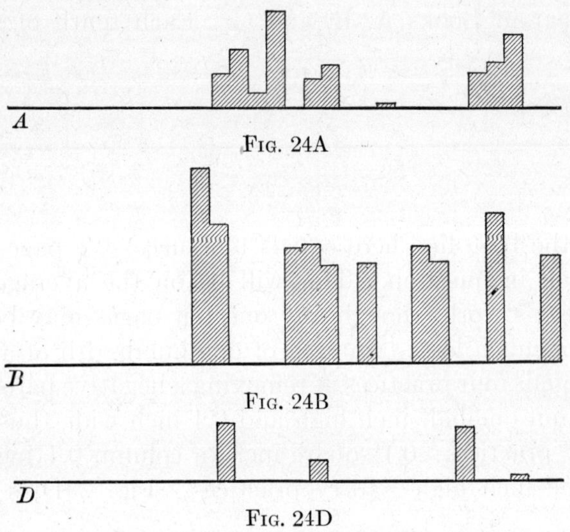

FIG. 24A

FIG. 24B

FIG. 24D

Figs. 25A, 25B, and 25D represent practices on "p^2-q^2 has $p+q$ and $p-q$ as its factors" in the same way.

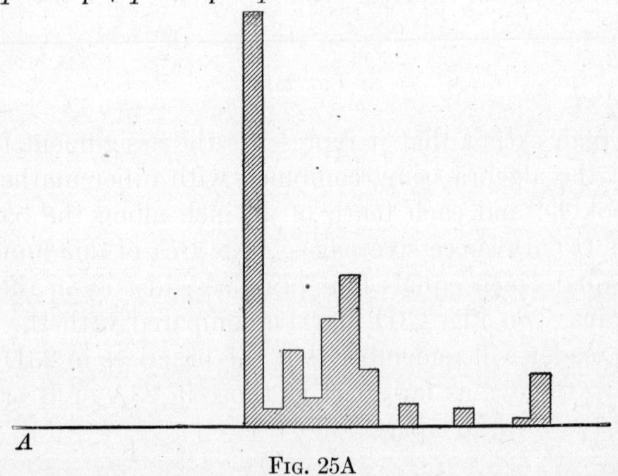

FIG. 25A

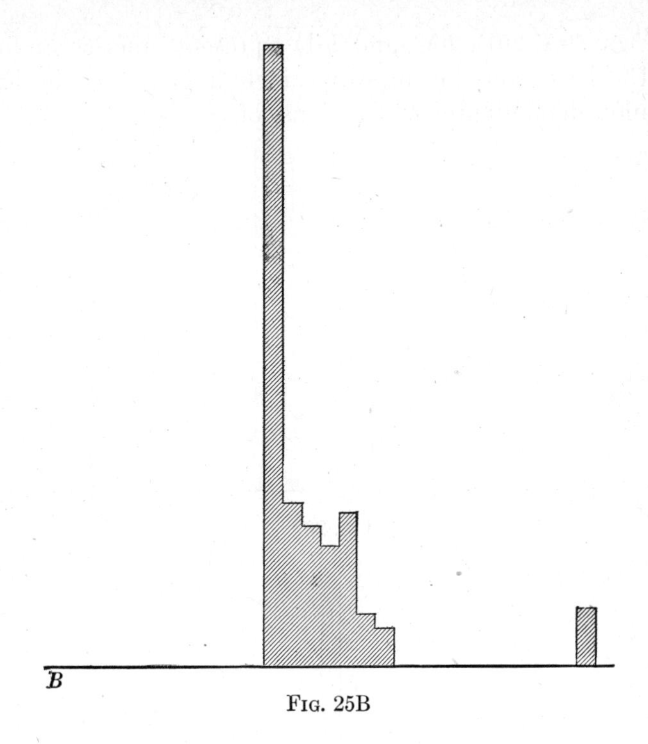

FIG. 25B

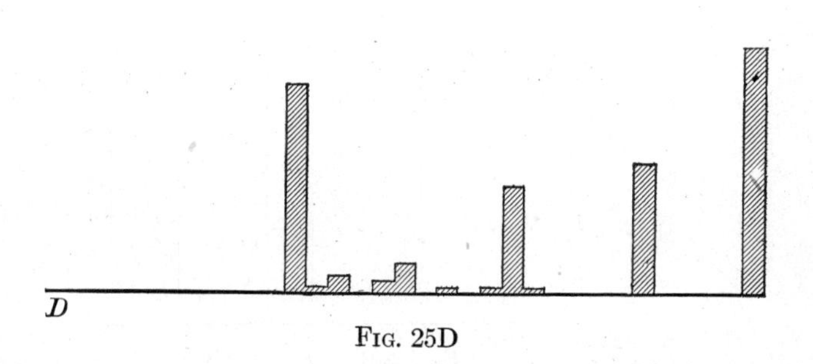

FIG. 25D

Figs. 26A, 26B, 26C, and 26D represent practices in finding the least common multiple (including finding the least common denominator of fractions, of course).

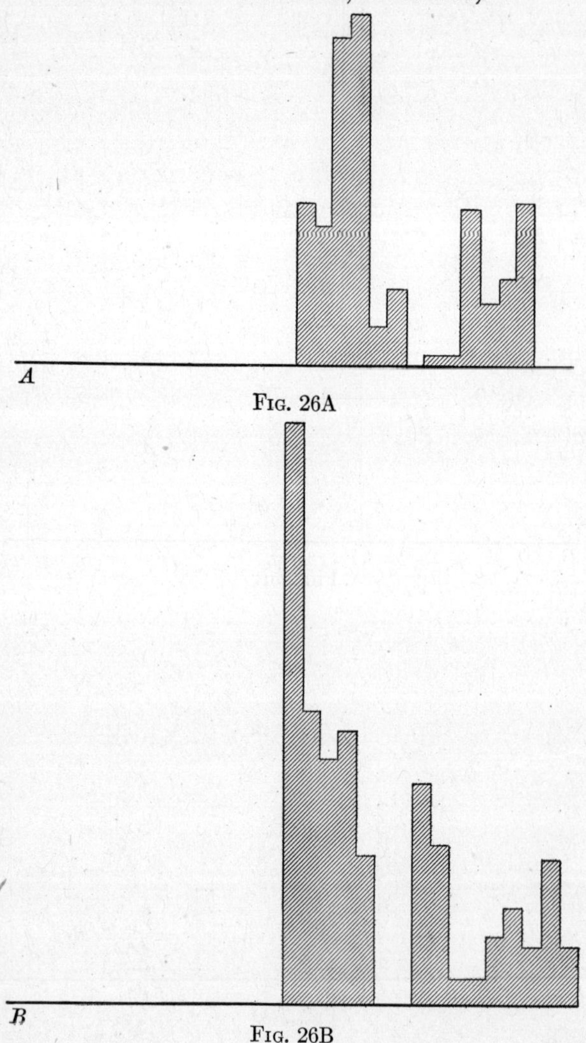

Fig. 26A

Fig. 26B

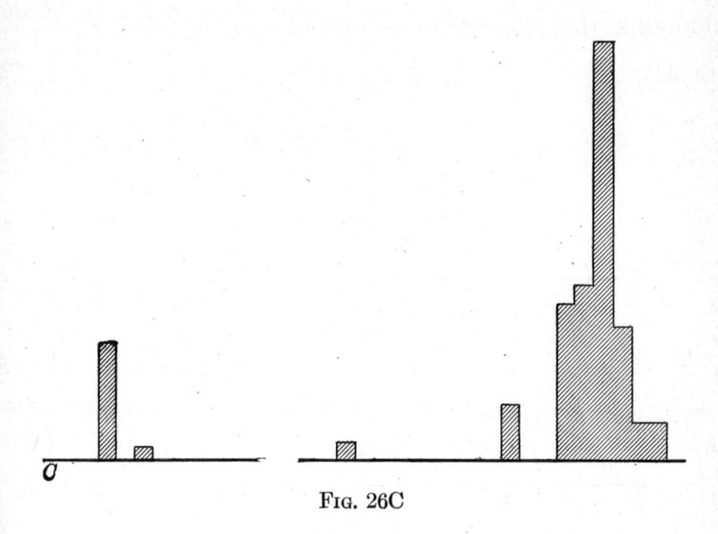

Fig. 26C

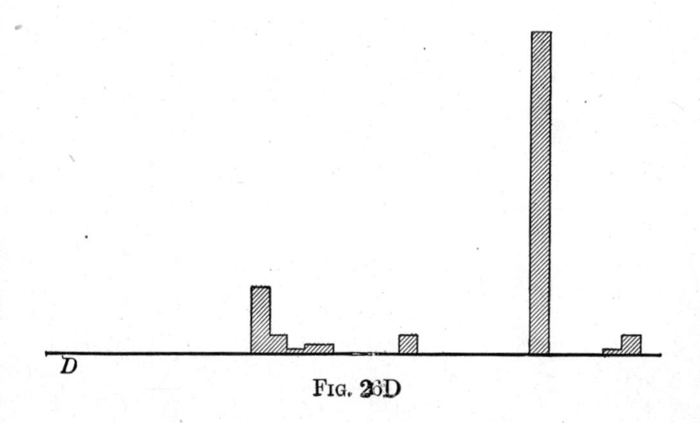

Fig. 26D

Figs. 27A, 27B, 27C, and 27D represent practice in changing the subject of a formula (e. g. solving $Q=\frac{dg_1}{M^2}$ for M).

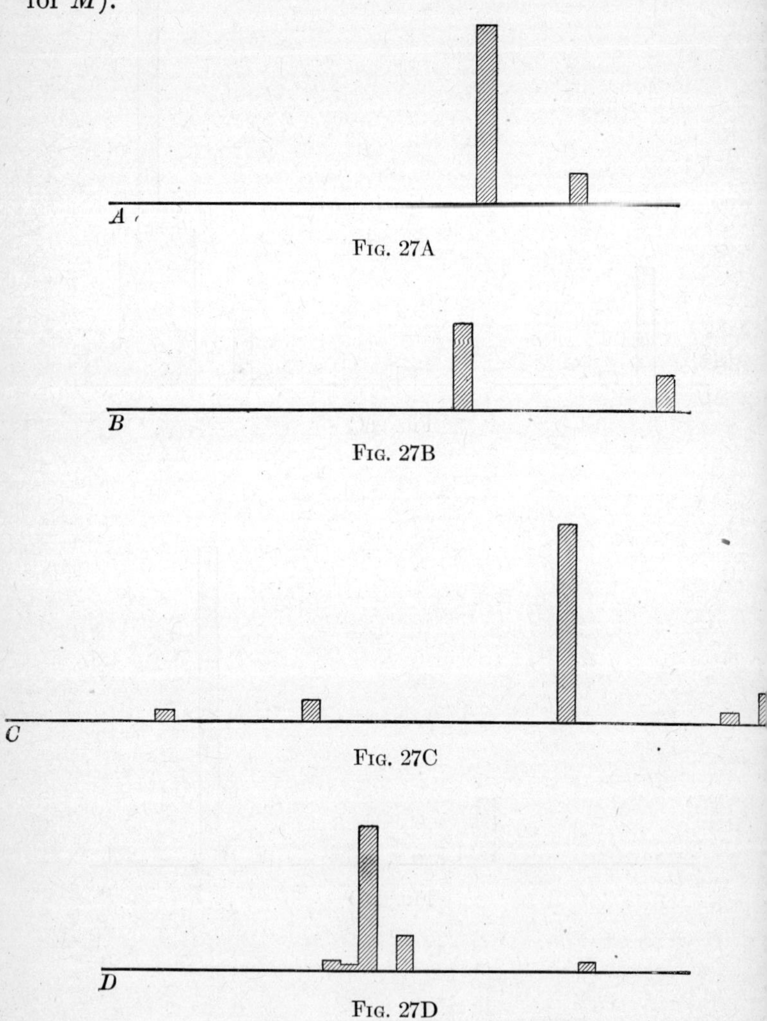

Fig. 27A

Fig. 27B

Fig. 27C

Fig. 27D

Figs. 28A, 28B, 28C, and 28D represent practice in factoring mx^2+px+q or x^2+px+q.

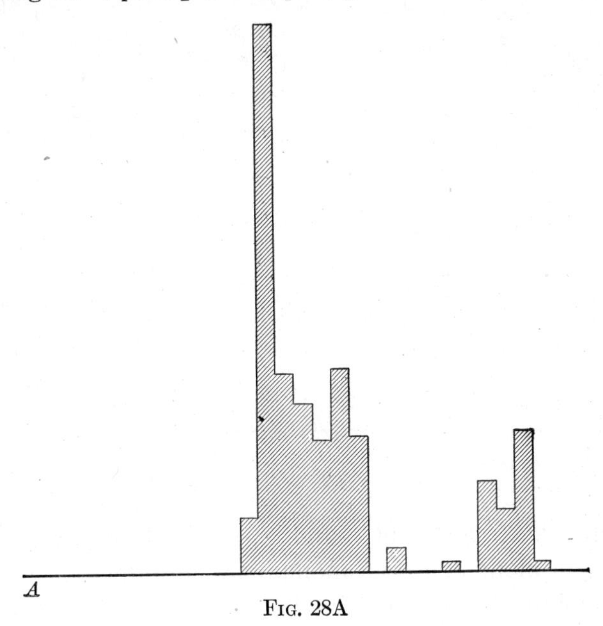

FIG. 28A

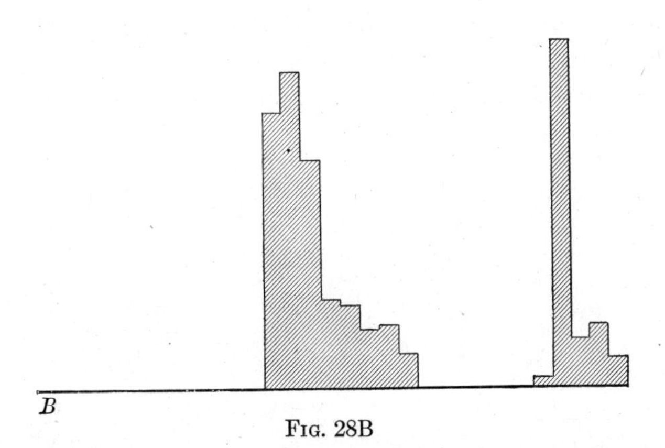

FIG. 28B

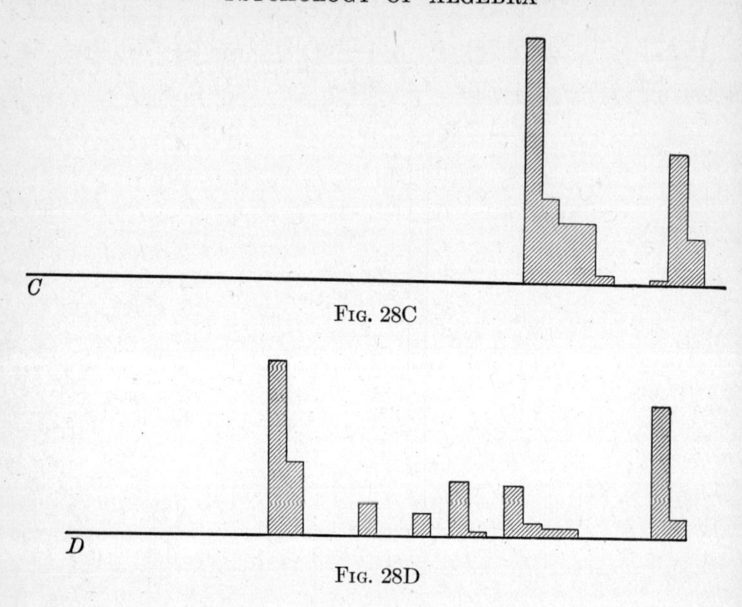

Fig. 28C

Fig. 28D

Figs. 29A, 29B, 29C, and 29D represent practice in dividing a monomial by a monomial containing at least one literal factor (exclusive of cases where the dividend and divisor are the same). In Figs. 29A, 29B, 29C, and 29D the area scale is different from that used in Figs. 17 to 28. Each hundredth of a square inch now equals twenty practices, a column one inch high and 0.1 wide equalling 200 practices.

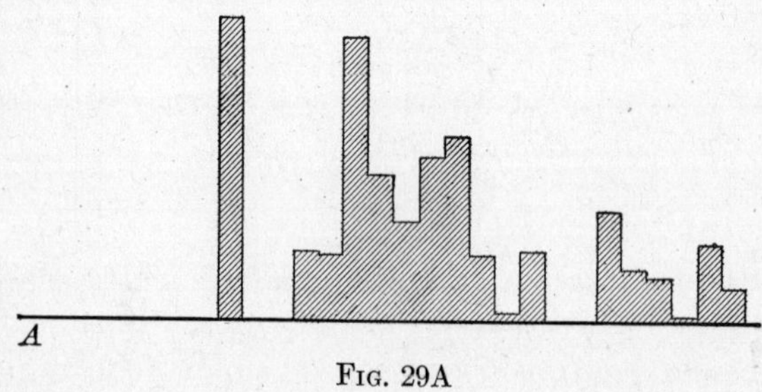

Fig. 29A

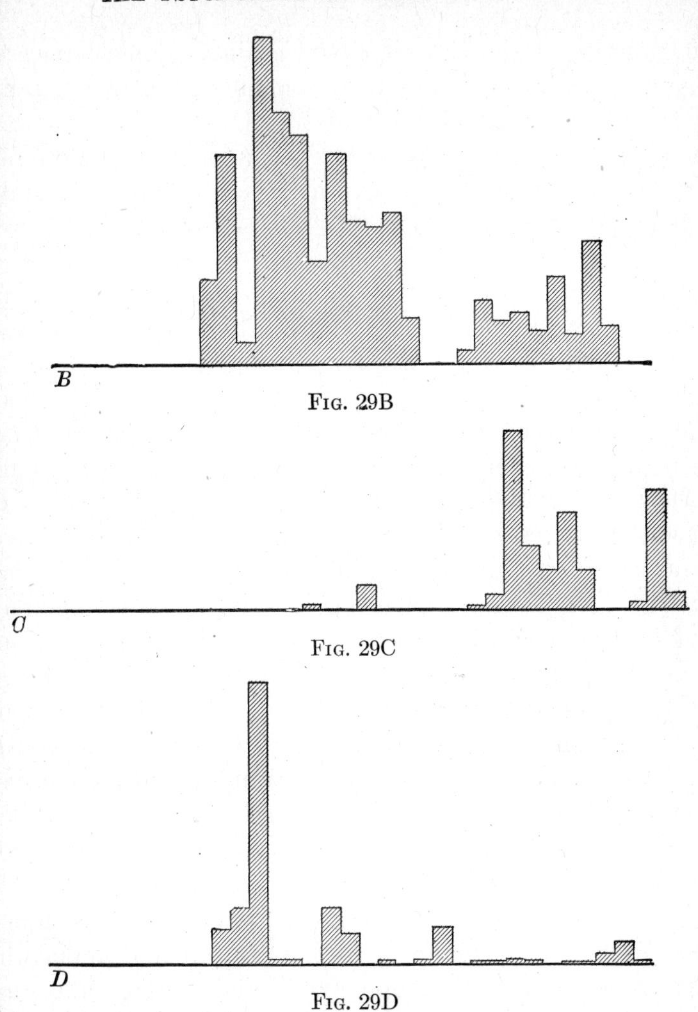

Fig. 29B

Fig. 29C

Fig. 29D

These distributions of practice do not show close correspondence with the psychological ideal. Nor would anybody probably regard them as the best possible distributions on pedagogical grounds. Some of them have too much of the practice at the time of first learning. Some have too long gaps without any practice; some have too great resemblance to a haphazard distribution. One can hardly avoid the conviction that every one of them could be made more helpful to learning by having practices subtracted at certain points and added at others.

The same impression is obtained from an examination of the distribution of practice on other abilities. Except for certain very commonly used abilities like the axioms, laws of signs, and rules for multiplying monomials which have so much practice in connection with the acquisition of other abilities that there is no question of economy in distributing their practice, every ability has a distribution which is imperfect, to say the least. It seems certain that a series of review exercises could be constructed to supply obvious deficiencies with great advantage to the learner.

The fact is that an author, no matter how talented, cannot keep in mind the facts for the distribution of practice for a hundred and more abilities, and provide effectively for all as he goes along, without sacrificing other more important matters. The only practicable way to provide for them is to chart them all as we have these samples, and then adjust them and readjust them to a closer approximation to the ideal so far as this can be done at no cost to vitality, interest, continuity and other desiderata in the course.

It is only within recent years that psychology has shown the teacher of algebra how to analyze out and define the mental connections the pupil should make. To make such a list, to decide how much practice each should have, to arrange the learning of them in such an order that each will furnish a maximum of facilitation and a minimum of inhibition to the rest; to distribute the practice for each so that there shall be a minimum of waste from overlearning what is sufficiently known and from relearning what should not have been allowed to lapse — this is a task which no teacher of mathematics had even considered until very recently, much less attempted.

It is worth considering. If we can take a dozen practices from page 100 where they bore the pupil already surfeited with a certain technique, and insert them on page 120 where they will help to save that technique from being forgotten, at no cost to other desiderata, we have made so much sure gain. It may be a small gain in comparison to that made by some clearer explanation or better procedure or more interesting material, but it is a gain; and it has the merit of requiring no added skill in the teacher.

The commonest defect in the distribution of practice is, in the psychologist's judgment, too great concentration. Consider the following cases:

Removing a double parenthesis	23 practices, all in one week.
Addition ot polynomials	38 practices, 18, 13, 7, in three successive weeks.
Cube of a monomial	21 practices, 16, 4, —, —, 1, all but one in two successive weeks.
Polynomial × a polynomial (to be multiplied out in detail)	112 practices, 89 in one week.
Square root of a polynomial	Book I, 20 practices, all in one week. Book II, 22 practices, all in one week.
Division of fractions	Book I, 173 practices, 160 in two successive weeks. Book II, 62 practices, 60 in one week.
Powers of fractions	42 practices, 37 all in one week.
Solving a quadratic with no x term	42 practices, all in one week.
Using simultaneous equations (2 unknowns) in solving problems	59 practices, 52 in one week.
Dividing polynomial by polynomial	Book I, 71 practices, 70 in one week. Book IV, 33 practices, 26 in one week.
Like powers of equals are equal	30 practices, all in one week.
Simultaneous equations (2 unknowns) solved by adding or subtracting	132 to 145 practices, 132 in two successive weeks.
Graphs, plotting points	166 practices, all in two weeks.
Completing the square by supplying an omitted term	93 practices, 92 in two weeks, separated by an interval.

Addition and subtraction of fractions...90 practices, 88 in two successive weeks.
Cancelling polynomial factor in fractions. 77 or more practices, all in one week
Cube root of a monomial perfect cube...54 practices, 44 in one week.
Changing sign of one term of fraction together with change of sign before the fraction..............................13 practices, all in one week.
Literal equations, other than formulas...26 practices, all in one week.
Forming a proportion from the factors of two equal products................23 practices, all in one week.
Multiplication of radical by radical.....20 practices, all in one week.

Doubtless wise teachers will supplement such assignments by reviews, but many teachers will expect more or less justifiably that a standard instrument of instruction should provide for reviews as well as for first learnings. Furthermore there is a more or less justifiable suspicion that a procedure which is never required later in the course of algebraic learning may itself not be worth learning.

In general also we can hardly expect teachers to plan for the distribution of learning as well as authors who have presumably much more ability and surely much more time. Observation of class work will show that the majority of teachers follow the assignments of the textbook rather closely. When they do not follow it, they do not as a rule improve upon it. In reviews they often go over the same tasks in about the same order and with about the same relative distribution.

These cases of very concentrated practice are of interest from another standpoint. One common defect in the distribution of practice is not revealed in our diagrams or tables, because our unit is so long, ten pages or about a week. This is the too great concentration of the first learning. If a certain ability is to be given, say, 100 minutes of practice during the week of first learning, it is customary to give that 100 all in one dose, say as 40 minutes of class work followed by 60 minutes of school or home study. Someimes this is wise, but in many matters of rather simple abits (like multiplying with monomials or dividing with nonomials or reducing fractions to lowest terms or collecting erms), it may be better to distribute the practice as say 0 on day one, 35 on day two, and 20 on day four. This an often be done by learning together two or more abilities which reinforce each other. Thus multiplication and division with monomials, changing fractions to higher terms and o lower terms, powers and roots $(a+b)$ $(a-b)$, and factoring $^2-b^2$ might be taught in pairs.[1]

Also there seems no reason why a pupil when primarily oncerned in learning to divide with monomials should for n hour do nothing else. This, in fact, encourages him to ay no attention to the division sign, and to learn the process ut of relation to other processes. After a score of such ivisions the next twenty might well be mixed with additions, subtractions and multiplications, say in the proportions 3, 1, 1, 1.

[1] Other things being equal, we teach one thing at a time, but the good fects of comparison and contrast make these pairings desirable.

CHAPTER XV

The Interest of Pupils in Algebra in Comparison with Other School Subjects

Except for the study by Schorling, Kahler, and Mille[r] [1916] there seem to have been no published reports of th[e] answers of high-school pupils to such questions as "Which subject do you like best?" or "What subjects have yo[u] studied in high school? Rank them in the order of you[r] liking for them." Willett has shown that the answers var[y] when the question is repeated, but has not published th[e] actual reported likings. Such reports need not, of course represent the truth. In certain matters pupils may deceiv[e] us or themselves or both. When taken critically, however they are well worth the consideration of teachers. Pupil[s] probably know what they like in school work rather bette[r] than most things about it; and the votes of pupils form [a] useful starting-point in any discussion or study of their in-terests. The facts we have to present tell at least one very clear and emphatic story — that pupils think that they lik[e] Latin less than any other subject that they study in high school.

In our investigation we first secured reports from about six hundred boys and about seven hundred girls in Grade 1[2] of four public high schools in New York (Group I). Thes[e] pupils are in most cases taking the "academic" course, as will be seen from the number reporting on each subject

They used the blank shown below in reporting. It would have been more convenient for our statistical treatment to have forced an ordering of subjects beyond two at each extreme, but for practical reasons this seemed inadvisable.

INT. IIa

Write your full name, your age, and what year of high school you are in. Write "*last*" if you are in the last year of high school. Write *n. l.* if you are in next to the last.

....

First Name *Middle Initial* *Last Name* *Age* *Year of H. S.*

Make a cross after each of these subjects which you have studied in High School. Then think which subject you liked most and write M. after it. Write N. M. after the one you liked next best. Write L. after the one you liked least. Write N. L. after the one you liked next to least.

English	....	....	Physiology	....	....
French	....	....	American History	....	....
Latin	....	....	European History	....	....
Spanish	....	....	Ancient History	....	....
Algebra	....	....	Drawing	....	....
Plane Geometry	....	....	Manual Training	....	....
Solid Geometry	....	....	Shop Work	....	....
Trigonometry	....	....	Typewriting	....	....
General Science	....	....	Stenography	....	....
Biology	....	....	Bookkeeping	....	....
Chemistry	....	....	Cooking	....	....
Physics	....	....	Sewing	....	....

The median rating for each subject was computed by calling M. equal to 5; N. M., 4; no rating, 3; N. L., 2; and L., 1. These numbers were considered to be the mid-points of the intervals for which they stand, e. g., 3 means 2.5 to 3.5; and it was assumed that the ratings in each interval were evenly distributed. The resulting medians indicate the *rank order* of reported interest, and nothing more.

TABLE 45

REPORTED INTERESTS OF BOYS IN GRADE 12: NEW YORK CITY

M. = Most Interesting. N. M. = Next Most Interesting.
L. = Least Interesting. N. L. = Next Least Interesting.

Rank	Subject	Number of Cases	Median Rating	Per-cent M.	Per-cent N. M.	Per-cent in Middle Group	Per-cent N. L.	Per-cent L.
1	American History	561	3.29	17.3	16.8	62.2	2.1	1.6
2	English	591	3.20	16.7	14.9	60.8	4.2	3.6
3	Physics	325	3.17	15.1	15.1	60.0	5.5	4.4
4	Bookkeeping	155	3.15	20.6	6.4	64.5	5.8	2.6
5	Trigonometry	140	3.12	8.6	15.7	67.0	5.7	2.9
6	Economics	243	3.11	9.9	8.2	76.5	2.9	2.5
7	Chemistry	422	3.05	10.7	9.0	67.3	6.2	6.9
8.5	Biology	499	3.03	6.8	6.1	78.4	5.0	3.8
8.5	European History	590	3.03	4.6	8.0	79.2	6.3	2.0
10.5	Spanish	273	3.01	8.4	13.6	57.4	11.4	9.2
10.5	Typewriting	189	3.01	1.1	3.2	93.6	1.1	1.1
12	Solid Geometry	57	3.00	1.7	5.3	86.0	1.7	5.3
13	Algebra	573	2.98	7.9	6.5	68.8	8.4	8.6
14	Physical Training	86	2.96	2.3	1.2	86.0	9.3	1.2
15	Stenography	63	2.95	3.2	6.3	73.0	11.1	6.3
16	Plane Geometry	557	2.93	5.6	4.5	70.4	9.7	9.9
17	French	388	2.88	6.2	7.7	58.2	13.6	14.2
18	Drawing	573	2.87	3.7	1.2	71.9	11.0	12.2
19	Music	106	2.84	....	....	75.4	13.2	11.3
20	Ancient History	543	2.83	.6	1.1	72.6	12.6	13.0
21	Hygiene	139	2.81	....	.7	77.6	9.4	12.2
22	Latin	274	2.39	1.8	4.0	42.4	17.2	34.6

To measure the amount of reported interest, we should have to compute from the individual records the percents reporting a certain study as more interesting or less interesting or unspecified as to interest. We should then need to make some assumption about the distribution of liking for studies in the group of individuals, and by putting the two facts together estimate how much more interesting one subject was than the other. It seems unwise to do this in view of (1) the labor involved, (2) the insecurity of the assumption, and (3) the fact that we can learn most of what we need to know about the matter by simpler and safer means.

TABLE 46

REPORTED INTERESTS OF GIRLS IN GRADE 12: NEW YORK CITY

M. = Most Interesting. N. M. = Next Most Interesting.
L. = Least Interesting. N. L. = Next Least Interesting.

Rank	Subject	Number of Cases	Median Rating	Per cent M.	Per cent N. M.	Per cent in Middle Group	Per cent N. L.	Per cent L.
1	English..........	709	3.35	27.2	14.5	55.3	1.3	1.6
2	Bookkeeping.....	38	3.14	18.4	13.2	52.8	13.2	2.6
4	French..........	609	3.11	13.5	14.5	56.8	6.6	8.7
4	Stenography.....	126	3.11	10.3	13.5	66.6	4.0	5.6
4	Trigonometry....	27	3.11	4.5	9.1	81.9	....	4.5
6	Chemistry.......	258	3.09	10.1	12.4	66.8	6.6	4.3
7	Physiography....	149	3.08	12.8	8.0	69.1	8.0	2.0
8.5	Biology..........	657	3.07	10.2	7.9	74.1	5.5	2.3
8.5	Spanish..........	263	3.07	11.8	13.3	57.4	10.6	6.8
10.5	American History.	684	3.06	6.3	11.5	74.0	4.7	3.7
10.5	Physics..........	128	3.06	9.4	14.8	58.6	7.0	10.2
12	Economics.......	372	3.04	7.5	9.7	71.2	5.4	6.2
13.5	Cooking.........	136	3.00	1.5	8.1	80.9	7.4	2.2
13.5	Current Events...	127	3.00	....	....	100.0	....	....
15.5	Home Nursing...	82	2.99	....	....	98.8	1.2	....
15.5	Music..........	423	2.99	1.7	2.4	89.3	4.0	2.4
17	Typewriting......	169	2.98	....	3.0	90.0	4.1	3.0
18.5	European History	687	2.97	3.2	5.2	77.8	8.2	5.7
18.5	Physical Training.	472	2.97	3.4	2.8	82.9	5.5	5.5
21.5	Drawing........	696	2.95	4.9	2.4	78.2	5.9	7.6
21.5	Elocution........	163	2.95	....	....	90.8	4.9	4.3
21.5	Hygiene.........	157	2.95	....	....	91.8	7.0	1.3
21.5	Sewing..........	283	2.95	1.8	1.1	85.5	6.7	4.9
24	Ancient History..	575	2.88	.7	1.7	76.4	10.3	11.0
25	Algebra.........	675	2.87	5.3	4.4	64.0	12.6	13.6
26	Plane Geometry..	553	2.80	3.8	5.4	57.9	15.4	17.7
27	Latin...........	388	2.73	4.1	6.4	51.0	13.1	25.3

The medians, as just stated, give us the order. English is at the top and Latin is at the bottom. Twelfth-grade pupils in these schools in general put English above Latin by about as much as they put their third choice out of fourteen above their twelfth. From English to Latin is such a difference that a pupil in Grade 12 reporting on fourteen subjects will on the average feel about that amount of difference in interest between the subject ranked third and that ranked twelfth.

This is, of course, not an enormous difference. If the reader will list fourteen subjects that he took in high school and rank them in order for interest, he may even feel no absolute certainty that he really did like his No. 3 better than his No. 12. Pupils are far from unanimous in finding English more interesting than Latin. The count, individual by individual, of the boys who took both subjects gives 155 voting English more interesting, 24 voting it less interesting, and 81 putting both in the middle ten. If these last had been forced to rate these two subjects in comparison probably about 55 would have put English higher.[1] So approximately 81 percent prefer English. A similar procedure with 385 girls gives 80 percent.

The difference in interest between one subject and another may be expressed approximately[2] as a fraction of this English-Latin difference (call it "elg" in the case of girls and "elb" in the case of boys): that is, the difference between the medians is divided by .62 in the case of the girls and by .81 in the case of the boys. Thus, for girls: English-French is .39 elg; French-Biology is .07 elg; Biology-American History is .02 elg; American History-European History is .14 elg; European History-Algebra is .16 elg; Algebra-Latin is .22 elg. For boys, American History-English is .11 elb; English-Biology is .21 elb; and so on.

If we leave out English and Latin in the case of the girls and American History and Latin in the case of the boys, all the remaining subjects are within a range of .53 elg for girls and .48 elb for boys.

It would of course be unfair to use these results without consideration of the fact that the studies reported by the

[1] If we divide them as those are divided who made the distinction, 86.6 percent would favor English. If they represented real indifference, the percent would, of course, be 50. Taking an even compromise the percent would be 68.3 and $.683 \times 81 = 55$.

[2] Not exactly, since we do not know the form of distribution.

smaller numbers are more often elective, and so probably better liked by those who take them than they would be by the others if they were required of all. Trigonometry, for example, is rated high by the 27 girls who took it, and algebra is rated low by the 675 girls who took it, but if all of these had taken trigonometry there might have been no difference. We have, therefore, made separate tables (Table 47 and Table 48) for those subjects reported as having been studied by 80 per cent of the total number, that is, by 475 or more boys or 570 or more girls.

In order to learn whether these results are peculiar to academic students in New York City or are of fairly general significance we have secured similar data from a city in the central west, a city farther west, and a New York high school less devoted to academic and college preparatory work. The median ratings for the ten commonly taken subjects are shown in Table 49. A combined median rating for

TABLE 47

Boys' Interests In Subjects Reported as Taken by 80 Percent or More

Subject	Median Rating	Liked Most or Next to Most	In the Middle Ten	Liked Least or Next to Least	Approximate Successive Differences in Terms of the English-Latin Difference
American History.	3.29	34.1	62.2	3.7	.11
English..........	3.20	31.6	60.8	7.8	.21
Biology..........	3.03	12.9	78.4	8.8	.00
European History	3.03	12.6	79.2	8.3	.06
Algebra..........	2.98	14.4	68.8	17.0	.06
Plane Geometry..	2.93	10.1	70.4	19.6	.07
Drawing.........	2.87	4.9	71.9	23.2	.05
Ancient History..	2.83	1.7	72.6	25.6	.54
Latin (274 boys)..	2.39	....	....	....	....

TABLE 48

GIRLS' INTERESTS IN SUBJECTS REPORTED AS TAKEN BY 80 PERCENT OR MORE

Subject	Median Rating	Liked Most or Next to Most	In the Middle Ten	Liked Least or Next to Least	Approximate Successive Differences in Terms of the English-Latin Difference
English..........	3.35	41.7	55.3	2.9	.39
French..........	3.11	28.0	56.8	15.3	.07
Biology..........	3.07	18.1	74.1	7.8	.02
American History.	3.06	17.8	74.0	8.4	.14
European History	2.97	8.4	77.8	13.9	.03
Drawing.........	2.95	7.3	78.2	13.5	.11
Ancient History..	2.88	2.4	76.4	21.3	.02
Algebra..........	2.87	9.7	64.0	26.2	.23
Latin (388 girls)..	2.73	....	....	....	.....

TABLE 49

MEDIAN RATINGS FOR TEN COMMONLY TAKEN SUBJECTS: GRADE 12[1] IN THREE SCHOOLS

Subject	School D		School E		School F	
	Number of Cases	Median Rating	Number of Cases	Median Rating	Number of Cases	Median Rating
English...........	123	2.91	68	3.34	134	3.70
American History..	114	2.92	64	3.09	99	3.04
French...........	27	3.03	53	3.00	60	3.20
Biology..........	—[2]	—[2]	56	3.07	64[3]	3.10[3]
European History.	32	2.88	67	2.93	50	3.00
Drawing.........	60	2.79	69	3.06	44	3.16
Algebra..........	108	2.94	65	2.81	134	2.90
Ancient History...	....	....	60	2.96	74	2.95
Plane Geometry...	102	2.92	56	3.04	117	2.81
Latin............	65	1.92	43	2.53	100	2.07

[1] Some from Grade 11 were included in one school.
[2] Biology had only four students.
[3] Botany.

the three schools is shown in Table 51 beside similar facts for the New York group. As in the New York group, English is at the top and Latin is at the bottom. The differences between the others are very small, their entire range being only .18 of the English-Latin difference.

TABLE 50

MEDIAN RATINGS FOR TEN COMMONLY TAKEN SUBJECTS: GRADE 12 IN FOUR SCHOOLS[3]

BOYS

	School G		School H		School I		Average of Medians
	No.	Med.	No.	Med.	No.	Med.	
American History	58	3.08[1]	38	3.28	36	3.96	3.44
English	79	3.07	44	3.11	62	3.04	3.07
French	55	2.86	28	3.19	55	2.97	3.01
Biology	34	2.91	40	2.98	7	3.20	3.03
European History	58	3.08[1]	42	3.00	48	2.95	3.01
Algebra	83	2.85[2]	45	3.00	62	2.89	2.91
Plane Geometry	83	2.85[2]	45	3.02	61	3.07	2.95
Drawing	49	3.39	44	2.87	19	3.00	3.09
Ancient History	30	3.19	42	2.69	41	2.90	2.93
Latin	56	1.40	6	2.50	57	2.22	2.04

GIRLS

	School G		School J		Average of Medians
	No.	Med.	No.	Med.	
American History	29	3.57[1]	26	3.14	3.35
English	61	3.48	50	4.60	4.04
French	58	3.79	48	2.74	3.27
Biology	17	2.78	25	3.05	2.92
European History	29	3.57[1]	27	3.00	3.29
Algebra	62	2.69[2]	48	2.87	2.78
Plane Geometry	62	2.69[2]	45	3.00	2.85
Drawing	12	2.83	32	3.08	2.96
Ancient History	28	3.21	36	3.02	3.12
Latin	48	1.92	41	2.15	2.04

[1] This median is for European and American History together.
[2] This median is for Algebra and Plane Geometry together.
[3] Schools H and I are boys' schools; school J is a girls' school.

As a further check we have used two private schools in New York City, a private school in Chicago, and another public school in New York City, concerned chiefly with academic and college preparatory work. The median ratings are shown in Table 50, by schools and sex.

A combined median rating for the four schools is shown in Table 51, beside similar facts for the New York group and the group of three schools used as the first checks. English leads and Latin is last, as before.

TABLE 51

COMBINED MEDIAN RATINGS OF SUBJECTS

	Group I Four New York Schools	Group II Three Schools	Group III Four Schools
English	3.28	3.32	3.44
American History	3.18	3.02	3.41
Biology	3.05	3.09	2.98
French	3.00	3.08	3.18
European History	3.00	2.94	3.12
Algebra	2.93	2.88	2.86
Drawing	2.91	3.00	3.03
Plane Geometry	2.87	2.92	2.93
Ancient History	2.86	2.96	3.00
Latin	2.56	2.17	2.04

Book [1922] and Schorling, Kahler, and Miller [1916] have reported pupils' votes somewhat similar to these. Book's results differ from ours in putting mathematics much higher. The results reported by the Schorling Committee are in accord with ours. Book asked high-school seniors in Indiana to report their favorite study. The results appear in Table 52. The percentages are of course not to be taken at their face value, since the number of pupils who had studied agriculture, or botany, for example, was probably much smaller than the number who had studied mathematics or English.

TABLE 52

NUMBER OF STUDENTS SELECTING DIFFERENT HIGH-SCHOOL SUBJECTS AS THEIR FAVORITE STUDY

Favorite Subject	Cases	Per-cent of Total Group	Favorite Subject	Cases	Per-cent of Total Group
Mathematics	1156	20	Latin	196	3
English and Literature	1119	19	Manual Training	147	3
History	683	12	Chemistry	144	3
Commercial	561	10	Music and Art	143	3
Science	368	6	Agriculture	87	2
Physics	323	6	Botany	53	0.92
Domestic Science	292	5	Debating	44	0.77
Language	240	4	No subject selected	192	3

Schorling, Kahler, and Miller [1916, p. 612f.] report the votes of pupils in three Chicago high schools as to the "Degree of enjoyment received from their studies," their choice being amongst "Very much," "A little," and "Not at all." [1] We quote their results in Table 53.

TABLE 53

STATEMENT OF STUDENTS AS TO DEGREE OF ENJOYMENT RECEIVED FROM THEIR STUDIES

UNIVERSITY HIGH SCHOOL

Subject	I Very Much	II A Little	III Not at All	I+II	Number of Students
History	67.1	31.6	1.3	98.7	149
French	63.1	34.2	2.7	97.3	149
General Science	62.6	33.8	3.6	96.4	83
Other Sciences	68.2	27.2	4.6	95.4	151
Popular Arts	56.9	36.4	6.7	93.3	165
German	47.3	44.5	8.2	91.8	146
Mathematics	49.7	40.9	9.4	90.6	318
English	43.6	46.5	9.9	90.1	372
Latin	45.5	42.5	12.0	88.0	233
Music	51.0	32.7	16.2	83.7	49

[1] "The principals of three high schools were asked to submit a questionnaire to the students in which each one was asked to indicate the subjects he 'enjoyed very much,' those he 'enjoyed a little,' and those 'enjoyed not at all.'"

TABLE 53—Continued

HYDE PARK HIGH SCHOOL

Subject	I	II	III	I+II	Number of Students
English	62.8	33.2	4.0	96.0	1416
Mathematics	51.5	40.2	8.3	91.7	955
German	52.2	37.6	10.2	89.8	322
French	60.7	32.3	7.0	93.0	295
History	52.6	39.1	8.3	91.7	504
Physics	50.0	38.2	11.8	88.2	134
Chemistry	58.7	38.5	2.8	97.2	109
Latin	40.6	38.7	20.7	79.3	660
Botany	68.0	27.0	4.1	95.9	147
Zoology	47.1	38.4	14.5	85.5	104
Physiology	71.3	22.8	5.9	94.1	412
Physiography	60.1	33.9	6.0	94.0	153

OAK PARK AND RIVER FOREST HIGH SCHOOL

Subject	I	II	III	I+II	Number of Students
Manual Training	79.6	16.8	3.6	96.4	349
General Science	61.7	34.4	3.9	96.1	444
Music	70.7	25.0	4.3	95.7	48
Drawing	73.3	22.4	4.3	95.7	165
History	57.6	37.6	4.8	95.2	460
French	62.2	32.4	5.4	94.6	114
English	56.0	37.1	6.9	93.1	1024
German	50.0	41.0	9.0	91.0	234
Mathematics	45.7	42.9	11.4	88.6	745
Latin	38.5	47.6	13.9	86.1	483

In Table 54 we have summarized the votes for certain subjects for convenient inspection by counting a vote of "Very much" as +1, "A little" as −1, and "Not at all" as −3. This procedure is somewhat arbitrary, but the reader versed in mental measurements will realize that it is not unreasonable, and that any reasonable weighting of the three sorts of votes would give approximately the same relative positions to the studies. Equal weight is given in Table 54 to each of the three schools.

TABLE 54

A SUMMARY OF THE SCHORLING COMMITTEE'S DATA

SUBJECT	AVERAGE WEIGHTED RATING	NUMBER OF STUDENTS
English	− 5.6	2812
History	+ 8.6	1113
French	+13.9	558
German	−18.6	702
Average for Science	+13.8	?
Mathematics	−21.5	2018
Latin	−48.0	1376

We cannot systematically compare these results of Schorling's committee with ours because they used pupils in all the grades of high school, whereas we used only pupils in Grade 12 (plus a few in Grade 11 from one city). In only one of his three schools were any two subjects reported on by 80 per cent of the pupils. Taking the percentages at their face value, Latin is at the bottom in all three schools; English is at the top in one, next to the bottom in another, and at about the middle in the third. The history vote we cannot use, since so much depends on whether American History or Ancient History is the subject, and this is not stated. Mathematics is low in all three schools.

If we take the subjects reported on by at least 50 per cent of the pupils (including Latin for comparison) we have English, Mathematics, and Latin in that order, with Latin much farther below Mathematics than English is above it.

In general, it appears practically certain that in present schools with present teaching, most pupils who study English, Mathematics, and Latin will like them in that order, the interval between Mathematics and Latin being much wider than the other. If all the pupils who were required to take English and Mathematics had been required to take Latin also, Latin would presumably have had a still lower relative position. Algebra is liked a trifle better than plane geometry.

It is not within the province of this chapter to discuss the possible causes of the relative interests reported. Certain thinkers will consider difficulty as the chief force; others will find evidence that abstractness is the essential; others, more hopeful of youth, will argue that these preferences are for what is modern, vital, and serviceable. In any case, we need to know these interests, most of all perhaps if it is our aim to change or thwart them.

CHAPTER XVI

THE INTEREST OF PUPILS IN VARIOUS FEATURES OF ALGEBRAIC LEARNING

Certain matters concerning pupils' likings for various features of algebra are matters of rather general agreement, for example, their preference for numerical rather than literal equations. Others are matters of very diverse opinions, for example, the relative interest in elaborate equations and simplifications involving shrewd inspection of the data and ingenious arrangement and manipulation of the numbers. Very little evidence has been published concerning any of these matters.

Such evidence may be obtained from three sources. The first is expert observation of the behavior of pupils in school and at home when engaged in the study of algebra; the second is the testimony of pupils; the third is the testimony of teachers. The study reported here used the second source. The testimony of school pupils is beset by many influences tending to error, but is especially competent in matters of likes and dislikes. Moreover, the nature of the questions asked was such as to reduce the forces productive of constant errors. The variable errors do little or no harm in the present inquiry. Any one pupil had only the simple task quoted below, of examining eight samples of algebraic tasks and choosing the four sorts of work liked best, next best, next to least, and least.

INT. Ia.

Look at these samples of some of the things you learned to do in algebra. Think which sort of work you liked most and mark it M. Think which you liked next best and mark it N. M. Think which you liked least, or disliked most, and mark it L. Think which you liked next to least and mark it N. L.

1. $4x-7=5x-4-(3x-1)$

2. Subtract $\frac{a^2-b}{c}$ from $a-\frac{b}{c}$

3. $\frac{x}{a-b}+\frac{x}{a+b}-a^2c=b^2c$ Find what x equals.

4. Simplify $\dfrac{\left\{1+\dfrac{2}{3-\frac{4}{5}}\right\}+\dfrac{1}{11}}{4}$

5. $\left.\begin{array}{l}2x+y=2(a+b)\\4x+2y=a+4b\end{array}\right\}$ Find what x equals and what y equals.

6. An engine can propel a boat 12 miles an hour in still water. How long will it take the boat to go upstream 10 miles and back again if the current is 2 miles an hour?

7. Draw graph of $x+2$, $x+4$, and $x+6$

8. Simplify $2a^2-18-(a+3)\left\{\frac{a^2+3a-18}{a+16}\right\}$

INT. Ib.

1. $4x+a=2x-b-(x-a)$ Find what x equals.

2. $a=1$, $b=2$, $c=3$ What does $\frac{a^2}{b}-\frac{c}{a+b}$ equal?

3. $\frac{x}{4}+\frac{5}{x+3}=\frac{x-2}{3}$ Find what x equals.

4. Simplify $\dfrac{\dfrac{2a^2}{(a+b)^2-(a-b)^2}}{b}-\dfrac{a^2+2ab-3b^2}{a-b}$

5. $2x+\frac{x}{2}=3x-4$ Find what x equals.

6. What is the total cost of n articles at c per article, if e is paid for expressage and $.02nc$ for insurance?

7. Draw graphs of $y=2x$ and $y=x^2$.

8. Half of a number plus a third of the number plus a fifth of the number equals one more than the number. What is the number?

INT. Ic.

1. $x+9=2x-5$

2. Subtract $\frac{a^2-b}{c}$ from $a-\frac{b}{c}$

3. $a=1$, $b=2$, $c=3$ What does $\frac{a^2}{b}-\frac{c}{a+b}$ equal?

4. Simplify $\frac{a^4-b^4}{(a^2+b^2)\ (a+b^2)}$

5. $\left.\begin{array}{l} 2x+4y=11 \\ 4x+2y=\ \ 9 \end{array}\right\}$ Find what x equals and what y equals.

6. An engine can propel a boat 12 miles an hour in still water. How long will it take the boat to go upstream 10 miles and back again if the current is 2 miles an hour?

7. Half of a number plus a third of the number plus a fifth of the number equals one more than the number. What is the number?

8. Simplify $\frac{\left[a+\frac{b}{c-\frac{a}{b}}\right]-\frac{abc}{bc-a}}{b-a}$

One hundred and forty-six boys and two hundred and seventy-one girls, scattered through a number of schools, did this, using form Ia; 151 boys and 260 girls similarly made choices, but using the eight sample tasks of Ib; 133 boys and 248 girls similarly made choices, but using the eight tasks of Ic. The use of actual sample tasks frees these ratings from many of the objections to questionnaires of the ordinary type, and is in general preferable to asking children general questions. For rigorous and convenient quantitative treatment it would have been better to have forced a complete ordering of all eight items, but it was our plan to make any necessary sacrifices to increase the probability that each pupil looked at each of the eight tasks, and ranked them according to his liking so far as he could. To force him to report differences where he felt none, or to follow some elaborate instructions, seemed likely to disturb him. Also it was desired to avoid the slightest burden on the teachers, beyond having the blanks distributed and collected.

Our own labor could have been reduced and the treatment made simpler if each pupil had been required to rank sixteen or more tasks instead of eight. We kept the number down, first, to simplify the task of rating; second, to prevent any pupil from considering such general issues as *letters* versus *numbers;* and third, to discourage any tendency to regard this as in any sense a test of individual merit and so to copy from other pupils. Blanks Ia and Ib, or Ia, Ib, and Ic were distributed in rotation about the classroom.

It will be noted that blank Ic duplicates two of the samples from Ia and two from Ib, thus serving as a link to put all the samples in one order for interest. The best way to obtain such an order is to pair every sample with every other, asking for each pair, "How many preferred 1 to 2, and vice versa? How many put them both in the undifferentiated middle group?" This is, however, extremely laborious and not necessary for the purposes of this article, which is to obtain approximate, not exact, measures.

The method which we have used is as follows: Treat boys' and girls' ratings separately. Let L., N. L., X, N. M., and M. be rated arbitrarily as 1, 2, 3, 4, and 5 respectively, 1, 2, 3, 4, and 5 meaning 0.5 to 1.5, 1.5 to 2.5, 2.5 to 3.5, etc. Tabulate the ratings as shown in Tables 55 and 56, and compute the median value for the ratings of each sample. Express these median values as divergences from some one reference point (we use the interestingness of $a2$, identical with $c2$, as this reference point). This gives the general order for interest of Table 57. These deviation numbers remain arbitrary and will require definition.

The procedure is as follows: Starting with $a2$ for boys as our point of reference or zero point we have $a1$ for boys equal

TABLE 55

RELATIVE INTERESTS IN VARIOUS SORTS OF ALGEBRAIC WORK: BOYS

Frequencies of votes of "Liked Most," "Liked Next Most," etc., in percents.

Set Ia ($n=146$)

	M.	N. M.	X.	N. L.	L.
a1	24.0	17.8	52.7	4.8	.7
a2	.7	6.2	71.9	13.7	7.5
a3	2.7	11.6	67.1	13.7	4.8
a4	2.1	.7	43.1	18.5	35.6
a5	34.2	21.9	32.4	8.2	3.4
a6	16.4	12.3	40.4	12.3	18.5
a7	17.8	26.0	41.8	4.1	10.3
a8	2.1	3.4	50.6	24.7	19.2

Set Ib ($n=151$)

	M.	N. M.	X.	N. L.	L.
b1	13.9	7.3	65.5	10.6	2.6
b2	28.5	19.9	39.1	6.6	6.0
b3	9.3	11.9	68.2	8.0	2.6
b4	2.0	8.6	23.8	20.5	45.0
b5	11.9	19.2	64.9	4.0	0.0
b6	6.0	8.6	45.7	21.2	18.5
b7	15.2	11.9	46.3	15.2	11.3
b8	13.2	12.6	46.3	13.9	13.9

Set Ic ($n=133$)

	M.	N. M.	X.	N. L.	L.
c1	29.3	25.6	39.1	5.3	.8
c2	0.0	0.0	57.9	33.8	8.3
c3	5.3	9.8	74.4	6.8	3.8
c4	2.3	7.5	67.6	16.5	6.0
c5	42.1	30.8	23.3	3.8	0.0
c6	11.3	9.8	62.4	9.0	7.5
c7	9.0	11.3	59.5	15.0	5.3
c8	.7	5.3	15.8	9.8	68.4

TABLE 56

RELATIVE INTERESTS IN VARIOUS SORTS OF ALGEBRAIC WORK: GIRLS

Frequencies of votes of "Liked Most," "Liked Next Most," etc., in percents.

Set Ia ($n=271$)

	M.	N. M.	X.	N. L.	L.
*a*1	39.5	23.6	35.1	1.1	.7
*a*2	1.1	3.3	84.8	6.3	4.4
*a*3	4.8	13.6	70.4	7.4	3.7
*a*4	1.8	2.6	46.1	26.9	22.5
*a*5	25.1	30.3	32.8	7.7	4.1
*a*6	5.9	4.8	27.7	16.2	45.3
*a*7	21.0	19.6	42.1	8.5	8.9
*a*8	.7	2.2	60.9	25.8	10.3

Set Ib ($n=260$)

	M.	N. M.	X.	N. L.	L.
*b*1	19.2	16.9	55.4	8.1	.4
*b*2	25.8	21.5	46.2	4.6	1.9
*b*3	9.6	15.0	66.2	8.1	1.2
*b*4	3.1	5.4	35.0	27.7	28.8
*b*5	13.5	14.6	68.8	2.3	.8
*b*6	1.5	.4	35.7	23.1	39.3
*b*7	14.2	13.5	47.7	10.0	14.6
*b*8	13.1	12.7	45.0	16.1	13.1

Set Ic ($n=248$)

	M.	N. M.	X.	N. L.	L.
*c*1	31.4	25.0	43.1	.4	0.0
*c*2	.4	2.8	79.8	12.5	4.4
*c*3	6.9	15.7	71.0	5.2	1.2
*c*4	7.3	13.3	64.5	12.9	2.0
*c*5	45.2	25.4	25.0	2.8	1.6
*c*6	.4	.8	39.9	19.8	39.1
*c*7	7.7	15.3	54.0	19.3	3.6
*c*8	.8	1.6	22.6	27.0	48.0

TABLE 57

THE ORDER FOR INTEREST OF SAMPLE ALGEBRAIC TASKS

		BOYS	GIRLS	AVERAGE (EQUAL WEIGHT TO BOYS AND GIRLS)
$\begin{cases} 2x+4y=11 \\ 4x+2y=9 \end{cases}$	$c5$	1.60	1.40	1.50
$x+9=2x-5$	$c1$	1.05	.85	.95
$\begin{cases} 2x+y=2(a+b) \\ 4x+2y=a+4b \end{cases}$	$a5$	.91	.72	.82
$4x-7=5x-4-(3x-1)$	$a1$	.44	1.09	.77
Graph $x+2$, $x+4$, $x+6$	$a7$	.45	.32	.39
$a=1,\ b=2,\ c=3\quad \frac{a^2}{b}-\frac{a}{a+b}$	$\begin{cases} b2 \\ c3 \end{cases}$	.39 .39	.15 .15	.27 .27
$\frac{a^4-b^4}{(a^2+b^2)\ (a+b^2)}$	$c4$	.27	.13	.20
$2x+\frac{x}{2}=3x-4$	$b5$	.14	−.05	.05
$\frac{x}{a-b}+\frac{x}{a+b}-a^2c=b^2c$	$a3$	−.03	.09	.03
$4x+a=2x-b-(x-a)$	$b1$	−.01	.06	.02
Subtract $\frac{a^2-b}{c}$ from $a-\frac{b}{c}$	$\begin{cases} a2 \\ c2 \end{cases}$	.00 .00	.00 .00	.00 .00
Half of a number, etc	$\begin{cases} c7 \\ b8 \end{cases}$	.35 −.09	−.06 −.23	.15 −.16 } −.01
$\frac{x}{4}+\frac{5}{x+3}=\frac{x-2}{3}$	$b3$	.01	−.07	−.03
Graph $y=2x$, $y=x^2$	$b7$	−.06	−.16	−.11
$2a^2-18-(a+3)\left\{\frac{a^2+3a-18}{a+16}\right\}$	$a8$	−.28	−.23	−.26
Engine problem	$\begin{cases} a6 \\ c6 \end{cases}$	.07 .40	−1.21 −.86	−.57 −.23 } −.40
$\frac{\left\{1+\frac{2}{3-\frac{4}{5}}\right\}+\frac{1}{11}}{4}$	$a4$	−.62	−.45	−.54
Cost of n articles, etc	$b6$	−.35	−1.21	−.78
$\frac{\frac{2a^2}{(a+b)^2-(a-b)^2}}{b}-\frac{a^2+2ab-3b^2}{a-b}$	$b4$	−1.33	−.93	−1.13
$\frac{\left[a+\frac{b}{c-\frac{a}{b}}\right]-\frac{abc}{bc-a}}{b-a}$	$c8$	−1.41	−1.34	−1.38

to +.44 (that is, 3.34−2.90); *a*3 for boys −.03 (that is, 2.87−2.90). Starting with *a*2 for girls, we have *a*1 for girls equal to +1.09 (that is, 4.05−2.96); *a*3 for girls equal to +.09 (that is, 3.05−2.96), and so on for all the *a* samples. Since *c*2 is identical with *a*2 we have *c*1 for boys equal to +1.05 (that is, 3.69−2.64) and so on with all the *c* samples. *c*3 = +.39 for boys and +.15 for girls. Since *b*2 is identical with *c*3 we set *b*2 as +.39 for boys and +.15 for girls and rate the other *b* samples to fit these values for *b*2. We could have made the *a*, *b*, and *c* samples comparable *via* *a*6 and *b*8 instead of *via* *a*2 and *b*2. The reliability of the ranking is measured by the differences between the ratings for *c*7 and *b*8 (the same sample), which is 0.31, and that for *a*6 and *c*6, which is 0.34. Apparent differences of .20 or .30 obviously should not be taken very seriously, though the truth is of course more likely to follow the table than reverse it.

We may compare the facts for interest as evidenced by the pupils' reports with ratings made by expert teachers of

TABLE 58

Task	Rank by pupils	Rank by teachers
*c*5	1	3½
*c*1	2	1
*a*5	3	13
*a*1	4	2
*a*7	5	9
*b*2, *c*3	6	6
*c*4	7	8
*b*5	8	3½
*a*3	9	15
*b*1	10	7
*a*2, *c*2	11	11
*c*7, *b*8	12	11
*b*3	13	5
*b*7	14	11
*a*8	15	16
*a*6, *c*6	16	14
*a*4	17	19
*b*6	18	17
*b*4	19	18
*c*8	20	20

algebra. These teachers considered all twenty specimen tasks and put them in a rank order for interest to pupils in general. By the median judgment of twenty-two teachers the ranking is as in the last column of Table 58. The rank order by the pupil is given for comparison. The two orders are much the same, the correlation (by the Spearman "foot rule") being .83. The agreement as to the dislikes is extremely close.

So far the numbers and differences of Table 57 remain arbitrary. We may define them by the magnitudes of the minorities in each case, since even the extreme difference between *c*5 and *c*8 is not unanimous. Six boys out of 133 and twelve girls out of 248 rate *c*8 as better liked than *c*5. The facts are as follows:

1.50 to .95 (*c*5 to *c*1) represents	55.7% of boys rating $c5 > c1$
	6.8% of boys rating $c5 = c1$
	37.6% of boys rating $c5 < c1$
	55.8% of girls rating $c5 > c1$
	6.8% of girls rating $c5 = c1$
	37.4% of girls rating $c5 < c1$
.95 to .20 (*c*1 to *c*4) represents	63.2% of boys rating $c1 > c4$
	25.6% of boys rating $c1 = c4$
	11.3% of boys rating $c1 < c4$
	61.8% of girls rating $c1 > c4$
	21.3% of girls rating $c1 = c4$
	16.9% of girls rating $c1 < c4$
.20 to −.40 (*c*4 to *c*6) represents	25.6% of boys rating $c4 > c6$
	39.1% of boys rating $c4 = c6$
	35.3% of boys rating $c4 < c6$
	64.4% of girls rating $c4 > c6$
	20.9% of girls rating $c4 = c6$
	14.9% of girls rating $c4 < c6$
−.40 to −1.38 (*c*6 to *c*8) represents	80.4% of boys rating $c6 > c8$
	5.3% of boys rating $c6 = c8$
	14.3% of boys rating $c6 < c8$
	53.0% of girls rating $c6 > c8$
	4.8% of girls rating $c6 = c8$
	42.2% of girls rating $c6 < c8$

Table 57 makes certain conclusions highly probable. First, there is no evidence that pupils prefer applied problems to computation. The three applied problems are rated $-.01$, $-.40$, and $-.78$, and the one of them that is rated highest is the least "applied."

Second, there is abundant evidence supporting the common belief that numerical computation is better liked than literal computation. $a1$ is substantially identical with $b1$ save that 7, 4, and 1 replace a, b, and a. Their ratings are $+.75$ and $+.02$. $a4$ is like $c8$ except that it is numerical and somewhat less elaborate. Their ratings are $-.54$ and -1.38. $c5$ is substantially $a5$ except that it is numerical and a trifle less elaborate. In only one pair of tasks ($b3$ and $a3$) does the numerical task's rating not exceed that of the corresponding literal task. Their ratings are $+1.50$ and $+.82$. Also the problem with literal numbers is rated lowest of the three problems.

Third, fractions have in general the same prejudicial influence that literal numbers have. $b5$, for example, is rated .90 below $c1$. The only added difficulty is to multiply through by 2 or otherwise handle the fractional quantity. It seems that the experiences with arithmetic and algebra have left a generalized antipathy to fractions. Pupils probably feel a lack of mastery and confidence in dealing with them.

Fourth, there is evidence that elaborate simplifications are notably disliked. With the literal problem they form the low end of the list. Elaborateness is, of course, correlated with difficulty; and part of the dislike of it is due to the dislike of difficulty. Whatever be the factors contributing to the dislike, it is certain that these simplifications are very much disliked by pupils in general, even when they have had, as these pupils had, a year more or less of algebra.

On the other hand, they feel no aversion toward very easy equations, evaluations, and graphing. These are at the very top of the list.

Fifth, the work in evaluation by substituting numerical values for several variables seems to be attractive to pupils in spite of the fact that most courses in algebra pay relatively little attention to it and that most pupils do such work badly.[1] Our one sample, appearing twice, however, is a little above the middle of the list.

Sixth, the graphing of equation-lines is (if we may assume that the pupils understood *a*7 and *b*7) by no means so dreary an exercise as the mathematician thinks, or as it may well be to the mathematically gifted. When the task is easy it ranks high, and even when it seems hard because of x^2 with its suggestions of quadratic theory and computation, it is barely below the middle. It may be that the pupils were influenced favorably by the mere word graph, which may have been associated with interesting statistical graphs from the world of affairs.

Seventh, difficulty is disliked, ease being especially liked where a little thought or work seems to produce a considerable result. Thus at the top of the list is the solution of

$$\begin{aligned} 2x+4y &= 11 \\ 4x+2y &= 9 \end{aligned}$$

which the pupil responds to as a task where he can see his way clear, has a sense of mastery and, as it were, solves two equations by one quick stroke. Many teachers will interpret this as due chiefly to a reprehensible "softness" in pupils, a love of mental idleness, a distaste for application and effort. Doubtless this is true of some pupils often and of many

[1] See the results given in Chapter XII, page 321. The work of Rugg and Clark [1917a] and that of Hotz [1918] give evidence of this; and we have corroborated this abundantly in an inventory test given to about eight hundred pupils in ten schools.

pupils occasionally, but it is not the whole truth. On the contrary, human beings in general and on the whole prefer the richness, complexity, and subtlety which constitute mental difficulty so long as they can do the work fluently, without being thwarted and baffled, and with confidence and mastery. Most of us would prefer to spend an hour with tasks like *c*8 than with tasks like *c*5. Most high-school pupils would very much prefer the *c*5's. This is often interpreted to mean that we are admirable creatures who like what we should like, whereas they are reprobates who like what they should not. But probably we and they act from much more nearly the same than opposite motives, both liking the maximum richness, complexity and subtlety that we can master comfortably. We like *c*8 because it gives us the same sort of exercise that *c*5 gives the ordinary pupils.

Difficulty, elaborateness, "fractionness," and "letterness" are closely associated in the twenty samples used. Their separate shares in producing dislike cannot well be determined without the use of many more samples chosen to represent variations of each factor separately. It is to be hoped that students of the learning of algebra will be led to make such an analysis. Where such an experimental analysis is difficult the method of partial correlation may be used.

This is a method that should become widely known and used by students of education. So we have applied it to our data though they are really too scanty to justify its use, and are unsuitable in other ways. We have in Table 57 a ranking of the twenty samples for interest. They were also ranked by psychologists and teachers of mathematics in an order of difficulty, in an order of elaborateness, in an order of "fractionness," and in an order of "letterness." These orders all correspond somewhat, the correspondences being

TABLE 59

Correlations among Interest (taken negatively), Difficulty, Elaborateness, "Fractionness," and "Letterness," for the Twenty Samples of Algebraic Work.

$r_{id} = .60$			
$r_{ie} = .65$	$r_{de} = .88$		
$r_{if} = .65$	$r_{df} = .42$	$r_{ef} = .65$	
$r_{il} = .37$	$r_{dl} = .28$	$r_{el} = .31$	$r_{fl} = .56$

as measured by the coefficients of correlation of Table 59. To find $r_{id.e}$ (the correlation coefficient which would obtain between interest and difficulty, for samples of algebraic work which were identical in respect to elaborateness) we use

$$r_{id.e} = \frac{r_{id} - r_{ie}\, r_{de}}{\sqrt{(1-r_{ie}^2)(1-r_{de}^2)}}$$

This is .08 whereas r_{id} was .60; $r_{ie.d}$ is .33 whereas r_{ie} was .65.

The numerical values of these coefficients should not be taken very seriously. Their probable errors are very high, due to the scantiness of the data. They are given, as stated, chiefly to illustrate the method. Taken at their face value, they show that elaborateness irrespective of difficulty has more influence on these pupils' interests than difficulty irrespective of elaborateness.

CHAPTER XVII

Individual and Sex Differences in Algebraic Achievement

SEX DIFFERENCES IN ALGEBRAIC ABILITY

Teachers in general realize from their daily experiences that there is no notable difference in algebraic ability between boys and girls. Three studies of school marks confirm this. One finds 57 percent of males reaching or exceeding the median score for females in mathematics (high-school boys and girls). The second finds 45 percent (college men and women). The third finds 41 percent of high-school boys reaching or exceeding the median score for girls in algebra, and 53 percent doing so in geometry.

Measurements comparing the sexes in actual tests have been lacking. We present here the results found where boys and girls in the same classes had, under identical conditions, the 40-element test described on pages 179 to 186. The tests were scored by the credit system described on page 180.

Table 60 shows the distributions in six schools of the scores in this test. The outstanding fact from inspection of the table is the parallel character of the distributions of boys and girls: in the same school the boys and girls have very nearly the same range. The medians are as follows:

School	Medians: Boys	Girls	Differences
A	22.5	25.9	−3.4
B	29.5	27.5	2.0
C	19.4	20.9	−0.5
D	56.2	35.6	20.6
E	64.3	55.0	9.3
F	53.1	54.2	−1.1
G	78.2	79.7	−1.5

TABLE 60

SCORES OF BOYS AND GIRLS IN THE I. E. R. ALGEBRA TEST TO QUADRATICS

	SCHOOL A		SCHOOL B		SCHOOL C		SCHOOL D		SCHOOL E		SCHOOL F		SCHOOL G	
	BOYS	GIRLS	BOYS	GIRLS	BOYS	GIRLS	BOYS	GIRLS	BOYS	GIRLS	BOYS	GIRLS	BOYS	GIRLS
95–100														2
90–94									2	2			7	5
85–89							1		4	1			7	9
80–84							3	1	6			2	12	6
75–79							1	1	8		5	2	8	8
70–74			3				3		13	3	1	4	7	6
65–69			1				1	1	9	2	6	2	4	2
60–64			4	1			1	2	7	1	7	3	4	3
55–59	1		6	4	1	1	4		4	2	3	5	3	
50–54		1	1	4	2		3	4	5	2	6	9	1	2
45–49	1	3	7	3		2	1	7	10	2	9	2	2	1
40–44		1	10	8	2	1	1	2	3		1	2		
3 –39		1	6	6	1	5	3	4	4	1	4	4	2	
30–34	4	2	6	9	1	5	1	8	4	5	3	2		1
25–29	3	6	10	8	5	10	1	1	2			2		
20–24	4	3	15	7	5	12		7	1	1	3			
15–19	3	3	9	8	8	15	1	3	1		1			
10–14	4	3	6	10	5	9	1	2	2					
5–9		3	2	6	4	7								
0–4	2		4	4	1	1								
	22	26	90	78	35	68	26	43	85	22	49	39	57	45
Median	22.5	25.9	29.5	27.5	19.4	20.9	56.2	35.6	64.3	55.0	53.1	54.2	78.4	79.7

In three of the schools the median of the boys is above the median of the girls, in four of the schools the median of the girls is above that of the boys. Apparently there is no great superiority of either sex in the ability in question. While in any school there is great range of ability, on the whole the ability of the sexes is on a par.

Since there is such a great range in the test scores a more significant measure of the sex differences is the percent of one group (say the boys) which reaches or exceeds the median of the other group (the girls). This is a measure of the overlapping.

School	Number		Percent of Boys which reach or exceed median of girls
	Boys	Girls	
A	22	26	41
B	90	78	53
C	35	68	49
D	26	43	81
E	85	22	62
F	49	39	47
G	57	45	47

The median of these percents is 49 with a P. E. of 10 percent. Although there is not a consistent story from all of the schools, the evidence supports the conclusion that the sexes are of approximately equal ability. Thorndike's conclusion in 1914 that "the individual differences within one sex so enormously outweigh the differences between the sexes in these intellectual and semi-intellectual traits that for practical purposes the sex difference may be disregarded" is acceptable in the case of algebra.

SEX DIFFERENCES IN INTEREST IN ALGEBRA

When boys and girls in Grade 12 place algebra in an order of school subjects for liking, the boys place it somewhat higher than the girls do. Using the system of values described on pages 387 to 391, the median rating for boys is 0.12 higher than that for girls. This 0.12 may be further defined as one-eighth of the difference between English and Latin in respect to interest. Boys and girls like English better than Latin, the difference being 0.89, 0.62, 1.03, and 2.00 by the four available methods of measurement. The median of these is eight times the number representing the difference between boys' liking for algebra and girls' liking for it. Probably the realization by boys of the uses of algebra in the physical sciences and engineering is sufficient to account for this sex difference in interest, without invocation of any deeper or subtler causes.

The available data are not adequate to give evidence concerning differences between the sexes in respect to various features of algebraic learning. Such data as there are appear in Table 57, on page 405; but, as was stated in connection with the description of that table, small differences (.5 or less) should not be taken too seriously. So far as the facts go, they show girls liking all three of the problems less than the boys (differences of −.28, −1.27, and −.86; average −.80), the engine problem especially; and liking the semi-clerical tasks of removing parentheses, transposing and collecting terms somewhat better (average difference for *c*1 and *a*1, +.22). The girls appear also to like the elaborate simplifications somewhat better than the boys (the average difference for *a*4, *b*4, and *c*8 is +.21). The boys perhaps like evaluation better than the girls do.

The difference in the interestingness of problems deserves further study. The problem situations used in textbooks on algebra much oftener concern men's activities than women's and the vote on the neutral problems about "Half of a number" and the "Cost of *n* articles" may be due to a general prejudice due in turn to this fact. If, *per contra*, girls disliked the aggressive planning and selection required for solving verbal problems in general, it would be evidence for the once orthodox but now rather discredited doctrine of a general passivity of the feminine intellect. In any event the difference in the case of the engine-boat problem teaches a useful lesson. If the slight mechanical and sporting interest which problems bear (or which boys think they bear) makes so great a sex difference in the interest ratings, we may expect a substantial gain in interest for either sex from replacing a remote unappreciated situation by one that deals with things known and cared about.

TABLE 61

DISTRIBUTIONS OF SCORES IN THE HOTZ TESTS

Score	Addition and Subtraction	Multiplication and Division	Equation and Formula	Problem	Graph
25	...	...	4	...	...
24	4	...	12	...	...
23	25	6	23	...	...
22	28	10	44	...	...
21	40	37	78	...	...
20	54	101	93	...	...
19	71	118	125	...	...
18	88	136	134	...	...
17	117	207	135	...	...
16	124	192	148	...	...
15	111	162	115	...	...
14	104	147	139	7	...
13	135	111	127	30	...
12	155	99	106	33	...
11	146	58	106	74	3
10	86	47	72	96	38
9	50	33	58	139	102
8	35	19	30	182	129
7	28	12	18	194	140
6	21	1	14	174	122
5	5	7	10	161	113
4	9	1	6	122	73
3	5	1	1	69	30
2	3	...	4	18	10
1	...	...	...	6	5
0	3	...	...	...	2
Median	14.6	15.9	15.7	7.6	7.1
S. D.	4.1	3.3	4.1	2.6	2.0

INDIVIDUAL DIFFERENCES

Hotz [1918], using the tests described on pages 173 to 179 with pupils who have studied algebra nine months, finds variations as shown in Table 61.

The enormous variation shown in this table is reduced two-fifths if we take any one school of those tested in place of the total group. That is, about two-fifths of the variation is due to differences between schools either in the quality of the pupils they attract or in the amount of improvement they make in them, or in both. This cannot be shown from the Hotz data themselves, which are not given separately by schools, but may be inferred from the facts found and reported herewith in Table 62 for pupils tested with the test shown on pages 179 to 186. Table 62 shows the variation in scores (computed according to the scheme of credits of page 180) for all the schools together, and for each school separately. In Table 62 only pupils who had studied algebra at least a year were taken. The mean square deviation is 24.2 for all together, and only 15.4 on the average for schools taken separately.[1]

[1] The facts, estimated as best we can for the Hotz data, are as follows:

Hotz gave his tests in eighty-four high schools, and we may expect these schools to differ widely. In the equation and formula test Hotz found the average score at the end of nine months to be 15.7 with an S. D. of 4.1. We should expect separate schools to make averages somewhat as follows: 1 school, under 8; 5 schools, 9 to 11; 20 schools, 12 to 14; 32 schools, 15 to 17; 20 schools, 18 to 20; 5 schools, 21 to 23; 1 school, 24 or 25. That such is not far from the truth is evident from data Hotz actually finds in various cities [1922, p. 33]:

City	Average Score on Equation and Formula Scale Nine-Month Group
Wisconsin Cities (1918)	17.2
Wellington, Kansas (1919)	18.1
Wellington, Kansas (1920)	16.7
Wellington, Kansas (1921):	
Average Group	14.4
Slow Group	11.2
Elizabeth City, N. C	12.8
Andover, Mass	12.8

TABLE 62

Distribution of Scores in the I. E. R. Algebra Test, to Quadratics

Score	A	B	C	D	E	F	G	H	I	J	K	All Schools Combined
100	......	......	......	......	......	......	......	......	......	......	......	......
95–99	5	......	......	......	......	......	......	2	1	......	......	8
90–94	3	......	......	......	......	4	......	12	1	......	1	21
85–90	4	......	......	......	1	5	......	16	2	1	......	29
80–84	8	......	......	......	4	6	2	18	7	3	1	49
75–79	7	......	......	......	2	8	7	16	8	......	......	48
70–74	3	......	3	......	3	16	5	13	12	7	......	62
65–69	3	......	1	......	2	11	8	6	10	8	2	51
60–64	2	......	5	......	3	8	10	7	8	10	1	54
55–59	2	1	10	2	4	6	8	3	8	13	1	58
50–54	2	1	5	2	7	7	15	3	8	3	2	55
45–49	3	4	10	2	8	12	11	3	3	1	3	60
40–44	1	1	18	3	3	3	3	......	1	......	......	33
35–39	......	1	12	6	7	5	8	2	1	......	......	42
30–34	......	6	15	6	9	9	5	1	......	1	1	53
25–29	......	9	18	15	2	2	2	......	1	......	......	49
20–24	......	7	22	17	7	2	3	......	......	......	......	58
15–19	1	6	17	23	4	1	1	......	......	......	......	53
10–14	......	7	16	14	3	2	......	......	......	......	......	42
5–9	......	3	8	11	......	......	......	......	......	......	......	22
0–4	......	2	8	2	......	......	......	......	......	......	......	12
No cases..	44	48	168	103	69	107	88	102	71	47	12	859
Median....	74.7	25.0	30.8	22.5	44.8	59.4	53.8	76.4	66.7	63.7	59.2	49.37
S. D......	17.4	13.0	16.9	11.8	19.8	19.5	15.3	13.6	13.1	9.7	15.1	{15.36 24.6

Where a careful system of classification is in force, the variation in a class taught together would be lower than that for a school. It would, however, still be large. Some pupil in a class of thirty is likely to be able at the end of the year to do (with easy work) at least twice as much per unit of time as some other pupil, and to do tasks whose difficulty

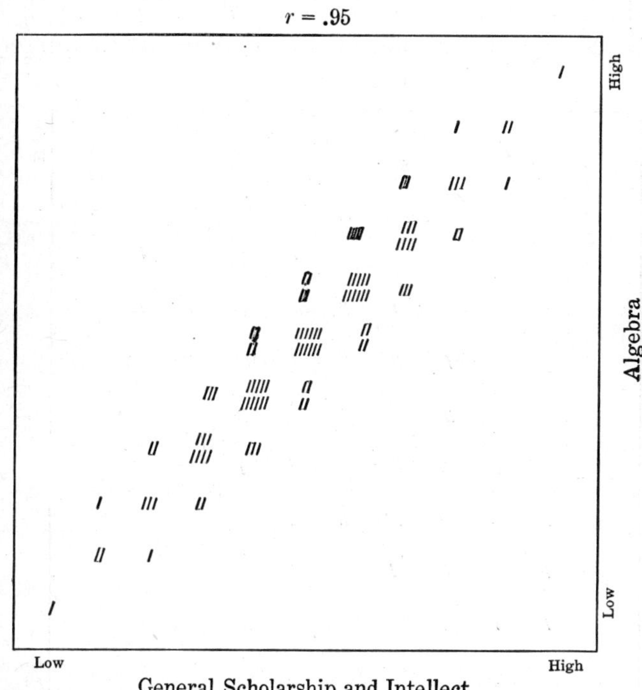

FIG. 30

is to the difficulty of those which the least competent pupil in the class can master as $\frac{3-2x}{(x-1)^3}+\frac{x+1}{(x-1)^2}-\frac{1}{x-1}$ is to $7x-$ $+6-4$ or as $\frac{6x-2}{x+3}-3=\frac{3x^2+13}{x^2-9}$ is to $7n-12-3n+4=0$.

Training is not, of course, absolutely equalized even fo pupils in the same class, but the larger part of the difference found amongst them is presumably due to differences in th capacities of the pupil.

CORRELATION BETWEEN ABILITY IN ALGEBRA AND GENERA INTELLECTUAL ABILITY

The differences in algebraic ability which are due t differences in personal capacity, are due in part to difference

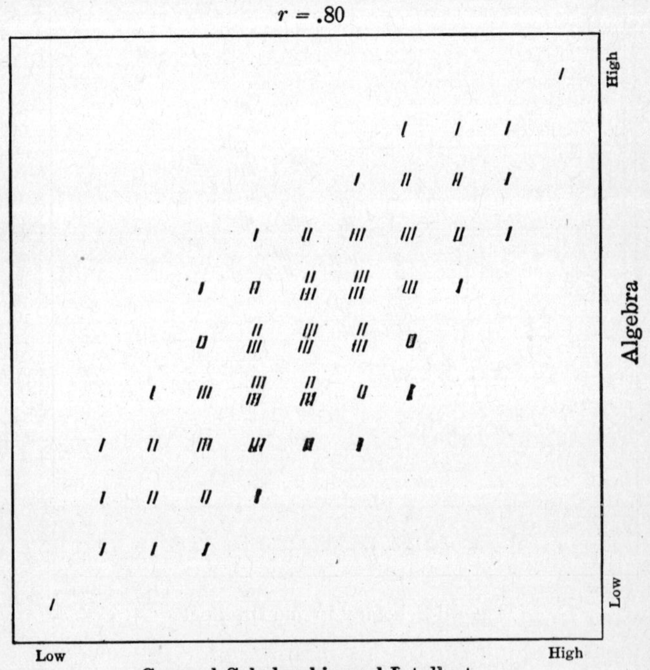

Fig. 31

in general capacity for abstract learning which make som pupils superior and others inferior in average academi ability. In small part, however, they are due to differenc in a special ability for algebra.

Assume there are available perfect measures of scholarship in general and of algebraic ability in particular, and that we have a hundred pupils each located by a little line representing his status, in the former by its distance to the right from Scholarship Low, and in the latter by its distance

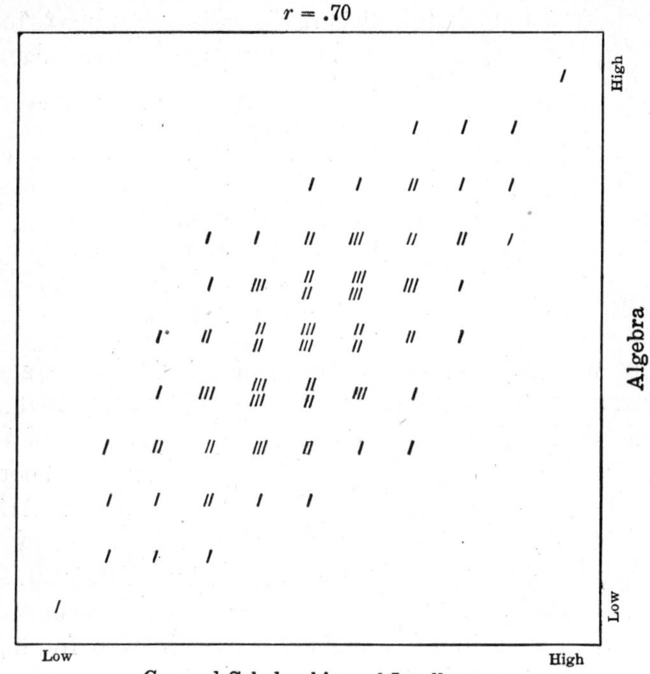

FIG. 32

upward from Algebra Low. The hundred pupils would not be located along the 45 degree line, nor very close to a line of nearly 45 degrees, as in Fig. 30, but approximately as shown in Fig. 31.[1] Some of those generally competent would be low in algebra and conversely.

[1] We are considering here a hundred taken at random from the male white population, supposing that all were carried on in school so as to give all an opportunity to learn algebra.

We may accept as a rough measure of general intelligence success in school work. Two investigators have obtained the correlation between algebra and a general average of school work. Weglein [1917] finds for a group of high-school graduates the correlation between first-year algebra and the general average (including algebra) of first-year subjects to be .64. Crathorne [1922] finds the correlation between algebra and the school average to be

School A	girls	.61
	boys	.58
School B	girls	.64
	boys	.64
School C	girls	.50
	boys	.62

But we may also use the result of a test of intelligence to give us the correlation. Crathorne [1922] finds that the correlation between algebra marks and an average of two intelligence tests is .50. Buckingham [1921] reports a correlation between algebra ratings and the score on Army Alpha of .38. Proctor [1921] reports a correlation of .46 between algebra grades and the Stanford-Binet I. Q. Thorndike [1922b] reports the following correlations between the algebra test in his intelligence examination for high-school graduates and the whole examination:

371 candidates at Columbia	.47
77 candidates at Columbia	.53
76 candidates at Columbia	.50
321 candidates at Columbia	.41
465 candidates at Columbia	.46
132 candidates at Columbia	.47
180 candidates at an engineering school	.50
97 women candidates	.51

The Mann Report of Engineering Education [1918] gives the correlation between a thirty-minute test in algebraic computation and an excellent criterion for intelligence as .62. The same test was given to Massachusetts Institute of Technology freshmen, but the intelligence criterion was not as good. The correlation in this case was .53.

These correlations as reported run all the way from .38 to .64. There are two factors which tend to displace correlations from comparable values. Both of these displacements are downward.

The first of these factors is the inaccuracy of the measurements. One source of such inaccuracy is in the tests themselves. If tests are made having too few scoring elements the tests do not show the real differences in ability. For instance, a test of five examples will show only five grades of ability, where a test of fifty examples might show fifty grades of ability. The same effect can be produced by having unequal steps in difficulty between the test elements. If the elements are bunched so that there are groups containing elements of nearly the same difficulty the result is the same as having a test with only a few elements. A second source of inaccuracy in the measurements is the variations in the situation — time of day, distractions, different examiner, etc. A third source of inaccuracy in the measurements is variations in the pupils taking the test. In a longer test, or in the same test spread out over various times, pupils are likely to show more nearly their true ability than in a short or single-session test.

The correction for this lowering of the correlation due to inaccuracy of the measurements may be estimated from Spearman's formula,

$$r_c = \frac{r_{xy}}{\sqrt{r_{x_1x_2} r_{y_1y_2}}} \text{ where}$$

$r_c = r_{xy}$ corrected for attenuation
r_{xy} = the obtained correlation
$r_{x_1x_2}$ = the reliability coefficient for x
$r_{y_1y_2}$ = the reliability coefficient for y

Using the following as reliability coefficients,

		Elementary School	High School	College
School Marks		.75	.70	.65
Standardized tests	5 minute test	.70	.65	.60
	10 minute test	.80	.75	.70
	20 minute test	.85	.80	.80
	40 minute test	.90	.85	.85

the correlation of Weglein becomes .86; those of Crathorne, .78, .86, .86, .67, .83, and .62; that of Buckingham, .49; that of Proctor, .60; those of Thorndike and the Mann Report become .62, .62, .70, .66, .54, .61, .62, .66, .67, .71, and .64.

The second factor tending to lower the correlations and to make comparison between them uncertain is the composition of the groups. It is well known that the correlation varies with the variability of the group. To make the correlations comparable they must be referred to some standard group; and in this instance the standard group chosen is the unselected adult population. The method by which the estimates for this correction are made is too complicated to describe here.[1]

Below are the results after making this correction:

Weglein	.90
Crathorne	.89, .87, .92, .80, .90, .78
Buckingham	.67
Proctor	.74
Thorndike	.78, .83, .81, .74, .78, .78, .81, .81
Mann Report	.83, .80

[1] The reader is referred to the monograph *Special Disability in Algebra*, by Percival M. Symonds, for a complete description.

The correlations obtained by using school marks are perhaps too high because the algebra marks were used in the average of studies, making the correlation in so far spurious. The best figures are those of Thorndike and those from the Mann Report. We conclude then that the correlation between algebra and intelligence is .80 when taken over the total range of the unselected adult population. For high school freshmen as a group this correlation would be in the neighborhood of .70.

The significance of these correlations is shown in Figs. 31 and 32, on pages 420 and 421, showing the "scattergrams" representing correlations of .80 and .70 respectively. Each "scattergram" contains one hundred individuals. These correlations are consistent with occasional larger discrepancies between general intellect and ability in algebra. Of the upper 22 in intelligence, in Fig. 31, one will be below the median in algebraic ability and two others in Fig. 31 will be near the median.

Fig. 32 represents the scatter of 100 pupils when $r = .70$, the figure suggested above as probable. Of the upper 22 in intelligence, two will be below the average of the group in algebraic ability and three about at the average. Such persons, we would say, had a special disability in algebra. Of course, in a group of a thousand pupils one or two would probably be found who stood very high in intelligence and very low in algebraic ability.

The converse to these statements is also true. Of the twenty-two stupidest pupils in a first-year high school of one hundred pupils, two will be above the average in algebra ability and three of about average ability. And out of a group of one thousand pupils it would be quite possible to find one or two very stupid pupils who stood well up in the group in algebraic ability.

Let not this emphasis on the extreme cases blind one to the main fact that, by and large, high intelligence means fine ability in algebra and low intelligence means poor ability in algebra. Of the twenty-two brightest pupils out of the hundred, seventeen will be above the average in algebra and of the twenty-two stupidest pupils, seventeen will be below the average in algebra.

SPECIAL DISABILITY IN ALGEBRA

Of special interest are those cases which form the border of the scatter diagram—and, in particular, those who are high in intelligence and low in algebraic ability. Such individuals do exist, and most teachers of mathematics are acquainted with the type. We know them in everyday life as the persons incompetent in mathematics who succeed along linguistic, literary, or artistic lines. But these cases are not nearly so numerous as popular impression would lead one to believe. This is evidenced both by the correlations and by the slight success of diligent search in four schools for such individuals. Talk with one of these intelligent and talented persons who have a horror at the sight of an algebraic formula and you will find that they "admit" doing very poor work in first-year algebra—they perhaps were lucky to have come off with a B or perhaps a C as their final grade. However, a B or a C does not by any means indicate that such an individual was a poor algebra student — he was undoubtedly in the upper quarter of his group. Because he easily won A's in his other subjects he feels that he is extremely poor in algebra, a feeling not borne out by comparison of his record with those of the others in his class.

In order to study the amount and nature of special disability in mathematics, the following investigation was undertaken. In the spring of 1921 a letter was sent to principals of a number of private schools in and around New York City stating the purpose of this study and asking permission to enter their schools to study whatever cases of special disability they had. In the fall of the same year arrangements were made to study whatever cases were available in the following schools: Horace Mann School for Boys; Columbia Grammar School; Friends School, Brooklyn; Brooklyn Polytechnic Preparatory Country Day School. Permission was obtained to study those pupils who during the previous year had received poor marks or had had trouble in mathematics, but were good in other studies.

A testing program was then arranged consisting of

- Army Alpha Intelligence Scale (form 7)
- Hotz Algebra Scales (series A)
  - Addition and Subtraction
  - Multiplication and Division
  - Equations and formulas
  - Problems
- Rogers' Tests of Mathematical Ability
- Thorndike-McCall Reading Scale (Form 1)
- Woody-McCall Mixed Fundamentals in Arithmetic. (Form 1)

The time necessary to give all of the above tests was a little under six hours. In each case, since the pupils were singled out from various classes, there was a little resentment at being "picked on," and fear that they were to be marked as "dubs." But when it was explained to them that as a group they were chosen as good students having special difficulty in mathematics, this attitude disappeared. The intelligence and reading tests were taken with good zeal, but it was noticeable that the mathematical tests were merely tolerated if not openly detested. However, all of the tests were given by Mr. Symonds in person, and every one of the scores recorded represents the result of a conscientious effort on the part of the one being tested.

The average T-score on the Thorndike-McCall reading test was 68. The norms given in the directions for using this scale place this score in Grade 12B, with a reading age of 16 years 9 months. There is, then, from the results of this scale, no difficulty in ability to read or in the intelligence necessary to do the tasks set by the reading scale.

The average score on the Woody-McCall Mixed Fundamentals in arithmetic is 29.1. This corresponds to the Grade 8 norm of 29.3, indicating that these students are much better in reading ability than in arithmetic computation as judged by grade averages.

To display the results on the other tests we have used the psychograph method. The purpose of the psychograph is to show a man's relation to the average of all men or of a class of men in certain traits. In these psychographs a vertical line near the middle of the chart stands for average ability in the traits which are displayed on the horizontal lines. If an individual is above the average in a trait he lies to the right of the average line in the graph. If he is below the average in a trait he lies to the left of the average line. Lines connecting an individual's position from trait to trait help to make these traits stand out better. There is no other reason for the existence of these lines,—they have no significance in themselves. Fig. 33 is the psychograph of the "special disability" group.

That these cases are actual cases of disability in algebra is abundantly testified. The average for the group on the test in algebraic computation in the Rogers series is 1.3 sigma below the average of Dr. Rogers' Horace Mann School Group—much lower in relative position than on any of the other Rogers tests. The averages on the Hotz tests are all below the nine-month norms. The averages on the addition and subtraction test and the equation and formula test are 1 sigma below the average; the average on the multiplica-

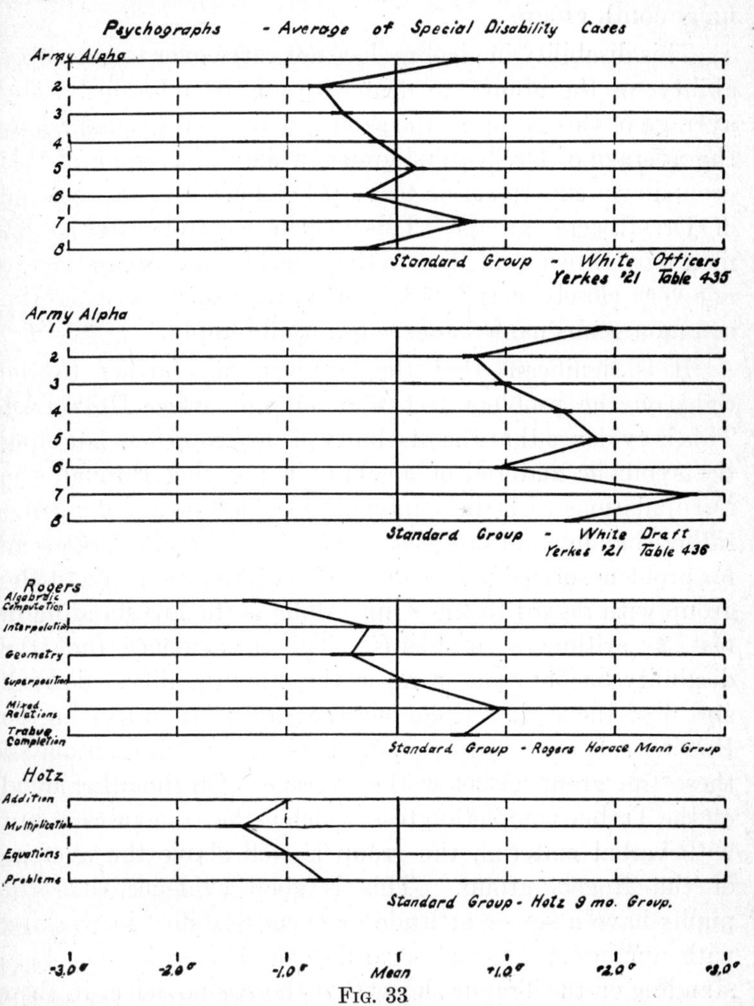

FIG. 33

tion and division test is 1.4 sigma below, but the P. E. of this average is larger than that of any of the others as there were not so many cases included in this group; the average on the problem test is 0.7 sigma below the average of Hotz's nine-month group.

This disability in algebra does not carry over to geometry ability, or the ability to deal with spatial relations. The average of this group in the geometry test is 0.4 sigma below the average of Dr. Rogers' group, while the average of this group in the superposition test is 0.1 sigma above the average of Dr. Rogers' group. This evidence corroborates statements by Brown and Rogers that algebra and geometry are not very closely correlated—not nearly so closely as the common genus name *mathematics* would imply.

It is significant that the group stands farther to the right on the problem test than on any other Hotz test. This is evidence that the disability of the group is in handling the symbolic material of algebra, in carrying through the manipulations with the combinations of letters and numbers, rather than in ability to reason and do the thinking necessary for problem solving in general. The striking thing about the group with regard to the Army Alpha is the low standing in test 2—arithmetic problems. This is evidence that the disability has its roots in an arithmetic disability. Tests 2 and 6 of the Army Alpha and the interpolation test on the Rogers scale contain numerical material, and in each one of these the group is below the average. On the other hand on the Trabue completion test, which is the same process but with verbal material, the group is well above the average of the Rogers group. This is good evidence that the pupils have a set or attitude or even disability in working with numerical material regardless of the form. The high standing on the Trabue shows that there is no defect in their intelligence, but that this intelligence is specialized with reference to the material on which it works.

The high standing on the mixed relations tests (Army Alpha 7 and Rogers Mixed Relations) is somewhat unexpected. Here is a test that has a distinctly mathematical form — the proportion. This is additional evidence that it is the material or content and not the form which presents difficulty to those who have special disability in mathematics. Dr. Rogers used this test in her sextet to predict mathematical disability. The reason this test was included in the Rogers sextet was because it correlated so highly with mathematics. Over the whole range of people, we have no better way to measure mathematical ability than by means of verbal, abstract intelligence. But when it is desired to make a close prediction over a more limited range, tests made up of numerical material seem to be more efficient in indicating disability in algebra. We get a poorer prediction of mathematical ability from a verbal test of mixed relations. A boy who gets a score of 30 on test 7 in Army Alpha on the average has more mathematical ability than a boy who gets a score of 22. But one gets a much poorer and less reliable estimate of mathematical ability from giving a test like the mixed relations test which is made up of verbal material than from an arithmetic problem test made up of numerical material. This is evident from Dr. Rogers' [1918] own correlations (see p. 74 of her monograph where she gives final corrected correlations with a criterion of mathematical ability). Algebraic computation correlates .81 with this composite criterion, the numerical completion test .78, arithmetic problems .74, whereas the mixed relations test correlates with it only .44, next to the lowest of all the seventeen tests which Dr. Rogers reports. If she had used the newer methods of obtaining weightings by regression coefficients, it is probable that the mixed relations test would not have been included in her group of six.

Figs. 34, 35, and 36 are psychographs of individuals Nos. 7, 13, and 20 who present decided cases of special mathematical disability. Individual 7 has studied algebra for 5 months, individual 13 for 1 year, and individual 20 for 1 year. That the characteristics of the average psychographs are not due merely to lumping together the scattered results of individuals is illustrated by the great similarity in the profile of these three individuals. Fig. 37, the psychograph of individual 39, who shows little or no evidence of special mathematical disability, gives on the other hand a profile of quite different characteristics with respect to intellective functions.

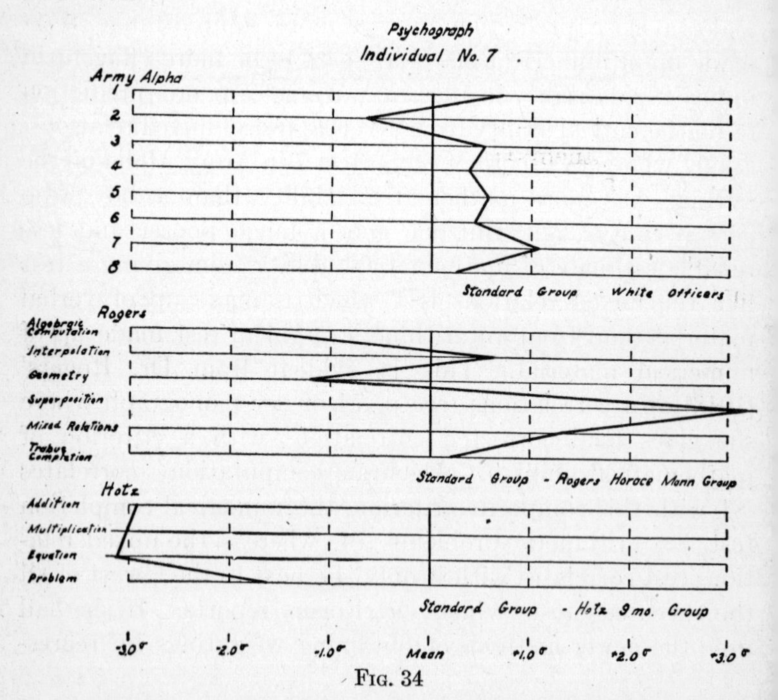

FIG. 34

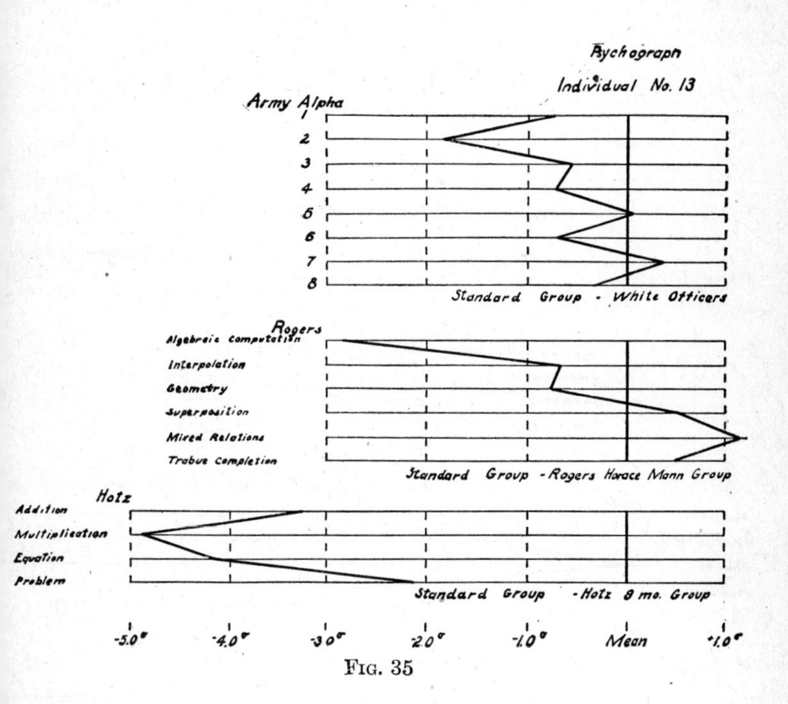
Psychograph
Individual No. 13
Army Alpha
1
2
3
4
5
6
7
8
Standard Group - White Officers
Rogers
Algebraic Computation
Interpolation
Geometry
Superposition
Mixed Relations
Trabue Completion
Standard Group - Rogers Horace Mann Group
Hotz
Addition
Multiplication
Equation
Problem
Standard Group - Hotz 8 mo. Group
-5.0σ
-4.0σ
-3.0σ
-2.0σ
-1.0σ
Mean
+1.0σ

Fig. 35

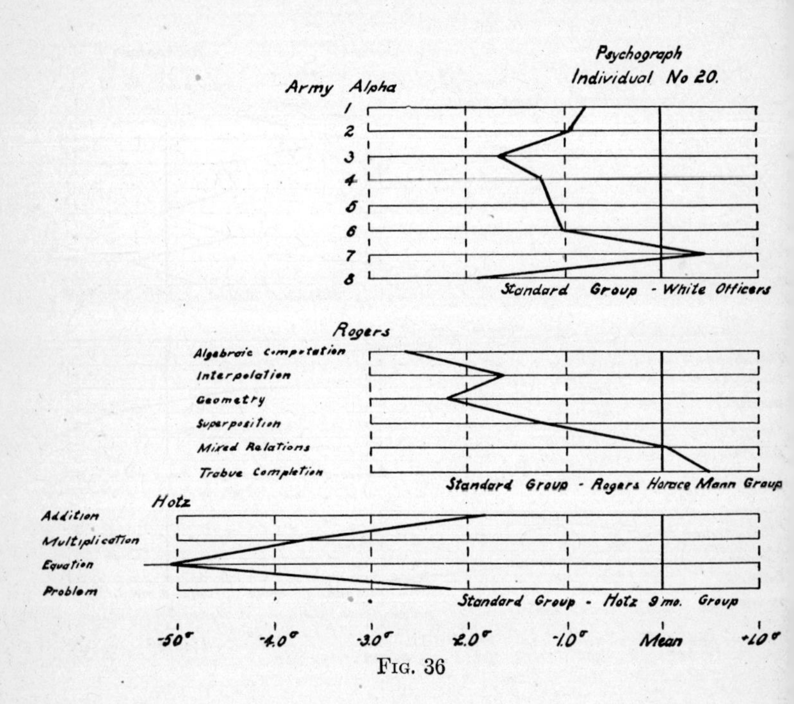

Psychograph
Individual No 20.
Army Alpha
1
2
3
4
5
6
7
8
Standard Group - White Officers
Rogers
Algebraic Computation
Interpolation
Geometry
Superposition
Mixed Relations
Trabue Completion
Standard Group - Rogers Horace Mann Group
Hotz
Addition
Multiplication
Equation
Problem
Standard Group - Hotz 9 mo. Group
-5.0σ
-4.0σ
-3.0σ
-2.0σ
-1.0σ
Mean
+1.0σ

FIG. 36

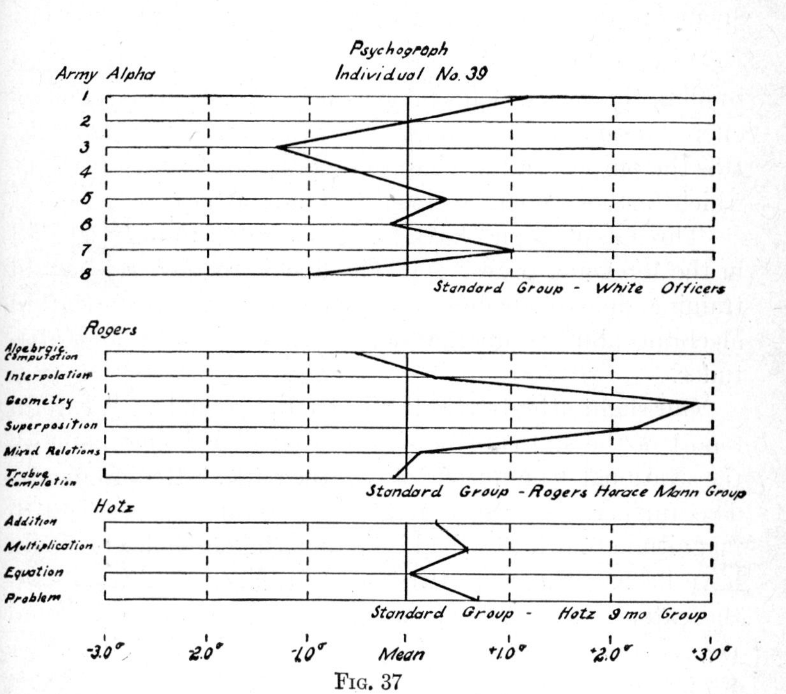
Psychograph
Individual No. 39
Army Alpha
1
2
3
4
5
6
7
8
Standard Group - White Officers
Rogers
Algebraic Computation
Interpolation
Geometry
Superposition
Mixed Relations
Trabue Completion
Standard Group - Rogers Horace Mann Group
Hotz
Addition
Multiplication
Equation
Problem
Standard Group - Hotz 9 mo Group
-3.0σ
-2.0σ
-1.0σ
Mean
+1.0σ
+2.0σ
+3.0σ

FIG. 37

CHAPTER XVIII

Suggestions for Research in the Psychology of Algebra

As was stated in the preface, the plan of our study was to fill in the various gaps in our knowledge of the psychology of elementary algebra, so far as our time and means allowed. It is the purpose of this chapter to note some of the questions which further researches may hopefully attack.

Three main topics have been somewhat scantily treated in the foregoing pages. (1) The first is the general mental training due to algebra, the influence of improvement in algebraic abilities upon wider abilities, interests and attitudes.

Our slight attention to this is due to two facts. A careful report of the status of present knowledge is being made by the National Committee on Mathematical Requirements. The further researches that were desirable were of such magnitude that if they had been undertaken, little else could have been. There are two promising lines of attack, first, the extension of Kelley's work noted in Chapter VI, and second, the comparison of the study of algebra with other studies and activities by means of repeated tests of selective and relational thinking, generalization and organization, and symbolism.[1]

(2) The conditions of algebraic learning, including methods and devices for increasing interest, aiding comprehension

[1] A study of the second sort is now being carried on by the Institute of Educational Research of Teachers College, Columbia University, with the aid of a grant from the Commonwealth Fund.

of principles, securing mastery of techniques, offer a wide field for psychological deductions and experiments. Especially needed are observations of pupils learning algebra in respect of what they find hard and easy, how they use time profitably and waste it, how much improvement they make from specified exercises. The facts which we have found and the applications of psychology which we have made regarding the constitution of algebraic abilities and the selection and elimination of mental connections, lead on to many questions concerning the detailed conditions of economical learning. These have not been reported on in this volume, but will be made the subject of later investigation by one of us.

(3) In Chapter XI we have considered the arrangement of algebraic subject matter or, in psychological terms, the order of formation of algebraic connections, and have applied psychology to several problems of arrangement. There are, of course, many other differences in present practice, and probably also many cases where the psychology of the learner suggests that we may devise an order better than any of those used in present practice. These will be made the subject of investigation and report later by one of us.

Besides these three topics, which will be investigated further by one or another of us, there are certain matters which we have noted as especially suitable for investigation and especially deserving of it. It seems proper to list these here with brief explanatory notes.

Mathematical Accuracy versus Clerical Accuracy

In experiments where gifted adults do algebraic work one notes a distinction in errors (or in precision, to look at the matter the other way about) which is not quite the same as the distinction between inability with principles and inability in computation, nor as the distinction between lack of knowledge and carelessness, though it is akin to both. It appears especially in work where operations are to be carried on with q elements in an order r; one operation is neglected or connected with a wrong element or done in the wrong order, or one element is neglected or taken in the wrong order, and the like. Until a certain ability is reached errors of this sort seem to testify to lack of mathematical ability. The pupil does not have the fundamental connections strong enough to operate them well in a new complex. After a certain ability is reached, however, they seem to testify rather to a lack of such system, care, and deliberateness as are qualities of a superior clerk rather than of a superior mathematician. This appearance may be illusory except after a degree of mastery which is seldom reached by high-school pupils, but the matter deserves study. The clerical side of algebraic ability, in so far as it has a clerical side, is a worthy ability, but it is not what the mathematician has in mind when he thinks of mathematical accuracy.

Mathematical Reasoning as Successful Guessing

To the psychologist it appears that the procedure of able pupils is, and that the procedure of all pupils should be largely guessing, or making an hypothesis, and modifying your guess or hypothesis until you guess right, in such cases as selecting the quotient figures in long division in arithmetic, factoring and simplifications in algebra, and finding proofs of originals in geometry. If this is so, teaching factorizations in algebra should be notably modified, becoming guidance in directing the guesses to likely ones rather than an imposing pretense at deducing with infallibility what the factors must be. One has the impression from textbooks that teachers of mathematics consider the mental process of making a successful hypothesis as a dubious adventure, indecorous if not immoral, and, in dignity, far below proving that a certain hypothesis is correct. Is not the contrary the case?

The Reliability and Significance of Instruments for Measuring Algebraic Ability and Achievement

One should not find fault with those who have done constructive work in devising and standardizing tests for algebra, because they have not done more. Much has been done, all within five years. The very men who have done the best work in this field, however, would be the first to admit that we do not know how "reliable" the tests are, that is, what the probable errors of determinations by the tests are;[1] and that we do not know what the tests signify, that is, what the correspondences are between them and various criteria of algebraic ability and achievement. The case is much worse with the ordinary tests and examinations improvised by examining boards and teachers. Surely teachers of mathematics should be eager to learn what their examinations really measure.

The investigations, though laborious, are straightforward, the chief difficulty being to secure groups of pupils who can be measured with enough different examinations and observations to ensure a composite rating which will be an adequate criterion by which to test the significance or validity of each of the separate examinations that are under consideration.

Given a test containing elements E_1, E_2, E_3, etc., which produces a certain result, say a correlation of .60 with a

[1] Usually measured in practice by the "self correlation" of the test, that is, the resemblance between the scores made in two independent trials with the test.

perfect criterion of the ability which it is designed to measure or prophesy, we may proceed to improve it, if it is improvable, or to discover that we are unable (with our present knowledge and ingenuity) to improve it. Our procedure will be to obtain the separate correlations of E_1, E_2, E_3, etc., with the criterion, and to find the best weights to attach to each by the partial correlations and regression equation. In proportion as the multiple correlation computed therefrom is above .60, we improve the test by using these weights in place of those originally used. In the course of this procedure certain of the E's may appear to be profitably replaceable by others to be invented, or even to be profitably eliminated (though the latter rarely occurs in a reasonably well devised test). What are wanted in such cases are new elements which correlate high with the criterion and low with the other E's. The most promising candidates are inserted in the test and it is tried in its new form.

Under the general heading of the detailed constitution of algebraic learning, almost all of the reconstructions now being tried or contemplated are suitable subjects for investigation. We note only a few, which seem either especially important or especially practicable.

The Use of Irregular Relation Lines, or "Statistical Graphs"

In Chapter III we advocated the use of only a few[1] statistical graphs for introduction to the regular or "mathematical graphs," and to lend interest to these. This conclusion was reached from as adequate a consideration of the psychological and educational facts as we could make.

[1] "A few" may be conveniently thought of as meaning about as many as are used by Nunn.

However, more facts are needed. The graphic presentation of facts is new in the world, and the psychology of reading with graphs and of composition with squared paper is largely unknown. We need careful observations of just what pupils gain from such treatments of graphs as those now given by, say, Rugg and Clark, or Schorling and Reeve, in comparison with what they gain from these treatments modified toward greater and toward less prominence of the irregular graph.

The Use of Derivations and Proofs of Formulas

These are not fashionable now in algebra and probably in their usual forms did encourage memoriter learning in most pupils. May it not be possible, however, to secure their merits without their defects? We do not wish pupils to attempt to understand generalizations which are beyond their abilities; on the other hand, we do not wish algebra to degenerate into a series of computations and solutions of particular equations.

It seems worth while to see what a more scientific and ingenious gradation, coupled with the avoidance of proving what is already perfectly known, will achieve.

The Geometry that Really Aids Algebra

In general the psychology of the so-called "combined" courses which are now the subject of so much discussion and experimentation is beyond the scope of this report; and we have hitherto refrained from considering even those cases where geometrical facts may be profitably taught for the sake of algebraic ability itself. These are probably few. It is doubtful, for example, whether many of the pupils who now study algebra are helped to understand irrational numbers by the geometrical facts that the length of a certain hypotenuse is exactly measured by $\sqrt{1+1}$ and vice versa, as much as they would be by the comparison of pairs like these: $\frac{1}{2}$ and $\frac{1}{3}$, 4 and 5, cost of a pound of sugar at 4 lbs. for 25 cents and cost of a pound of sugar at $4\frac{1}{2}$ lbs. for 25 cents. Some, however, there probably are, and it would be a very useful piece of work to make a list of candidates, inventory their merits and demerits, and assign the valuable ones to their exact posts.

The Standardization of Problem Material

According to the judgments of our psychologists and mathematicians there is a very wide variation in merit among the problems now in use. It would be a clear gain for teaching if the four or five hundred best problems now available were collected and arranged with a statement of the special service performed by each and of its approximate difficulty. To this basic list problems of local or temporary importance could be added. Textbooks and courses of study would, however, in general accept it, adding only such problems as had substantial claims to superior merit. The fantastic, incomprehensible and misleading problems would be driven from practice to somewhat the same extent that individual perversities have been driven from the spelling book by standard spelling lists.

The Standardization of Practice Material

A natural sequel to the investigations concerning the psychology of drill in algebra will be a series of exercises devised to form the desired mental connections with a maximum of facilitation and interest and a minimum of interference and drudgery. We find, for example, that we need much more practice with $\frac{a}{ax}, \frac{x}{ax}, \frac{ax}{bx}$ and the like than is now usually given; that we need practice with $\frac{ax \pm x}{x}$, $\frac{x}{ax \pm x}$, $\frac{ax \pm ay}{ax}$, $\frac{ax}{ax \pm ay}$ and the like, soon after work in the division of monomials has been begun, to prevent the pupil from canceling terms instead of factors. A series of exercises which will take account of all such matters as these and provide practice specified to produce certain abilities in place of the somewhat impressionistic and indiscriminate practice now in vogue would deserve and soon obtain wide adoption.

The Value of Explanations

There have been and are many ways of explaining the principles and techniques of algebra. In the case of negative numbers, for example, we find the following used: temperature above and below zero, money owned and debts, force of gravity and the pull of a balloon or spring, distance north and south, distance east and west, distance above and below sea level, movement of an elevator up and down. The following might well be used and probably are by some teachers: marks expressed as deviations from the average of the class or from some standard for an age or grade, deviations from norms in gymnasium tests, credit and penalty scores.

In the course of time, by a process of trial and observation of results, the better forms of explanation probably outlive the worse, but there is need of a systematic comparison, which will expedite this selective process and, we may hope, increase our knowledge of the general principles of effective explanation and illustration.

In the case of algebra two matters specially need investigation—the use of explanation where what is really needed is familiarity, and explanations by analogy with space relations. The former needs investigation because we have all probably underestimated the dependence of thought and reasoning upon the content with which we think and about which we reason. The latter needs investigation because it seems probable that space relations are far more interesting and illuminating to mathematicians than they are to non-mathematicians of equal intelligence. Consequently they seem likely to be overworked in explanations of algebra. Nunn's use of space analogies in explanation of the laws of signs and of special products and factors is an excellent case to investigate.

The Values of Various Aids to Interest

The chief appeal of algebra has been, and probably always will be, to the love of thought for thought's sake by those who can play the game of thought well. If we arrange the learning process so that pupils can learn algebra readily they will be in so far inclined to like it. In addition, teachers are trying one or another means of enlisting the common human interests in action, rivalry, mastery, sociability, approval, and so on. These means need to be surveyed critically in respect to their psychological effects, and to have their actual classroom results measured. For example, the two that have been perhaps the most prominent in the last five years are the use of pictures and biographies of great mathematicians, and the making of algebraic ability instrumental in the service of personal projects outside algebra rather than an end in itself. It does not seem wise to take either of these upon faith.

Concerning the former the psychologist is emphatically of the opinion that the pictures of these queer-looking old gentlemen, and most of what is said about them, will act negatively upon interest. He agrees with the suggestion of G. A. Miller [1915, p. 809] that these pictures and biographies make algebra seem a dead science, antiquated and behind the times, and that they should be replaced, or at least supplemented, by pictures of twentieth-century mathematicians and tales of what is being done now. The difference of opinions needs actual experimentation for settlement.

Concerning the latter we all may appreciate the force of a plea to utilize the pupil's individual purposes as well as the purpose of society to make him a certain sort of boy or girl. We may also appreciate the theoretical importance of other personal projects than the project of learning the algebra which wise people think it is desirable that you should learn. But in practice what are these projects and purposes to which algebra is to be instrumental; and how far and whither will they lead one in the acquisition of algebraic ability? Here again experimentation is needed.

The same is true of algebra matches, mathematical entertainments, mathematical clubs, applications to technology and sociology, and many others of the means advocated for increasing interest in algebra.

These suggestions for productive researches in the psychology and teaching of algebra might be increased by many others and extended by the specification of details. They will, however, serve as samples.

APPENDIX I

The Abilities Involved in Algebraic Computation and in Problem Solving [1]

There is a tendency to compare algebraic computation unfavorably with problem solving in respect to the intellectual demands which it makes and the functions which it trains. The former is said to be "mechanical," or to be a matter of mere memory, or to be "formal," unlike the real thinking needed in life.

Such statements, though true to a certain extent, are misleading. It may be possible for a pupil to learn to manipulate coefficients, exponents, radicals, signs, parentheses, numerators, and denominators, by habituation to certain fixed rules like "In the product of two numbers like signs give plus and unlike signs minus," with very little genuine understanding of what he is doing or why he does it. It is, however, extremely hard for a pupil to learn to operate algebraically in this way. If he understands algebra, each habit usually helps the formation and retention of the others. Each habit is like a meaningful sentence in a story which has a general unity of plot and a few consistent characters. Just as the context helps a person to read each sentence in such a story, so the plan and content of algebra, when understood, help the pupil to remember who Mr. Coefficient is, what you do to $x+14=2x+8$ to find the value of x, and

[1] Reprinted, with some additions, from *School and Society*, vol. 15, pp. 191-193.

whether or not $-\sqrt{x}$ is the same as $\sqrt{-x}$. In proportion as the pupil does not understand algebra, but merely says its rules like a parrot and performs its operations as a trained dog does his tricks, he has the task of memorizing many unrelated facts which are of such nature as to become confused very, very readily. No parrot or dog could in fact be taught in twenty years to do the algebraic "tricks" which a bright child will easily master in twenty hours. Rules about what is done in multiplication and division with the other factors as distinguished from the coefficients are enormously difficult to apply if the pupil does not understand the general nature of literal notation.

What commonly happens is not that pupils lacking insight learn to compute mechanically, but that they *do not learn to compute.* If a pupil does learn to compute, he probably has gained some insight into algebra.

As Upton says [1923], "To understand the principles by which an equation is cleared of fractions, by which a formula for the area of a circle may be changed to a formula for the radius of the circle in terms of the area, by which $\sqrt{\frac{1}{3}}$ is reduced to its simplest form, by which we know that $(a^2)^3$ is a^6 and not a^5, by which we relate the work in exponents to logarithms, by which we show the close relation between exponents and radicals, by which we know why we change signs when we 'transpose' in an equation, or, better, by which we understand that we really never do 'transpose' any term from one side of an equation to the other, and by which we know that $3x^4+2x^2+1$ can be put into the form of a quadratic expression — to understand all of these is to my mind as important an outcome in the teaching of algebra as the solution of verbal problems. The principles of equivalent equations, as a further example, and their use in explaining those mysterious 'extraneous' roots, are to me quite as important and quite as thought-provoking, and often far more interesting to the pupil, than many verbal or 'clothed' problems, as they are called abroad. Certainly, the opportunities for logical thinking in handling the above topics are as numerous as they are in any other part of the algebra course. There is a great danger, therefore, that all of these opportunities may be lost or subordinated in our endeavor to get the fundamental processes speeded up to the habit level as soon as possible."

The case of computation *versus* problem solving is in this respect somewhat like that of knowledge of words and simple sentences *versus* ability to follow legal arguments or pass a difficult completion test. The latter taxes understanding somewhat more and requires a wider and more ingenious organization of one's mental forces, but the former is usually far from being a matter of "mere memory." So far is it, indeed, that even knowledge of single words is one of our best quick indices of intellect.

The argument from analysis and *a priori* considerations could be carried further and still be instructive, but the whole matter is perhaps better settled by evidence *a posteriori*, showing that in fact algebraic computation involves very much the same abilities that problem-solving does. This evidence may be had in the correlations of each of them with general intellectual ability in verbal, mathematical and other abstract and symbolic situations (which we may call I).

The common derogatory comments concerning computations leave the reader with the impression that computation correlates with I little above 0, but that problem-solving correlates with it nearly 1.00. This is very far from the truth. Seven years ago I gave to groups of freshmen in engineering schools the test in computation and the tests in problem solving shown below. These were parts of an elaborate set of examinations, the total weighted average of which is an excellent measure of I.[1] The correlations were as shown below.

Correlations with I (General Abstract Intelligence)

	Arithmetical Verbal Problems	Algebraic Problems	Algebraic Computation
34 students, Columbia College....	.63	.80	.63
41 students, Mass. Inst. Tech....	.43	.53	.44[2]

It should be noted that if we could arrange to have a hundred or more children selected at random and all given (or forced to take) education including algebra until they were eighteen, and if we then tested these hundred children with algebraic computation, problem solving, and the total I examination, both correlations would be very much higher than the .53 and .66 which were the averages for the Columbia and M. I. T. groups. They would probably be about .85 and .95, respectively. Consequently, if we wish to speak of the comparative amounts of I involved in computation and in problem solving, it is truer to speak of .80 and .90 than of .53 and .66.

Algebraic Problems

1. If $2-\dfrac{\frac{x}{a}-1}{\frac{2}{a}}=0$, what does x equal?

2. A man has a hours to spend riding with a friend. How far can they ride together, going out at the rate of b miles an hour, and just covering the return trip at the rate of c miles an hour?

3. How much water must be added to a pint of "alcohol 95% pure" to make a solution of "alcohol 40% pure"?

[1]In one group the criterion for I included also high-school records, college marks, ratings by instructors, and ratings of one another by the students themselves.

[2] The higher correlations for the Columbia group are due partly to the more accurate criterion, partly to the greater range of ability, and partly, probably, to chance.

4. Given that $2x-3$ is less than $x+5$ and that $11+2x$ is less than $3x+5$, to find the limits (*i. e.*, the values) between which x lies.

5. A cube containing eight inches was plated with copper. The difference in the weight of the cube before and after the plating was 0.139 lbs. One cubic inch copper weighs 0.315 lbs. Form an equation from which the approximate thickness of the copper plating could be calculated. State whether the approximate estimated thickness by your equation would be less or more than the exact thickness.

Arithmetical Problems

1. A boy is tested with a series of 16 problems in algebra. He did nothing at all with six of them; he did one correctly except for a mistake in changing signs; he did two with many mistakes in each; he did the others perfectly. He finished the work in 100 minutes. What was his total credit, supposing that he is given a credit of 8 for each example right, a credit of 3 for each example right except for changing signs, and a penalty of 1 for each minute spent over an hour and a half?

2. Suppose the following to be true: Out of every ten thousand men in an army 920 will probably be exposed to typhoid fever. Out of every thousand exposed to typhoid fever 350 will probably fall sick. Out of every hundred who will fall sick 20 will probably die of the fever. How many men will probably die of the fever in any army of one hundred and fifty thousand men?

Algebraic Computation

1. If $a=6$ and $b=1$, what does $2ab-ab^2$ equal?

2. If $x-2a+b=2x+2b-4a$, what does x equal?

3. If $a=6$ and $b=3$, what does $\sqrt{a}\sqrt{2b}$ equal?

4. If $\frac{6x+7}{5}-\frac{2x-1}{10}=4\frac{1}{2}$, what does x equal?

5. If $\frac{1}{a}-\frac{1}{x}=\frac{1}{x}-\frac{1}{b}$, what does x equal?

6. If $\frac{x+a}{x-a}-\frac{x-a}{x+a}-\frac{x^2}{a^2-x^2}=1$, what does x equal?

7. If $x=\frac{a+b}{2}$, what does $\left(\frac{x-a}{x-b}\right)^3-\frac{x-2a+b}{x+a-2b}$ equal?

To supplement the data given above I have computed the correlations of algebraic computation (10 min. test shown above) with the I score obtained in the Thorndike Intelligence Examination for High School Graduates of which it is a part, and have done the same for the sum of two five-minute tests with arithmetical problems, also contained in that examination. The facts are as follows:

	Correlations of Total Score in the Intelligence Examination with	
	Score in Algebraic Computation	Score in Arithmetical Problems
371 candidates at Columbia	.47	.64
77 candidates at Columbia[1]	.53	.51
76 candidates at Columbia	.50	.66
180 candidates at an engineering school	.50	.61
97 women candidates	.51	.70

A weighted average correlation is about .50 for the computation and about .13 higher for the problems.

These correlations also would both be much higher for a random sampling of the total population, carried on in school work, including algebra, to the age of eighteen, probably about .8 and .9.

Algebraic computation as actually found is then emphatically an intellectual ability. It is not so indicative of intellect as problem solving, partly because it involves less abstraction, selection, and original thinking, partly because it involves only numbers, not numbers and words. It is, however, far above the reproach of being a mechanical routine which can be learned and operated without thought.

[1] In this case the correlations were not with *Total Score*, but with Score in Part I plus Score in Part II minus the score in Algebraic Computation (or Arithmetical Problems).

APPENDIX II

THE PERMANENCE OF SCHOOL LEARNING[1]

It is common, in discussions of educational values, to use the alleged fact that we soon forget most of what we learn in school.

We have been able to make a rough measurement of the loss in the case of algebra. The subjects were 189 college graduates, all now first-year students in a law school. The conditions were such that detailed information could not be had concerning their study of algebra, but approximately it dated back at least four or five years in the great majority of cases. All of them had studied algebra, it being presumably a required subject for entrance to the colleges whence they were graduated.

They spent ten minutes without any previous notice that a test in algebra was to be given, half of them upon the five questions of A and half of them upon the five questions of B. The A and B sets are of about equal difficulty and will be treated alike in all that follows:

A

1. $a=1$, $b=2$, $c=3$, $d=4$. What does $a^2b^2c^2-abc$ equal?

2. $ax=bx+1$. What does x equal?

3. $a=-4$, $b=3$, $c=-2$, $d=1$. What does $b^2-(2c)^2$ equal?

4. $\frac{b}{x-1}=1$. What does x equal?

5. If $5t=7$ and $3b=4\frac{1}{2}$, what does $2bt^2-\frac{b}{2}$ equal?

[1] Reprinted with some changes from *School and Society*, Vol. 15, pp. 625-627 and 649.

B

1. $k=\frac{1}{2}$, $e=\frac{1}{3}$, $m=\frac{1}{4}$. What does $(2k^2)+(3e^2)-m$ equal?
2. $x-(c+d)-5=2x-(a+d)+3$. What does x equal?
3. $x=a$, $y=4$, $a=y^2$. What does $x+2y$ equal?
4. $\frac{x}{c}+1=d$. What does x equal?
5. What is the sum of $(a+b)^2$, $(a-b)^2$ and $(a+b)\ (a-b)$?

Eight and one-half percent of them did no one of the five correctly; 14.8 percent did only one correctly; 19 percent did two correctly; 21.7 percent did three correctly; 23.8 percent did four correctly; 12.2 percent did all correctly. The median ability was thus three right.

At first thought this may seem to be evidence of extreme forgetting. If, after four years or more, half of these men of excellent abilities cannot solve $\frac{b}{x-1}=1$ or $\frac{x}{c}+1=d$, one is tempted to conclude that only a small fraction of the algebraic ability once acquired has persisted. How small the fraction is, depends, however, on how large the original ability was. We must estimate how well they would have done in a similar test at or near the time when they were studying algebra. This we are able to do approximately since 10 minutes out of 60 spent on Part II of the Thorndike Intelligence Examination for High School Graduates is given up to a set of six algebraic tasks of which the first five are equal in difficulty to A or B above; and many hundreds of records of college freshmen are available. Under the conditions of the examination these freshmen could spend some extra time upon this algebra test if they did not use all the allotted time upon the other tests. This may be regarded as about balancing the addition of the sixth task. It should be noted also that, with few exceptions, both the college graduates and the freshmen complete five of the algebra tasks; they do them, but make errors. There is one more special consideration. College graduates are a selection of the abler freshmen in respect to scholarship; and college graduates who study law are perhaps a still abler selection.

In view of tnese facts the most probably fair comparison seems to be with the top half of freshmen as rated by the total score in the intelligence examination. The facts for 200 of them in the algebra test were as shown in the table below (Table 63) in comparison with the facts for the college graduates: $1\frac{1}{2}$ percent of them did no example correctly; $2\frac{1}{2}$ percent of them did only one example correctly; 6 percent did two correctly; $16\frac{1}{2}$ percent did three correctly; 22 percent did four correctly; $51\frac{1}{2}$ percent did at least five correctly.

TABLE 63

ABILITY IN ALGEBRA IN THE CASE OF COLLEGE GRADUATES AND COLLEGE ENTRANTS OF APPROXIMATELY EQUAL ABILITY

Number Correct	Freqencies: In Percents	
	College Graduates	Top Half of Freshmen at Entrace
0	8.5	1.5
1	14.8	2.5
2	19.0	6.0
3	21.7	16.5
4	23.8	22.0
5	12.2	51.5 { 33.0
6		18.5

The effect of the four years or more from college entrance in September to April of the first year in law school is thus approximately a reduction from the ability to do four or five of such tasks to the ability to do three. Or, we may say that at entrance to college the median of these graduates could have done tasks as hard as:

$\frac{b}{x-1}=1.$ What does x equal?

$\frac{x}{c}+1=d.$ What does x equal?

What is the sum of $(a+b)^2$, $(a-b)^2$ and $(a+b)\ (a-b)$?

If $5t=7$ and $3b=4\frac{1}{2}$, what does $2bt^2-\frac{b}{2}$ equal?

whereas, four years or more later he could do only tasks as hard as:

$ax=bx+1$. What does x equal?

$x-(c+d)-5=2x-(a+d)+3$. What does x equal?

What does $b^2-(2c)^2$ equal if $a=-4$, $b=3$, $c=-2$, and $d=1$?

What does $x+2y$ equal if $x=a$, $y=4$, and $a=y^2$?

Or, we may express the change as follows: After the interval most of those who at entrance could do one or two correctly could do none; most of those who could do three at entrance could do only one; most of those who could do four at entrance could do only two; those who could do five at entrance could do three or four.

Consider now the loss from the time of completion of the study of algebra. We can do little better than speculate about this, since we do not know the ability of such a group as our college graduates when they completed their study of algebra in Grades 9 or 10. The available material includes the data given by Hotz and records which we have obtained from about 800 pupils in high schools taken during their study of algebra or within a few months after they had finished it. The test they took was not like A or B above, and we cannot estimate with surety how well they would have done with a ten-minute test like A and B. Estimating as best we can, without an undue expense of time, we obtain percents corresponding to those found for college graduates and college freshmen as shown in Table 64, for a group of students in preparatory schools of high grade who are perhaps somewhere near the grade of ability represented by our college graduates when these were at that stage. According to this very imperfect determination, the median man of our group of college graduates was able to do all or all but one of these tasks when he was studying algebra, and has immediate mastery now of only three. How long it would take him to recover the ability he had six or seven years ago, we do not know.

TABLE 64

ABILITY IN ALGEBRA IN THE CASE OF COLLEGE GRADUATES AND POSSIBLY COMPARABLE GROUPS AT ENTRANCE, AND AT THE TIME OF STUDY OF ALGEBRA

	College Graduates	Top Half of College Entrants	Top Half of Pupils in Excellent Preparatory Schools, While Studying Algebra
0	8.5	1.5	..
1	14.8	2.5	..
2	19.0	6.0	..
3	21.7	16.5	5
4	23.8	22.0	36
5	12.2	51.5	59

In connection with these data the following facts may be of interest. Three of the Hotz scales, Series A (Equation and Formula, Problems and Graphs) were used with 37 graduate students. Eighteen of these were students in a course on Educational Tests; the others were almost all students of psychology or education, without special interests in quantitative sciences.

The results were:

EQUATION AND FORMULA

Task	Attempts	Rights	Correct Method, Wrong Answer
1	37	37	..
2	37	35	..
4	37	34	1
6	37	32	1
8	37	29	1
11	37	34	..
14	37	25	..
18	37	23	5
19	37	15	2
23	37	13	2
24	37	24	2
25	37	18	[illegible]

PROBLEMS

TASK	ATTEMPTS	RIGHTS
1	37	36
2	37	33
4	37	34
5	37	36
7	37	33
8	37	33
9	37	28
12	37	15
13	37	20
14	37	20

GRAPHS

TASK	ATTEMPTS	RIGHTS	CORRECT METHOD, WRONG ANSWER
1	37	37	..
3	36	34	..
4	36	36	..
5	35	26	..
6	36	32	..
7	32	21	..
10	32	15	1
11	34	21	5

The work in graphs is perhaps largely irrelevant to our problem, since such work was not often studied by these individuals when in high school, and on the other hand may have been studied directly or indirectly in recent years.

The facts reported here are hardly more than hints, but they may at least prevent very exaggerated notions of the loss of learning.

APPENDIX III

The Effect of Changed Data Upon Reasoning[1]

The older psychology, perpetuated in current educational doctrines and practices, regarded reasoning as a force largely independent of associative habits, which worked back to correct or oppose them. Our present psychology finds that the mind is ruled by habit throughout, the correction or opposition being of certain more simple, thoughtless and coarse habits, by others which are more elaborate, selective, and abstract. It defines reasoning as the organization and coöperation of habits rather than as a special activity above their level; and expects to find "reasoning" and habit or association working together in almost every act of thought.

One interesting and rather important consequence of this view is the theorem that "Any disturbance whatsoever in the concrete particulars reasoned about will interfere somewhat with the reasoning, making it less correct or slower or both." It is the purpose of this article to give illustrations of and evidence for this theorem, drawn from a simple experiment which can be easily given to any class, and which is perhaps deserving of inclusion in a list of group experiments in psychology.

Consider the two sets of tasks in algebra, printed below. Each pair of tasks demands the application of the same principle, but the concrete situation in the one case is that

[1] Reprinted, with additions, from the *Journal of Experimental Psychology*, Vol. 5, pp. 33-38.

with which our ordinary associative habits have been made, whereas in the other case the concrete particulars are somewhat altered. The alterations vary from slight ones, such as using p instead of x, or b_1+b_2 instead of $x+y$, to a change from a very customary statement of a set of relations to a very rare statement, as in No. 9.

Ha.

Habitual

1. What is the square of $x+y$?
2. What is the square of a^2x^3?
3. Simplify $4ac+\left(\frac{b^2}{c^2}\times\frac{c}{d^2}\times\frac{d^3}{b}\right)$.
4. What are the factors of x^2-y^2?
5. Multiply x^a by x^b.
6. Simplify $ac-[a(b+c)]$.
7. Solve
$$x+y+z=15,$$
$$2x+y+3z=22,$$
$$x+2y+z=25.$$
8. $e^2+ef=\frac{g}{x}$. What does x equal?
9. There are two numbers. The first number plus 3 times the second number equals 7. The first number plus 5 times the second number equals 11. What are the numbers?

Ch.

First Changed Form

1. What is the square of b_1+b_2?
2. What is the square of $r_1^3 r_{11}^2$?
3. Simplify $5a^2b+(b^2\div c^2)\,(c^2\div d^2)\,(d^3\div b)$.
4. What are the factors of $1/x^2-1/y^2$?
5. Multiply 4^a by 4^b.
6. Simplify $p_1p_3-[p_1(p_2-p_3)]$.
7. Solve
$$c_1+c_2+c_3=15,$$
$$2c_1+c_2-3c_3=22,$$
$$c_1+2c_2+c_3=25.$$
8. $e^2+ef=\frac{g}{p}$. What does p equal?
9. $y=ax+b$. When $x=3$, $y=7$. When $x=5$, $y=11$. What does a equal? What does b equal?

The probability of error and delay in the necessary reasoning is clearly increased when the task is of the changed sort (Ch.) rather than the habitual sort (Ha.).

The subjects which I used were ninety-seven graduate students, divided at random into three groups of 34, 32 and 31. Group A did tasks 1 and 7 of Ha., tasks 4, 5, 6, 8 and 9 of Ch. and two other tasks as follows: "What is the square of k^2p^3?" and "Simplify $3ab+[b^2cd^3/bc^2d^2]$." Group B did tasks 3, 6 and 9 of Ha., tasks 1 and 2 of Ch. and four other tasks as follows: "What are the factors of $x-y$?" "Multiply x^4 by x^3." "Solve $x+y=13$, $y+z=12$, $x+z=5$." "$e^2+ex=g/a$. What does a equal?" Group C did tasks 2, 4, 5 and 8 of Ha., tasks 3 and 7 of Ch. and three other tasks as follows: "Multiply (b_1+b_2) by (b_2-b_1)." "Simplify $mn-[2n(m-p)]$." "The cost of a season ticket is b dollars plus k times the cost of a month ticket. When the cost of a month ticket is \$6, the cost of a season ticket is \$15. When the cost of a month ticket is \$15, the cost of a season ticket is \$33. What is the value of b? What is the value of k?" [1]

The relative abilities of the groups may be estimated by comparing the scores in tasks estimated to be of equal difficulty, such as:

$(x+y)^2=$? and Solve $e^2+ef=g/p$ for p, for A,
$x^4\times x^3=$? and Solve $e^2+ex=g/a$ for a, for B,
Factor (x^2-y^2) and Solve $e^2+ef=g/x$ for x, for C.

Using these combinations, the percent of right answers was 71 for Group A, 84 for Group B, and 63 for Group C. There is thus an approximate equalization of ability of subjects between the *Habitual* and the *Changed* series.[2]

All three groups worked by the same time-schedule, which was: "0, begin; 1 min. 30 sec., even if you have not finished 1, begin on 2; 2:30, even if you have not finished 2, begin on 3; 4:30, even if you have not finished 3, begin on 4;" and

[1] The fractions in the above and throughout were all presented in the usual "horizontal" form, e. g. $\frac{g}{a}$.

[2] The B group is probably really somewhat more superior than these figures show, and the differences shown in Table 65 are somewhat greater than would be the case with perfect equalization. However, there can be no doubt that a substantial difference would remain.

so on with 6:30 for proceeding to 5; 8:30 for 6; 9:30 for 7; 12:30 for 8; 13:30 for 9; 16:30 to stop. The subject could thus use time saved on one task for another, absolute uniformity being had only in respect to the 16 min. 30 sec. spent on the entire set of nine tasks.

The percents wrong or incomplete for the nine tasks where customary associations were favored and for the nine where some change was made were, in order:

TABLE 65

	Customary	Changed	Nature of Change
1.	6	28	$(x+y)$ to (b_1+b_2)
2.	34	47	a^2x^3 to $r_1^3 r_{11}^2$
3.	37½	64½	fraction to ÷ and × to parenthesis form
4.	22	41	x^2-y^2 to $\frac{1}{x^2}-\frac{1}{y^2}$
5.	55	70½	x^2 to 4^2
6.	25	53	ac, etc., to p_1p_3, etc.
7.	62	61	x, y, z, to c_1, c_2, c_3
8.	52	53	x to p
9.	16	70	form of problem

In all but two cases (7 and 8) there is a substantial interference with thought by the change. In No. 9 the amount of novelty introduced is much greater than in the other cases, and is perhaps more truly called a change in the principles and operations used than a change in the concrete particulars reasoned about. In No. 6 the use of a minus sign in place of the plus may have added a little difficulty over and above the change from literal to subscript distinctions.

We may now examine certain facts from the accessory tasks which were used partly to conceal the special point of the experiment from the participants and partly to secure additional data.

"Multiply (b_1+b_2) by (b_2-b_1)" is not strictly comparable with "What is the square of $x+y$?" but the rise in errors from 6 percent for the latter (Group A) to 61 percent for the former (Group C) would probably have been much less if "Multiply $(x+y)$ by $(y-x)$" had been used.

The change from "What is the square of a^2x^3?" (Group C) to "What is the square of k^2p^3?" (Group A) raises the errors and omissions from 34 percent to 47 percent.

Simplify $3ab+[b^2cd^3/bc^2d^2]$ produces more errors than the same task when arranged in $+(\times\ \times)$ or in $+(\div\ \div)$ form (71 percent in Group A). This may seem to be in contradiction to our general theorem, but an inspection of the errors makes it probable that the parentheses were a protection against two habits, one of canceling upper and lower numbers seen in proximity, the other of clearing of fractions any expression that contains one. The former leads to canceling the b of $3ab$ with the b of bc^2d^2, which is done by one of the thirty-four subjects. The latter adds an operation in which errors are made by four of the subjects, and makes the needed cancellations more difficult. It also encourages the error of dropping out the common denominator, which is done by four more, and permits the worker to feel that he has done his duty by the task, though he has not cancelled. Finally, there are further erroneous manipulations of the denominator, as in $(3ab+b^2cd^3)/bc^2d^2$, from which the parentheses and signs of multiplication and division were perhaps a protection in the other forms of the task.

The question "What are the factors of $x-y$?" evokes responses which show that when a novel situation is met, one's associative habits do not retire while reason attacks. On the contrary, the forces of habit are nowhere more evident than in the treatment of novel situations. Which habits will act depends on which elements or features of the situation are given weight and upon the amount of weight given to each. With our subjects, the features that this is the difference not of two particular numbers but of any number from any number and that "any number" is as truly the square of "$\sqrt{\text{any number}}$" as "(any number)^2" is of "any number," are usually neglected or underweighted, there being only four answers of

$$(\sqrt{x}+\sqrt{y})\ (\sqrt{x}-\sqrt{y})\ \text{or}\ (x^{1/2}+y^{1/2})\ (x^{1/2}-y^{1/2})$$

The underweighting of these features permits the habits of not factoring prime numbers in arithmetic, and of not factoring expressions like $x-y$ in algebra, to act acceptably, answers of $1(x-y)$ being the commonest given (11 out of 32).

"Multiply x^4 by x^3" is, consistently with our general theorem, easier than x^a by x^b.

On the whole, it seems certain that even such slight changes as from the customary a, b, x, and y, to k and p or to p_1, p_2, and p_3 or to p_1, p_{11}, p_{111}, impede thought, and that the general theorem does hold that "Any disturbance whatsoever in the concrete particulars reasoned about will interfere with reasoning."

The results of the experiment described above have been checked and corroborated in general by a second experiment performed with 118 undergraduate and graduate students. As before, three sets (A, B, and C) of nine tasks were used, but in this experiment three of the nine tasks were identical in all six sets. These three were:

Divide a^2bc^3 by ab^2c^2.
Multiply $x+3$ by $x-2$.
Multiply $3ax^2-bx$ by $\frac{b}{x}$.

Using the records for these three, the group having Set A had 47 percent of right answers; the group having Set B had 48 percent of right answers; and the group having Set C had 38 percent of right answers. Enough low-scoring individuals were then dropped from the C group to make a

"Modified C" group with 47 percent of right answers. We may then treat the A group, B group and "Modified C" group as of closely equal ability in algebraic computation.

So treating them, we find the following percents wrong or incomplete for the six tasks where there was a comparison of customary with changed data:

TABLE 66

Customary	Changed	Nature of Change
18	71	$(x+y)^2$ to $(b_1+b_2)^2$
56	76	$(a^2x^3)^2$ to $(r_{11}^{3} r_{111}^{2})^2$
3	39	Factor x^2-y^2 to factor $\frac{1}{x^2}-\frac{1}{y^2}$
58	71	Multiply x^a by x^b to multiply 4^a by 4^b
55	80½	$ac[a(b+c)]$ to $p_1 p_3 [p_1 (p_2+p_3)]$
67	76	$e^2+ef=\frac{ge+gf}{x}$. Solve for x, to
		$e^2+ef=\frac{ge+gf}{p}$. What equals p?

Changes from $x+y$ to $g+h$, from a^2x^3 to k^2p^3, from x^2 and y^2 to h^2 and k^2, and from "Solve for x" to "What does x equal?" did not seem to cause interference in this group of students. Changes from $x^a x^b$ to $A^a A^b$, and from *abc* to *mnp* in the task of removing parentheses and collecting terms, did seem to do so. Neither the absence of the former nor the presence of the latter effect is at all certain, however. The general effect of the six changes especially experimented with is sure.

BIBLIOGRAPHY OF REFERENCES MADE IN THE TEXT

Book, W. F. . . . . . . '22 *Intelligence of High-School Seniors.*

Boughn, E. T. . . . . . . '17 "A Mathematical Contest;" *School Science and Mathematics*, XVII, 329 f.

Buckingham, B. R. . . . . '21 "Mathematical Ability as Related to General Intelligence;" *School Science and Mathematics*, XXI, 205–215.

Byrne, Lee . . . . . . '22 "How Much Education Have the American People;" *School and Society*, XV, 289–292 and 327-331.

Childs, H. G. . . . . . . '17 "The Measurement of Achievement in Algebra;" *Bulletin of the Extension Division*, Indiana University, II, No. 6.

Commission on the Reorganization of Secondary Education Appointed by the National Education Association '20 *The Problem of Mathematics in Secondary Education* (United States Bureau of Education Bulletin No. 1).

Commission Appointed by the Board of Review of the College Entrance Examination Board . . . . '21 *Definition of the Requirements in Elementary Algebra, Advanced Algebra, and Plane Trigonometry.*

Committee of the Central Association of Mathematics and Science Teachers (by Mabel Sykes) . . . . '19 "Final Report of Subcommittee on Content of Course in First-year Mathematics;" *School Science and Mathematics*, XIX, 259–264.

Counts, G. S. . . . . . . '15 *A Study of the Colleges and High Schools in the North Central Association* (U. S. Bureau of Education Bulletin No. 6).

CRATHORNE, A. R. . . . '22 "A Critical Study of the Correlation Method Applied to Grades;" to be chapter 10 of *The Reorganization of Mathematics in Secondary Education*, a summary of the report of the National Committee on Mathematical Requirements.

DALMAN, M. A. . . . . . '20 "Hurdles, A Series of Calibrated Objective Tests in First-year Algebra;" *Journal of Educational Research*, I, 47–63.

DAVIES, C. . . . . '59 *New Elementary Algebra*, revised edition; date of preface, 1859, date of publication, 1866.

DOUGLASS, H. R. . . . . '21 "The Derivation and Standardization of a Series of Diagnostic Tests for the Fundamentals of First-year Algebra;" *University of Oregon Publications*, I, No. 8.

HEDRICK, E. R. . . . . '22 "Functionality in Mathematical Instruction in Schools and Colleges;" *The Mathematics Teacher*, XV, 191–207.

HOTZ, H. G. . . . . . '18 "First-Year Algebra Scales;" *Teachers College, Columbia University, Contributions to Education*, No. 90.

HOTZ, H. G. . . . . . '22 *Teachers' Manual for First-Year Algebra Scales.*

INGLIS, A. J. . . . . . '15 *Principles of Secondary Education.*

KELLEY, T. L. . . . . . . '20 "Values in High-School Algebra and Their Measurement;" *Teachers College Record*, XXI, 246–290.

KOCH, E. H., and SCHLAUCH, W. S. As yet unpublished. *A Study of the Opinions of College Professors and Leading Men of Commerce and Industry upon the Relative Importance as well as the Desirability of Including or Excluding Certain Mathematics from a List Suggested for the Secondary School Syllabus.*

KOONS, G. J. . . . . . . '17 "Vocational Distribution of High-School Graduates and of Pupils Leaving School before Graduation;" *Educational Administration and Supervision*, III, 358–360.

MADSEN, I. N., and SYLVESTER, R. H. . . . . '19 "High School Students' Intelligence Ratings According to the Army Alpha Test;" *School and Society*, X, 407–410.

MANN, C. R . . . . . . '18 "A Study of Engineering Education;" *Bulletin No. 11* of the Carnegie Foundation for the Advancement of Teaching.

MILLER, G. A. . . . . . . '15 "Historical Notes in Textbooks on Secondary Mathematics;" *School Science and Mathematics*, XV, 806–809.

MITCHELL, H. E . . . . '14 "Distribution of High-School Graduates in Iowa;" *School Review*, XXII, 82–90.

MONROE, WALTER S . . . '15 "A Test of the Attainments of First-Year High-School Pupils in Algebra;" *School Review*, XXIII, 160–171.

MONROE, WALTER S. . . . '15 "Measurement of Certain Algebraic Abilities." *School and Society*, I, 393–395.

National Committee on Mathematical Requirements . . '20 *The Reorganization of the First Courses in Secondary School Mathematics* (U. S. Bureau of Education Secondary School Circular, 1920, No. 5).

National Committee on Mathematical Requirements . . '21 "College Entrance Requirements in Mathematics." *The Mathematics Teacher*, XIV, 224–245.

NUNN, P. . . . . '14 (a) *The Teaching of Algebra, including Trigonometry*, (b) *Exercises in Algebra*.

PROCTOR, W. M. . . . . . '21 "Psychological Tests and Guidance of High-School Pupils;" *Journal of Educational Research Monographs*, No. 1.

ROGERS, A. L. . . . . . . '18 "Tests of Mathematical Ability and their Prognostic Value;" *Teachers College, Columbia University, Contributions to Education*, No. 89.

RUGG, H. O., and CLARK, J. R. '17 *Standardized Tests in First-Year Algebra*.

RUGG, H. O., and CLARK, J. R. '17a Data given on the Instruction and Scoring Sheets which accompany the Rugg-Clark Tests.

Rugg, H. O., and Clark, J. R. '18 "Scientific Method in The Reconstruction of Ninth-Grade Mathematics;" *Supplementary Educational Monographs*, II, No. 1.

Rugg, H. O., and Clark, J. R. '18a *Fundamentals of High-School Mathematics.*

Schorling, R. (with Kahler, F. A., and Miller, O. M.) . '16 "The Place of Mathematics in the Secondary Schools of Tomorrow;" *School Science and Mathematics*, XVI, 608–616.

Shallies, G. W. . . . . . . '13 "The Distribution of High-School Graduates after Leaving School;" *School Review*, XXI, 81–91.

Smith, D. E. . . . . . . '00 *The Teaching of Elementary Mathematics.*

Strayer, G. D., and Thorndike, E. L. . . . . . . '13 *Educational Administration: Quantitative Studies.*

Stromquist, C. E. . . Undated *University of Wyoming Algebra Test.*

Symonds, P. M. . . . . . '22 "The Psychology of Errors in Algebra;" *Mathematics Teacher*, XV, 93–104.

Thorndike, E. L. . . . . . '14 "An Experiment in Grading Problems in Algebra;" *The Mathematics Teacher*, VI, 123–134.

Thorndike, E. L. . . . . . '22 *The Psychology of Arithmetic.*

Thorndike, E. L. . . . . . '22a "Instruments for Measuring the Disciplinary Values of Studies;" *Journal of Educational Research*, V, 269–279.

Thorndike, E. L. . . . . . '22b "The Abilities Involved in Algebraic Computation and in Problem Solving;" *School and Society*, XV, 191–193.

U. S. Bureau of Education. . '17 Report of the Commissioner of Education for 1916.

U. S. Bureau of Education. . '18 *The Elyria Survey.* Bulletin No. 15.

U. S. Bureau of Education. . '20 *Private High Schools and Academies, 1917–1918.* Bulletin No. 3.

U. S. Bureau of Education. . '20a *Statistics of Public High Schools, 1917–1918.* Bulletin No. 19.

U. S. Bureau of the Census . . . Thirteenth Census, Vol. I.

Upton, C. B. . . . . . . '23 Chapter XIII in "The Reorganization of Mathematics in Secondary Education." *Report of the National Committee on Mathematical Requirements.*

Weglein, D. E. . . . . . '17 "The Correlation of Abilities of High-School Pupils;" *The Johns Hopkins University Studies in Education,* No. 1.

Wells, W., and Hart, W. W. '12 *New High School Algebra.*

Wentworth, G. . . . . . '06 *Elementary Algebra* (revision of 1906).

Wood, B. D. . . . . . . '21 *The Reliability and Difficulty of College Entrance Board Examinations in Algebra and in Geometry.* (A pamphlet issued by the College Entrance Examination Board.)

Yerkes, R. M. (Editor) . . '21 "Psychological Examining in the United States Army." *Memoirs of the National Academy of Sciences,* XV.

Young, J. W. A. . . . . . '07 *The Teaching of Mathematics in the Elementary and the Secondary Schools.*

'10 *Centralblatt für die gesamte Unterrichtsverwaltung in Preussen,* Ergänzungsheft, 27.

INDEX